SAUNDERS GOLDEN SUNBURST SERIES

EXPERIMENTAL METHODS IN MODERN BIOCHEMISTRY

GEORGE RENDINA, Ph.D.

Chemistry Department, Bowling Green State University

W. B. SAUNDERS COMPANY Philadelphia—London—Toronto

W. B. Saunders Company: West Washington Square
Philadelphia, PA 19105

1 St. Anne's Road
Eastbourne, East Sussex BN21 3UN, England

1 Goldthorne Avenue
Toronto, Ontario M8Z 5T9, Canada

Listed here is the latest translated edition of this book together with
the language of the translation and the publisher.

Spanish (1st Edition) – NEISA, Mexico 4 D.F., Mexico

Experimental Methods in Modern Biochemistry   ISBN   0-7216-7550-6

Print No.:   9   8   7   6

To Irma

ACKNOWLEDGMENT

The author has had the advantage of participating in the design and revision of student biochemistry laboratory experiments at several universities. He also has been fortunate to have access to unpublished laboratory manuals and experiments used in departments at other universities. From these beneficial experiences it became possible to put together a manual which takes on some of the form and the content of these many sources. Therefore, there is a great debt to the staff scientists at Kansas, Johns Hopkins, and Wisconsin for their dedication to teaching. In the course of rewriting and organizing, some of this material has undergone extensive selection, modification, and revision with the view of enhancing the learning experience of beginning students of biochemistry methodology. Acknowledgment is extended to those students who groped with earlier versions, and who by their critical comments have rendered valuable service by making the manual more meaningful for succeeding students.

The Department of Chemistry Staff of Bowling Green University, and particularly its chairman, Professor W. Heinlen Hall, should be commended for financial aid and encouragement during the years it took to complete this manual.

There is also great debt to Mrs. Marjorie Motten, whose patience with an incomprehensible script has made this manual a reality.

CONTENTS

CHAPTER 3

ABSORPTION AND ANALYTICAL SPECTROPHOTOMETRY

CHAPTER 4

THE PREPARATION, CHEMICAL, AND PHYSICAL PROPERTIES, AND QUANTITATIVE DETERMINATION OF PROTEINS

CHAPTER 5

THE ELECTROPHORETIC SEPARATION OF SERUM PROTEINS

CHAPTER 6

THE IDENTIFICATION OF *N*-TERMINAL GROUPS OF PROTEINS BY SANGER'S METHOD

CHAPTER 7

THE CATALYTIC PROPERTIES OF AMYLASE, CATALASE, AND UREASE

CHAPTER 8

THE ISOLATION OF DNA AND RNA

CHAPTER 9

THE DISTRIBUTION OF NUCLEIC ACIDS IN SUB-CELLULAR PARTICLES

CHAPTER 10

THE MEASUREMENT OF HYPERCHROMIC SHIFTS AND VISCOSITY CHANGES IN PURE DNA

CHAPTER 11

CHEMICAL PROPERTIES OF CARBOHYDRATES

CHAPTER 12

THE USE OF POLARIMETRY FOR CONFIGURATIONAL AND QUANTITATIVE ANALYSIS OF CARBOHYDRATES

CHAPTER 13

THE ISOLATION OF GLYCOGEN AND THE DETERMINATION OF THE DEGREE OF BRANCHING BY PERIODIC ACID OXIDATION

CHAPTER 14

THE SEPARATION AND PHYSICAL AND CHEMICAL PROPERTIES OF LIPIDS

CHAPTER 15

THE PURIFICATION OF EGG WHITE LYSOZYME (MURAMIDASE)

CHAPTER 16

THE KINETICS OF EGG WHITE LYSOZYME AND HEXOKINASE

CHAPTER 17

THE DETERMINATION OF THE MOLECULAR WEIGHT
OF PROTEINS BY GEL FILTRATION

CHAPTER 18

THE TURNOVER NUMBER OF TRYPSIN AND CHYMOTRYPSIN

CHAPTER 19

CATION REQUIREMENTS OF NUCLEOTIDE PHOSPHOHYDROLASES

CHAPTER 20

THE RATE OF GLUCOSE ABSORPTION IN RAT INTESTINE (DETERMINATION OF THE CORI COEFFICIENT)

CHAPTER 21

GLYCOLYSIS IN CELL FREE EXTRACTS; THE EMBDEN-MEYERHOF CYCLE

CHAPTER 22

THE ISOLATION OF LIPIDS AND THE ISOTOPIC DETERMINATION OF LIPID TURNOVER IN BRAIN

CHAPTER 23

ENZYMES OF OXIDATION-REDUCTION REACTIONS IN MAMMALIAN TISSUE

CHAPTER 24

OXIDATIVE PHOSPHORYLATION IN RAT LIVER

INTRODUCTION TO BIOCHEMICAL ANALYSIS

I. FUNDAMENTALS OF QUANTITATIVE BIOCHEMICAL ANALYSIS

A. General Comments

The experiments described in this manual have been designed to illustrate both major principles and modern techniques of biochemistry. Some of these techniques will involve very complex biochemical concepts which may have been only briefly discussed in a lecture course. Through supplementary reading, as suggested by the references given at the end of certain experiments, the more difficult aspects of both the technology and the principles of these experiments may be clarified and made more meaningful. Presumably, such additional effort should arise from natural inquisitiveness and from the need to better understand the subject or phenomenon under investigation.

It is impossible to illustrate many biochemical concepts by means of experiments which an inexperienced student can perform by himself because very often special skills and elaborate and costly equipment are required. As far as is possible, however, demonstration of advanced research methods should be presented to supplement the basic experiments performed by the student. On the other hand, the experiments included in this manual have varying skill and equipment needs. Some of the experiments are fairly difficult to perform and are intended for graduate students or seniors; others are simpler and more basic in character, intended for less advanced students. However, there are enough experiments to allow planning of a laboratory course suited to the needs of undergraduate or graduate students of biochemistry.

Many students develop an interest in biochemistry via descriptive biology, and are therefore unprepared to think in terms of quantitation and molecular

1

models. In these circumstances, the principles of physics and chemistry are frequently disregarded. This can be a fatal error for those students arriving at biochemistry by this path. Part of the purpose of this course therefore is to develop useful scientific attitudes by showing such students that they are missing a great deal of excitement and challenge in biology by persisting in glossing over a mastery of these more difficult and abstract disciplines. At the same time, it is important to point out to biology students that descriptive biology has yielded to conceptual biology, a rather more dynamic representation of biology embracing all living organisms and the environmental changes they are subjected to, i.e., from the ecological to the subcellular levels. All properties attributed to life—reproduction, growth, aging, death, and decay—are processes involving change, both chemical and physical; thus the only constancy in nature is an endless spatial arrangement and rearrangement of molecules with time. It is a phenomenon which uniquely characterizes living things, and it is one which offers both vitality and continuity. The modern biologist must realize this very early during his educational training and he should respond accordingly.

The experiments described in this manual are designed to teach the student to perform a number of quantitative analytical measurements. Modern biochemistry is dependent on exact measurement of metabolic components in tissue fluids, and the serious student will find that a rapid mastery of quantitative skills will be extremely useful. Moreover, recent scientific progress, particularly in biochemistry, has been both dramatic and exceedingly rapid. In order to be able to understand, if not only merely to cope with these advances, it is mandatory to develop proficiency in theoretical and quantitative aspects of physics and chemistry, particularly in those areas which are directly applicable to biochemistry. For this reason, and because some students may have no analytical and physical chemistry background, and also because there is a trend toward eliminating a formal course in quantitative analysis in chemistry curricula, this introduction to the manual provides a concise treatment of the basic tools and concepts of quantitative analysis and physical chemistry as they pertain to the needs of modern biochemistry. In the text of each experiment there is a stress on each of these disciplines in order to expose the student, as much as is practically possible, to more sophisticated biochemical techniques and metabolic concepts. The successful performance of a number of the more difficult experiments described in this manual ultimately will provide the student with the basic arts and skills required to enhance his understanding of life processes. In turn, they will also allow him to utilize these newly acquired skills in ancillary fields of endeavor in the ever widening demands for technological services in medicine, biology, and chemistry, or to utilize them in the more exciting and active frontiers of research.

B. Instructions

Although each student will individually carry out most of the experiments described, some are to be done with a partner or in groups. In addition, a number of "special" experiments are included in the manual. These involve somewhat more advanced apparatus and techniques. They will be carried out by groups of students, under close supervision. Nearly every student should have participated in at least one of the "special" experiments during the course.

Before attempting to carry out any experiments, study the directions carefully and plan your work in advance before coming into the laboratory. A "cook book" or "dry lab" approach will be a complete waste of time.

C. Recording Data

You are expected to keep your own notebook. It is suggested that you purchase a standard loose-leaf notebook. As you do each experiment, you can insert the tables, graphs, and descriptive information you find important. You are expected to turn in reports for each experiment. After examination, they will be returned to you so that they can be incorporated with the typed description of the experiment provided for you in the laboratory manual.

A suggested format for keeping notes is as follows:

I. Statement of Purpose—a sentence or two which summarize the main idea of the experiment.

II. Data—includes graphs, tables, drawings, calculations, and so forth. Organize these as well as possible and as neatly as possible.

III. Concise discussion of data and conclusion.

Mention possible *sources of error*. Briefly discuss results in terms of expectations; outline reasons for "unexpected" results. Be *precise and clear*. Provide a single statement of conclusions from the experiment. Your skill in concise expression, rather than verbosity, will be regarded highly.

Use the questions and directions in the text of each experiment as a guide for write-up, but do not repeat any part of the descriptive portion or the instructions for setting up the experiments as part of your report. Spend the best part of your time organizing and completing your data. Do not leave any details of your data or calculations out.

D. Stoichiometry and Concentration

A *molar solution* contains one gram molecular weight of a substance in one liter of *solution* . Examples: (a) The molecular weight of NaOH is 40.01. A molar solution of NaOH contains 40.01 g in one liter. (b) The molecular weight of NAD^+, the coenzyme nicotine adenine dinucleotide, is 664.44; hence a 1×10^{-4} M solution contains 66.44 mg of NAD^+ per liter.

A *normal solution* contains one gram *equivalent* weight of a substance in one liter of *solution* . The equivalent weight of the substance depends on the nature of the reaction for which the solution is used. In acid-base analysis the equivalent weight of an acid or base is the weight in grams which yields or combines with exactly 1.008 grams of H^+ (the weight of one gram atom of H^+).

Examples:

(1) One mole of NaOH can combine with one mole of H^+; hence the equivalent weight of NaOH is 40.01 g and a normal solution of NaOH has the same concentration as a molar solution. Acetic acid has a molecular weight of 60.05. A normal solution of HAc contains 60.05 g of HAc per liter of solution since one molecule of HAc yields one H^+. However, succinic acid (m.w. 118.09) contains two ionizable hydrogen atoms. Its equivalent is therefore the gram molecular weight divided by 2, or an equivalent weight of 59.045 g, and a normal solution of succinic acid contains 59.045 g per liter.

(2) In oxidation-reduction reactions the equivalent weight of the substance is that weight in grams which yields or combines with *one electron equivalent*, which is 1.008 g of hydrogen. Since oxidation-reduction reactions involve the loss and gain of electrons, *a change of one electron is the basis of stoichiometric calculations.*

When, for example, the ferrous form of cytochrome *c* is oxidized to the ferric form in biological oxidations, one electron is transferred to a suitable oxidizing substance, in this case another cytochrome. The half equation is

$$\text{cyto}_c\text{Fe}^{+2} \rightleftharpoons \text{cyto}_c\text{Fe}^{+3} + 1e.$$

The equivalent weight of the Fe^{+2} form is therefore equal to the gram molecular weight, since one Fe^{+2} atom loses one electron. However, in the case of the reduction of the coenzyme FAD, flavin adenine dinucleotide, by oxidizable substrates in the presence of certain dehydrogenase enzymes, two electrons are transferred to FAD. The half equation is: $FAD + 2e + 2H^+ \rightleftharpoons FADH_2$. In this case the equivalent weight for FAD is, therefore, one-half its molecular weight.

All oxidation-reduction reactions involve an electron donor or acceptor species of chemical participants. When equations of half reactions of an oxidation-reduction process are balanced, the same number of electrons transferred must be involved in each equation. Thus, during respiration when $FADH_2$ is acting as a reducing agent and is itself being oxidized, the overall equation for the transfer of electrons from $FADH_2$ to Fe^{+2} cytochrome *c* may be represented as the sum of the reverse forms of these two half reactions. The electrons gained and lost are cancelled out.

$$\text{FADH}_2 + 2\,\text{cyto}_c\text{Fe}^{+3} \rightleftharpoons 2\,\text{cyto}_c\text{Fe}^{+2} + \text{FAD} + 2H^+.$$

It is always necessary to know the exact nature of the reaction in oxidation-reduction stoichiometry, particularly because a given oxidizing or reducing agent may have more than one equivalent weight depending on the reaction.

E. Calculations of Stoichiometry

The practical unit for stoichiometric calculations is the *milliequivalent,* which is one-thousandth of the equivalent weight. One milliliter of a 1.0

normal solution contains one milliequivalent of the solute. This is a convenient and practical unit because one can calculate the number of milliequivalents of reagent present in a given solution simply by multiplying its normality by its volume (in milliliters). For instance, 25.0 ml of 0.0800 N NaOH contain 25.0×0.0800 or 2.00 milliequivalents. Since one milliequivalent of acid is chemically equivalent to one milliequivalent of base, the following relationship is obvious.

$$Normality_1 \times volume_1 = milliequivalents = Normality_2 \times volume_2$$

$$N_1 \times ml_1 = N_2 \times ml_2.$$

The following types of problems may be solved with the above equation:

(**1**) 25.0 ml of 0.0800 N NaOH was exactly neutralized by 32.0 ml of a solution of HCl. What is the normality of the HCl solution?

$$N_1 \times ml_1 = N_2 \times ml_2$$

$$0.080 \times 25.0 = N_2 \times 32.0$$

$$N_2 = 0.0625 \text{ N}$$

(**2**) A sample of 1.201 g of pure acetic acid (CH_3COOH) was dissolved in H_2O and was found to require 40.0 ml of a NaOH solution for exact neutralization. What is the normality of the NaOH solution?

Molecular weight acetic acid = 60.05

Equivalent weight acetic acid = 60.05

1.201 g acetic acid represents $\dfrac{1.201}{60.05} = 0.0200$ equivalents or 20.0 milliequivalents, and since $N_1 \times ml_1 =$ milliequivalents,

$$N_1 \times 40.0 = 20.0$$

$$N_1 = \frac{20.0}{40.0} = 0.500 \text{ N}$$

F. Standard Solutions

In volumetric analysis the concentration of solute in an unknown solution is often determined by measuring that volume of the unknown solution which reacts exactly with a given volume of reagent of an accurately known concentration. For example, the concentration of NaOH in a solution can be determined by finding what volume of HCl of accurately known concentration will exactly neutralize a given volume of the NaOH. The normality of the unknown NaOH solution is then calculated by the relationships described above.

A solution of a reagent of which the concentration is very accurately known and which is, therefore, useful in determining the concentration of an unknown solution is called a *standard solution*. Standard solutions of acids, bases, salts, and oxidizing and reducing agents may be prepared for use in volumetric analysis. It is necessary that the substance used in the preparation of a standard solution be stable, so that the concentration of this reagent will not change rapidly with time. Solutions of NaOH, HCl, NaCl, I_2, $Na_2S_2O_3$, $AgNO_3$, and other compounds are frequently used as standard solutions.

In some cases, standard solutions can be made up directly by weighing the solid substance in question and diluting the solution to an exact volume. This is possible with certain stable compounds having constant composition, such as NaCl, $AgNO_3$, potassium acid phthalate and tris (hydroxymethyl) aminomethane (TRIS). Such reagents are known as *primary standards*. However, with other substances such as NaOH, HCl, and $Na_2S_2O_3$, this approach is not practical because the reagent in question is not available in sufficiently pure and constant form to allow direct weighing. For instance, NaOH is commonly contaminated with variable amounts of Na_2CO_3 and is highly deliquescent; HCl is a gas at room temperature; and $Na_2S_2O_3 \cdot 5H_2O$ is efflorescent (i.e., it loses water of crystallization when exposed to air and hence is of uncertain composition). In these cases it is necessary to make up a solution to approximate the desired concentration and then determine its concentration accurately by *standardization* against a *primary standard solution* of some reagent with which it reacts quantitatively and which can be weighed out with certainty. McGuiness and Clarke (J. Chem. Ed. **45** 740, 1968), however, report on a simple azeotropic method of distillation of HCl that has distinct advantages over titrometric procedures. TRIS can serve conveniently as a standard base and potassium acid phthalate as a standard acid because they can be obtained as stable, pure reagents.

G. Serial Dilution

Serial dilution is often used to obtain solutions whose concentrations of solute vary by some constant amount. This technique is frequently employed to establish a calibration curve in spectrophotometry. For example, one could begin with a stock solution of 1 mg/ml of material to be measured and set up a series of tubes in which the dilution factor (as it refers to the stock solution) progressively changes by one-half. In this illustration, the diluted tubes would be 1:2, 1:4, 1:8, 1:16, and so on. This could be accomplished by pipetting twice the volume of stock solution needed in the first tube, e.g., 2.0 ml, then pipetting one-half that volume (1.0 ml) of suitable diluent (H_2O) into each of the other tubes. The next step involves pipetting one-half the stock solution into the second tube, containing 1.0 ml of H_2O, thus diluting the stock by half. The contents are mixed well by pipetting alternately up and down several times. If the stock concentration was 1 mg/ml, the second tube will now have a solute concentration of 0.5 mg/ml. One-half of the total volume in the second tube is then pipetted into the third tube and the entire mixing procedure repeated. When completed, the third tube will have a solute concentration only one-half of that of the second tube or 0.25 mg/ml. The procedure can be continued for as many repetitions as desired or needed, with one-half the volume of the last tube being discarded to ensure equal volume in all tubes. Although an initial volume of 2.0 ml of stock and 1.0 ml of diluent in each subsequent tube would obviously be the most convenient for the student, various procedures often call for different final volumes and must therefore be taken into account. If the spectrophotometer tubes require 3.0 ml to obtain a reading, then the first tube would have to contain 6.0 ml of stock and the dilution tubes each would contain 3.0 ml of diluent.

Although this technique is quite workable for general use, it is very important to emphasize the necessity for extreme care in pipetting. Each successive dilution will "magnify" a mistake, so each volumetric measurement must be done accurately.

H. Aliquot

You will often meet this term in analytical work. It means "part of the whole." It refers to the practice of taking a small measured fraction of the volume of a given solution for analysis. If a 5.0 ml sample of one liter of a reagent is taken for analysis, the total reagent present in the liter of solution can be calculated from the content of the 5.0 ml sample by simple multiplication. The 5.0 ml sample is called an "aliquot."

I. Dilution Techniques and Calculations

Dilution provides a simple and accurate method for changing the concentration of a solution. It is also a method of indirectly "weighing" a solute whose weight is below the usual accuracy limits of an analytical balance.

In order to understand dilution methods more fully, a knowledge of basic nomenclature is necessary. How would you prepare a 1:10 dilution? The answer to this question depends on which method is employed, but fundamentally a 1:10 dilution means a solution in which the ratio of solute to solvent is one part in 10 total parts. Since grams or milligrams are used as units of weight and liters or milliliters as units of volume, the problem is simplified—we can assume that 1.0 ml of water (the most common solvent) weighs 1.0 g.

There are three ways of preparing a 1:10 dilution. In the *weight-to-weight* system (w:w) you would dissolve 1.0 g of solute in 9.0 g of water, giving 10 total parts by weight of which one part is solute. The *weight-to-volume* (w:v) system is most commonly used. A graduated cylinder is filled with ½ the desired volume of solvent. The solute is added, followed by further dilution to exactly 10.0 ml. In this preparation, one part (by weight) is dispersed in 10 total parts (by volume). If one gram of NaCl is dissolved in 10 ml of water, the solution will weigh 11.0 g but the volume will remain essentially unchanged at 10 ml. Since most solutions are very dilute, the error involved is acceptable.

If the amount of solute added is large, the volume will increase measurably, so the practice in the w:v procedure is to bring the solution to the desired volume with solvent after dissolving the solute in a volume of solvent less than the final volume.

Finally, the third way to prepare a 1:10 solution would be encountered when the solute is a liquid. Suppose, for example, it is desired to prepare a 1:10 solution of pure ethyl alcohol. The usual method would be to add 1.0 ml of ethanol to 9.0 ml of water, resulting in a ten part solution in which alcohol was one part. This would be called a *volume-to-volume* system (v:v). Alternatively, we could mix 1.0 g of ethanol in 9.0 g of water and achieve almost the same solution on a weight-to-weight basis, since alcohol is a little lighter than water. *In any case, the method of preparing a solution should be clearly indicated so as to avoid confusion.*

Per cent concentration means *parts per hundred*, and is commonly employed to indicate solution concentration. Since one part in 10 is the same as 10 parts in 100, the above solutions could be described as "10 per cent" solutions. Each of them differs slightly in actual concentration, so again, clarity is assured only when solution (a) is designated 10 per cent (w:w), solution (b) as 10 per cent (w:v), and solution (c) as 10 per cent (v:v).

J. Changing Concentration

The biochemist is often confronted with the problem of converting a solution of one concentration into another. Obviously the only practical direction of such a conversion would be from a more concentrated solution to one more dilute. Assume that you have 10.0 ml of a 1:5 (20 per cent) solution and you wish to make it 1:20 (5 per cent). One way to approach this problem would be to determine how many times the 20 per cent solution must be diluted to make a 5 per cent solution:

$$\frac{1/5}{1/20} = 4.$$

The new concentration will be achieved by diluting the original 10.0 ml four times, i.e., adding water to a total volume of 40 ml. A 1:5 solution would contain 20 g/100 ml (or 2 g/10 ml) and a 1:20 solution would contain 5 g/100 ml (or 2 g/40 ml).

Examples:
   (1)   Convert 10 ml of a 1:7 dilution of NaCl to a 1:22 solution.
      (a) Calculate dilution factor:

$$\frac{1/7}{1/22} = 3.14.$$

      (b) 3.14 × 10 = 31.4.
      Thus add water to the 10 ml of 1:7 NaCl until the volume is 31.4 ml.

(2) You have a stock solution of 1:20 NaCl and you wish to prepare 2.0 ml of a 1:400 dilution. By using the equation $Vol_1 \times \% Conc_1 = Vol_2 \times \% Conc_2$ a simpler solution of these problems can be achieved.

(a) $\frac{1}{20} \times 100 = 5\%$

$\frac{1}{400} \times 100 = 0.25\%$

(b) $2.0 \times 0.25 = Vol_2 \times 5$

$Vol_2 = 0.1$ ml

Therefore, diluting 0.1 ml of 5 per cent NaCl to 2.0 ml will give the desired concentration.

(3) Prepare 50 ml of 70 per cent ethanol from 95 per cent ethanol.

(a) $50 \times 70 = Vol_2 \times 95$

$Vol_2 = 36.9$ ml.

(b) Therefore, add 13.1 ml of water to 36.9 ml of 95 per cent ethanol to make 50 ml of a 70 per cent solution.

(4) To what volume must 25.0 ml of 0.0800 N HCl be diluted to yield a concentration of 0.05 N HCl?

$$N_1 \times ml_1 = N_2 \times ml_2$$
$$0.080 \times 25.0 = 0.0500 \times ml_2$$
$$ml_2 = 40.0 \text{ ml.}$$

K. Principles of Logarithms

Stoichiometric calculations may be done with the slide rule, but since the ordinary 10 inch slide rule allows calculations to only three significant figures, whereas most operations with standard solutions are carried out to four figures, it is convenient to use logarithms for such calculations. The use of logarithms in calculations will be briefly explained with some examples. Logarithm tables are found in the Appendix.

General

Any number may be expressed in an exponential form. For example: $100 = 10^2$, and 100 may be expressed logarithmically as $\log_{10} 100 = 2$. This states that 2 is the logarithm to the base 10 of the number 100. In general, if $10^x =$ some number y, then $x = \log_{10} y$. If x is given, its antilogarithm y may be found by looking up in a table the number which has x for its logarithm, that is, the antilogarithm y is the number obtained by raising 10 to the x power.

The Characteristic and Mantissa

Logarithms are divided into two parts: the characteristic which comes to the left of the decimal point, and the mantissa which comes to the right. The mantissa determines y, the number representing the antilogarithm, and the characteristic determines the position of the decimal point of the number. For example: In the logarithm 2.5933, 0.5933 is the mantissa and 2 is the characteristic. The antilogarithm of 0.5933, as found in the table, corresponds to the figure sequence 3920. The characteristic indicates that the number belongs in the hundreds, and therefore, the antilogarithm is 392.0.

The rule is that a characteristic of zero corresponds to antilogarithms between 1 and 10, and for each increase of one in the characteristic, the number has one more significant figure at the left of units place. Similarly, for each decrease of one in the characteristic, the first significant figure is moved one place to the right of the decimal point.

Application of Logarithms

The use of logarithms should be clear from the following:

$$\log \quad 3.0 = 0.4771 \quad \text{(see log tables in the Appendix)}$$

$$\log \quad 30.0 = 1.4771$$

$$\log 300.0 = 2.4771$$

$$\log 0.3 \quad = 9.4771 - 10 \quad \text{or} \quad = \overline{1}.4771$$

$$\log 0.03 \quad = 8.4771 - 10 \quad = \overline{2}.4771$$

$$\log 0.003 = 7.4771 - 10 \quad = \overline{3}.4771$$

$$\text{antilog of } 0.3010 = 2.0$$

$$\text{antilog of } 1.3010 = 20.0$$

$$\text{antilog of } 2.3010 = 200.0$$

$$\text{antilog of } 9.3010 - 10 = 0.2$$

$$\text{antilog of } 8.3010 - 10 = 0.02$$

$$\text{antilog of } 7.3010 - 10 = 0.002$$

Problem: What is the logarithm of 1052? It will be seen that the table gives the values of $\log 1050 = 3.0212$ and $\log 1060 = 3.0253$. The desired log must be interpolated from these two figures. A "proportional part," which represents the appropriate fraction of the difference between 3.0212 and 3.0253, is found from the table. The log of 1052 is therefore $3.0212 + 0.0008 = 3.0220$.

Problem: It is found that 10.91 ml of 0.0945 N NaOH are exactly neutralized by 13.42 ml of HCl. What is the normality of the HCl?

$$N_1 \times ml_1 = N_2 \times ml_2$$

$$N_1 = \frac{0.0945 \times 10.91}{13.42}.$$

To multiply two arithmetic numbers, add their logarithms and find the antilog of the sum. To divide, subtract the log of the denominator from the log of the numerator and find the antilog of the difference.

$$\log N_1 = \log 0.0945 + \log 10.91 - \log 13.42.$$

This multiplication and division can be carried out as follows:

$$
\begin{array}{lll}
\log 0.09450 & = & 8.9754 - 10 \\
\log 10.91 & = & 1.0378 \\
\hline
\text{Sum} & = & 10.0132 - 10 \\
\text{Subtract} \quad \log 13.42 & = & 1.1277 \\
\hline
\text{Difference} & = & 8.8855 - 10
\end{array}
$$

Antilog $8.8855 - 10 = 0.07682$ N, the desired normality.

Negative Logarithms

In chemical calculations, negative logarithms are frequently encountered. These cause problems. Logarithms are exponents, and they may be either positive or negative. The mantissas as given in logarithmic tables are always positive. The decimal point between the characteristic and the mantissa *always signifies addition*, as it does in ordinary decimal numbers. When the minus sign is written above the characteristic, it applies only to the characteristic, but when it is written in front of the characteristic, it applies to both mantissa and characteristic. If the logarithm, therefore, is written -3.4777, this means that both characteristic and mantissa are negative. In order to obtain the antilog of this negative, you must first convert the mantissa to a positive value. By adding 1 and subtracting 1 we can perform this conversion.

$$
\begin{array}{l}
\bar{4}.0000 + 1.0000 \\
\qquad\quad - 0.4777 \\
\hline
\bar{4}.0000 + 0.5223 = \bar{4}.5223
\end{array}
$$

This is the correct expression of the logarithm -3.4777.

L. Accuracy and Precision of Measurements and "Significant Figures"

The terms *accuracy* and *precision* are used to denote two different aspects of the validity of a measured quantity. A method may be accurate but not precise, and a method may be precise but not accurate. Ideally a method should be both accurate and precise.

A method is *accurate* when the average of a large number of determinations agrees very closely with the theoretical value of the quantity measured. A method in which single determinations can be very closely duplicated is *precise*. Frequently, an accurate method is not precise. Although the *average* of a large number of determinations may be a correct measure of the quantity involved, the individual determinations may deviate widely from the average. On the other hand, a given method may be very precise, yet be inaccurate because, although readily reproducible, the determination itself does not give a true measure of the quantity in question. In this circumstance, a new methodology should be developed.

In performing the calculations of analytical chemistry, it is helpful to know something about the limits of precision of the various steps involved in the analytical method. Often much time can be saved if it is realized that it is possible to be "over-accurate" at one stage of a quantitative determination. This statement is not intended to give license to carelessness but to point out that an analytical method cannot be more accurate and precise than its *least* accurate and precise step. A particular spectrophotometer in use, for instance, may have an error of 3 per cent. It is pointless to try to obtain a precision of 0.001 per cent in some other step of the procedure, since the whole method can be no more precise than 3 per cent.

The precision of a measurement may be indicated by the number of "significant figures." A measurement expressed as 5.2 units (two "significant figures") indicates that the method is repeatable to about one part in 52 or about 2 per cent. If the measurement is expressed as 5.20 units (three "significant figures"), it indicates that it is reliable to one part in 520 or about 0.2 per cent, and so on. A measurement is best expressed in the number of "significant figures" to which the method is reliable in order to convey the precision of a measurement. Since any given analytical method can be no more reliable than its least precise and accurate step, it is permissible to "round off" all quantities to the same number of significant figures as the least reliable step permits. For instance, it is possible to weigh 100 grams to three decimal places on the analytical balance (100.000 g) giving six significant figures (precision of 1 part in 100,000 or 0.001 per cent). If a solution prepared from this material is then subjected to spectrophotometric analysis with an error of 1 per cent (1 part in 100, three significant figures) the result should be expressed in three significant figures only, all other quantities and calculations being "rounded off" at three significant figures. It is obvious that one need not have weighed out the sample in this case to 0.001 gram. A weighing to 0.1 g would have been sufficient. In "rounding off" numbers, add one to the last significant figure if the rejected figure in the next position is 5 or greater. For instance, 16.557 becomes 16.56 and 16.554 becomes 16.55.

M. Analysis of Statistical Reliability of Data

A number of experiments conducted in biochemistry are subject to analysis by statistical methods. There are many ways of interpreting a measurement. In practical research, statistics are used to give as accurate an answer as possible as to whether a series of measurement is reliable. Because of time limitations, it is difficult to conduct student experiments on a statistically valid basis. As a general rule, the more samples or trials employed, the more nearly correct the data will be, provided the method of measurement is in itself reasonably accurate.

Despite the existence of this time problem, some exposure to simple statistical analysis may be helpful. Most students are familiar with the statistical concept of the *average* (mean). In a series of determinations involving, for example, the height of the individual members of the class, the sum of all heights divided by the total number of people in the class would give an average or mean height (often symbolized as m or $\overline{X}$). If we use the symbol $\sum$ to signify "the sum of," x to signify any experimental value, x_n to represent all values of x, and N to represent the number of trials of values of x, the following formula can be used to calculate the mean $\overline{X}$:

$$\overline{X} = \frac{\sum x_n}{N}.$$

However, the average value does not always signify that most individual measurements have that value. If there are equal numbers of men and women, and all men are six feet tall and all women are five feet tall, obviously no one is the average value of five feet six inches. In order to show the relative close-ness of the average to the values of the measurements taken, the *standard deviation* (σ, sigma) is calculated. Basically the standard deviation describes by how much *average* variation all the sample values differ from the mean, and is therefore an *average deviation* or variation from the average. It may be calculated in the following manner, using some experimental data. The figures given in Table 1 were fluorometric determinations of serum levels of phenyl-alanine in normal individuals.

Table 1

Trials (N)	Value (x)	x^2	$(\overline{X} - x)$	$(\overline{X} - x)^2$
1	2.00	4.00	0.01	0.0001
2	1.82	3.31	0.17	0.0289
3	1.51	2.28	0.48	0.2304
4	3.93	15.44	1.94	3.7636
5	2.20	4.84	0.21	0.0441
6	0.55	0.30	1.44	2.0736
7	1.93	3.72	0.06	0.0036
$\sum$	$\sum = 13.94$	$\sum = 33.89$	$\sum = 4.31$	$\sum = 6.1443$

$$\overline{X} = \frac{13.94}{7} = 1.99 \qquad \text{average error} = \frac{\sum (\overline{X} - x)}{N} = \frac{4.31}{7} = 0.62.$$

The average serum phenylalanine level is expressed as $= 1.99 \pm 0.62$ within the limits of the average error.

A more meaningful expression is given in terms of the standard deviation. At ± 1 standard deviation (σ), 66 per cent of the determinations should fall between its limits; at $\pm 2\sigma$, a 95 per cent confidence level is expected.

Calculation of the standard deviation: The standard deviation is the square root of the variance or σ^2, where

$$\sigma^2 = \frac{\sum (\overline{X} - x)^2}{N - 1}$$

and N = number of samples,

$\overline{X}$ = the average or mean, and

x = individual sample values or measurements.

In this particular case,

$$\sigma^2 = \frac{6.1443}{6}$$

$$\sigma^2 = 1.024$$

and

$$\sigma = \sqrt{1.024} = \pm 1.01.$$

Now the mean is expressed as 1.99 ± 1.01, where 66 per cent of the values should fall within these limits.

The standard deviation can be used to calculate the standard error of the mean, σ_m, which gives some indication of the validity of the experimental values (the expected per cent error).

$$\sigma_m = \frac{\sigma}{\sqrt{N}} = \frac{1.01}{\sqrt{7}} = \frac{1.01}{2.64} = 0.38 \quad \text{or} \quad 38 \text{ per cent error of the mean.}$$

It can be seen that the more samples (N) taken, the lower the expected per cent error. The error is also dependent upon precision; the more precise the determination, the narrower will be the limits of σ, hence the lower the error.

There are many ways to calculate the standard deviation. The method illustrated above is summarized by the following formula.

$$\sigma = \sqrt{\frac{(\overline{X} - x)^2}{N - 1}}.$$

A shorter method where $N < 25$ uses the formula

$$\sigma = \sqrt{\frac{N \sum x^2 - (\sum x)^2}{N(N - 1)}} = \sqrt{\frac{\sum x^2 - \frac{(\sum x)^2}{N}}{N - 1}} = \sqrt{\frac{33.89 - \frac{(13.94)^2}{7}}{6}}$$

$$= \pm 1.01.$$

For $N > 25$ use $\dfrac{1}{N} \sqrt{N \sum x^2 - (\sum x)^2}$.

In the genetic disorder, PKU (phenylketonuria), which leads to mental retardation, the serum phenylalanine is much above the normal level. The question arises: Is this difference statistically significant? In order to evaluate this possibility, data obtained from such individuals must be analyzed. For instance, if it is found that in PKU, $\bar{X} = 27.26$, $\sigma = 7.28$, and $\sigma_m = 1.63$ by analysis of variance and applying the T test, it is possible to evaluate the probability of statistical significance between the two sets of data.

Calculation of the standard error of the mean difference, σ_{md}, gives

$$\sigma_{md} = \sqrt{(\sigma_{m_1})^2 + (\sigma_{m_2})^2} = \sqrt{(1.63)^2 + (0.38)^2} = 1.67.$$

Applying the T test,

$$T = \frac{\bar{X}_1 - \bar{X}_2}{\sigma_{md}} = \frac{27.26 - 1.99}{1.67} = 15.2.$$

Any value for T over 2.7 is regarded as significant. Looking up the probability of this difference in suitable tables, for this value of T, the probability ≤ 0.001, i.e., there is one chance in a thousand that these differences are not statistically significant.

PROBLEMS

1. What is the per cent concentration of (a) 1 M solution of NaCl, (b) 0.15 M Na_2CO_3? *Ans.* (a) 5.854%, (b) 1.59%.

2. To what volume must 25.0 ml of 0.1198 N HCl be diluted to produce a concentration of 0.0900 N? *Ans.* 33.27 ml.

3. Analytical reagent grade HCl can be purchased as a 37 per cent solution. What is the normality? The specific gravity of concentrated HCl is 1.18. *Ans.* 11.9 N. If you assume that the concentration of HCl is on a weight volume basis, you will get an answer of 10.15 N which is incorrect.

4. How would you go about making 1 liter of 1 N HCl from concentrated HCl? The specific gravity of HCl is 1.18. *Ans.* Dilute 83.7 ml of concentrated HCl to 1 liter.

5. The label on a bottle of commercial adenosine triphosphate (ATP), an important coenzyme, indicates that it is 98 per cent pure, contains 3.5 moles of water of hydration per mole of ATP, and is in the disodium salt form. How

would you go about making 10 ml of a 0.03 M solution? The molecular weight of free ATP is 507.2. *Ans.* Dissolve 187 mg in 10 ml of H_2O.

6. In determining the concentration of protein in tissue fluids, a standard solution of pure albumin is used for reference. How would you go about making up 100 ml of 1 mg/ml of albumin in 0.9% NaCl? *Ans.* Add 100 mg of albumin to 100 ml of water containing 900 mg of NaCl, since the volume change would be insignificant.

7. In assays for the formation of acetoacetate by β-hydroxybutyrate dehydrogenase enzyme, a standard solution containing 0.1 μmole per ml of freshly distilled ethyl acetoacetate is converted to the potassium salt. How would you go about making a stock standard of 100 ml to contain 10 μmoles/ml of potassium acetoacetate? The density of ethyl acetoacetate is 1.025 g/ml. *Ans.* 0.127 ml of ethyl acetoacetate is treated with base and diluted to 100 ml.

8. Enzyme assays are carried out either by measuring the rate of appearance of product or the disappearance of substrate (reactant). Succinic dehydrogenase catalyzes the oxidation of succinic acid to fumaric acid. The specific activity of an enzyme is defined in this case as the μmoles of succinate disappearing per minute per mg of enzyme at 37° C. The specific activity of mammalian liver enzyme is 120 μmoles per min/mg of enzyme at 37°. Starting with 40 μmoles of succinate, how many μmoles of substrate will remain after one hour in the presence of 1 μg of enzyme? *Ans.* 32.8 μmoles.

II. BUFFERS AND pH

A. The Concept of pH

Hydrogen ion concentration is of considerable significance for all living organisms. Small changes in the hydrogen ion concentration of blood, for instance, are accompanied by marked changes in the chemical composition of the blood, by marked changes in physiological processes such as respiration, and also by unpleasant subjective changes in behavior. Small changes in the hydrogen ion concentration can affect the efficiency of enzymes as catalysts. As another example, of which there are many, most microorganisms require a rigid control of hydrogen ion concentration in order to sustain an optimum rate of cell division and growth.

That hydrogen ion concentration cannot be measured by ordinary titrimetric procedures is shown by the following example. Equal volumes of 0.1 N acetic acid and 0.1 N HCl give the same titration values with 0.1 N NaOH when phenolphthalein is used as an end-point indicator. In other words, the "hydroxyl ion combining capacity" of 0.1 N acetic and of 0.1 N HCl are identical. However, the hydrogen ion concentration of 0.1 N HCl is approximately 0.1 N, while that of 0.1 N acetic acid is only about 0.001 N. This difference in hydrogen ion concentration is attributable to the fact that

0.1 N HCl is almost completely dissociated while 0.1 N acetic acid is only about 1 per cent dissociated.

Experimental evidence indicates that acetic acid dissociates reversibly according to the following equations:

(1) $$HA + H_2O \rightarrow H_3O^+ + A^-$$

(2) $$H_3O^+ + A^- \rightarrow HA + H_2O.$$

According to the mass action law, the velocities of reactions such as (1) and (2) are proportional to the product of the concentrations of the reacting substances.

Accordingly, we assign v_1 for the velocity of reaction (1) and v_2 for reaction (2).

(3) $$v_1 = k_1[HA][H_2O]$$

and

(4) $$v_2 = k_2[H_3O^+][A^-].$$

At equilibrium the velocities of (1) and (2) must be equal, hence $v_1 = v_2$ and:

$$k_1[HA][H_2O] = k_2[H_3O^+][A^-].$$

From this, we obtain on rearrangement:

(5) $$\frac{k_1}{k_2} = \frac{[H_3O^+][A^-]}{[HA][H_2O]} = K$$

where K is the thermodynamic equilibrium (ionization) constant. This equation may be simplified by a convenient convention. Water is generally present in aqueous systems in much higher concentration than any of the other constituents; pure water is 55.6 molar. Because of its high concentration this figure is essentially unchanged and equation (5) reduces to:

(6) $$K_a = \frac{[H_3O^+][A^-]}{[HA]}$$

where K_a is a new constant which combines the acid ionization constant with that of water. Solving for $[H_3O^+]$ we obtain:

(7) $$[H_3O^+] = K_a \frac{[HA]}{[A^-]}.$$

Take the logarithm of both sides of the equation:

$$(8) \qquad \log[H_3O^+] = \log K_a + \log \frac{[HA]}{[A^-]}.$$

Multiply each side by -1.

$$(9) \qquad -\log[H_3O^+] = -\log K_a + \log \frac{[A^-]}{[HA]}.$$

We define $-\log[H_3O^+]$ as pH and $-\log K_a$ as pK_a. The equation now becomes:

$$(10) \quad pH = pK_a + \log \frac{[A^-]}{[HA]} \quad \textbf{(Henderson-Hasselbalch Equation)}.$$

This equation has been derived by assuming that there is incomplete dissociation of the weak acid, and that for certain substances, depending upon the conditions, the concentration of ionic species is affected by the degree of dissociation. The term activity (α) refers to the effective or apparent concentration of an ionic species, and γ is the activity coefficient of that ion, or the fraction of the concentration that is effective in a given reaction. The product of the concentration of a given species and the approximate activity coefficient consequently is equal to the activity of the species under consideration. The Henderson-Hasselbalch equation can be written as:

$$(11) \qquad pH = pK_a + \log \frac{[A^-] \cdot \gamma_{A^-}}{[HA] \cdot \gamma_{HA}}.$$

Activities of individual ions cannot be measured but concentrations can; accordingly, pK_a and $\log \dfrac{\gamma_{A^-}}{\gamma_{HA}}$ are combined into a new term pK_a' so that Equation (11) reduces to

$$(12) \qquad pH = pK_a' + \log \frac{[A^-]}{[HA]}$$

where K_a' is now the *apparent* equilibrium constant. Experimentally, it is found that the activity of a substance approaches its actual concentration in solution as its concentration approaches zero; that is, as solute becomes more dilute, the activity coefficient approaches unity. It is evident that pK' approaches pK (and that K' approaches K) as the concentrations of the constituents A^- and HA approach zero.

The constant K_a' may differ quite significantly from K_a, and the extent of the difference depends in large part on the total concentration of ions in the solution. Ions tend to attract about them ions of opposite charge. Around any given ion there is an "ion atmosphere" consisting of ions of opposite charge. Thus pK'_a will vary with ionic strength. (Ionic strength is designated by $\Gamma/2$. It will be discussed in more detail in a later section.) In addition, most ionic species are hydrated. These effects are major ones in determining the values of the activities and concentrations and the difference between K_a and K'_a. Further, both K_a and K'_a are temperature dependent. Accordingly, a value of K'_a determined under one set of experimental conditions may differ significantly from the value determined under another set of conditions. Thus in determining K'_a all conditions must be clearly specified.

B. Dissociation of Water; the pH Scale

In the Brönsted-Lowry definitions an "acid" is a species having a tendency to lose a proton and a "base" is a species having a tendency to gain a proton. Examination of Equations (1) and (13) shows that according to these definitions water can act both as a base (proton acceptor) and as an acid (proton donor). (In Equations (1) and (2) HA, the acid, is also a proton donor and A^-, the anion, is a proton acceptor.) According to this definition the ionization of water takes place between conjugate acid-conjugate base pairs.

$$(13) \qquad HOH + HOH \rightleftarrows H_3O^+ + OH^-$$

$$(14) \qquad K_i = \frac{[H_3O^+][OH^-]}{[H_2O]^2}.$$

The ionization constant can be calculated for pure water: $[H^+] = 10^{-7}$ M $= [OH^-]$; therefore, $K_i = \dfrac{10^{-14}}{[55.6]^2} = 3.24 \times 10^{-18}$. A new constant, the ionization product of water, K_w, may be obtained by combining K_i and $[H_2O]^2$:

$$(15) \qquad K_w = K_i \times [H_2O]^2 = 1 \times 10^{-14}.$$

With the conventional notation for the log of the reciprocal we have:

$$(16) \qquad pK_w = pH + pOH \quad \text{since} \quad K_w = [H_3O^+][OH^-].$$

Since $K_w = 10^{-14}$, $pK_w = 14$. At pH = 7, pOH = 7. Expressed in terms of concentration (activity is assumed equal to concentration because of the low concentration) these values are equivalent to 0.0000001 (or 1×10^{-7}) moles per liter. At pH = 13, pOH = 1; in terms of concentrations, $[H_3O^+] = 0.0000000000001$ M or 10^{-13} M, and $[OH^-] = 0.1$ M. The convenience of the pH scale is evident. Most physiological processes occur under conditions where $[H_3O^+]$ is about 0.0000001 M (pH 7). For instance, in severe acidosis, the serum pH of a patient can change from 7.5 to 7.2. This represents a doubling of H^+ concentration, and causes severe illness. A change in pH from 7.5 to 6.5, a ten-fold change, is fatal.

Can you verify the magnitudes of these changes by calculation?

C. Buffer Action

Consider now in more detail the Henderson-Hasselbalch equation (12):

$$pH = pK_a' + \log \frac{[A^-]}{[HA]} \cdot$$

Let $pK_a' = 7.0$ and assign the following values to the ratio $\frac{[A^-]}{[HA]}$: 0.05, 0.1, 0.2, 0.3, 0.4, 0.5, 0.6, 0.7, 0.8, 0.9, 0.95. Calculate pH for each of these values as ordinate with the corresponding value of the ratio as abscissa. Draw a smooth curve through the points. Note that the slope, i.e., the change of pH for a unit change of $\frac{[A^-]}{[HA]}$, is least when $pH = pK_a'$. It will be helpful to estimate directly from your graph the actual change of pH for unit changes in $\frac{[A^-]}{[HA]}$. This small change in pH when $\frac{[A^-]}{[HA]}$ varies is the essence of buffer action. When a weakly dissociable acid is titrated with base, $HA \rightarrow A^-$, and when $[HA] = [A^-]$, at the 50 per cent titration point, $\log \frac{[A^-]}{[HA]} = \log 1 = 0$, and $pH = pK_a'$. This means that the pK_a' of an acid is the pH at which it is titrated by half.

   Turn in your calculations and graph.

A numerical example will illustrate this point further. Suppose a solution contains 0.05 moles of HPO_4^{-2} per liter and 0.05 moles of $H_2PO_4^-$ per liter. (Here $[A^-] = [HA]$, and $pH = pK_a$.) The pK_a' is 6.8 for phosphate ions.

$$pH = 6.8 + \log \frac{0.05}{0.05} = 6.8 + \log 1 = 6.8.$$

Now add to one liter of this solution 10.0 ml of 1 M HCl. It is found that a quantity of $H_2PO_4^-$ equivalent to the added acid, i.e., 0.01 mole, is converted to HPO_4^{-2}. Our equation is now

$$pH = 6.8 + \log \frac{0.04}{0.06} = 6.62.$$

A change of approximately 0.20 pH unit occurs. Now suppose the solution contains 0.02 moles of HPO_4^{-2} per liter and 0.08 moles of $H_2PO_4^-$. Calculate the change of pH occuring when 10.0 ml of 1 M HCl are added to this solution. It will be found that the change in pH is now much greater than 0.20. (Note that if 10.0 ml of 1 M HCl had been added to pure water of pH 7.0, the pH would have decreased to about 2.0.) *The useful* pH *zone for a given buffer system may be set arbitrarily at about* ± 1.0 *unit about the value of the* pK_a' *of the dissociable group.*

The importance of the concentration of the buffer system is sometimes forgotten. A buffer with total concentration of [HA] + [A⁻] of 0.1 will be more "efficient" as a buffer than one with a lesser total concentration.

A *buffer system* can be defined as a solution of free acid and salt (or anion) in water in such proportions that addition of an acid or a base results in a smaller change of pH than would occur if the buffer system were not present. (In the case of the very weak acid ammonium ion, NH_4^+, the species NH_3 corresponds to the anion, e.g., to CH_3COO^- in the acetate system.) It is obvious that a judicious choice of buffer system must be made if solutions are to be "buffered" at different pH and it should be apparent that a system's "capacity" to buffer is determined by *the relative concentrations of the buffer components and the acid or base added.*

D. The Measurement of pH of Weak Acids, Bases and Salts

pH of Weak Acids and Bases

The equation for dissociation of weak acids is

$$HA \rightleftharpoons H^+ + A^-$$

$$K_a = \frac{[H^+][A^-]}{HA} .$$

If all of H^+ and A^- is presumed to come from HA, then $[H^+] = [A^-]$

$$K_a = \frac{[H^+]^2}{[HA]} ; \qquad [H^+]^2 = K_a \times [HA]$$

$$[H^+] = \sqrt{K_a[HA]}$$

or in logarithimic terms:

$$\log [H^+] = \log \sqrt{K_a[HA]}.$$

Changing signs and simplifying:

$$-\log [H^+] = \tfrac{1}{2}(-\log K_a - \log [HA])$$

$$\boxed{pH = \frac{pK_a - \log [HA]}{2}}$$

The equation for dissociation of weak bases is:

$$pOH = \frac{pK_b - \log [BOH]}{2}$$

(Attempt to derive this equation
as above for a weak acid.)

pH of Partially Ionized Salts

NH_4Cl, the salt of *a strong acid and a weak base*, dissociates in water.

$$NH_4Cl + H_2O \rightleftarrows NH_4OH + HCl.$$

Since HCl is dissociated to a greater extent than is NH_4OH, this salt solution will be acidic in aqueous solution.

This reaction can be expressed partially as the hydrolysis constant for NH_4^+.

$$NH_4^+ + H_2O \rightleftarrows NH_4OH + H^+.$$

Therefore, $K_h = \dfrac{[NH_4OH][H^+]}{[NH_4^+]}$. H_2O again can be dropped out by assuming that its concentration is large and does not change, and hence for all practical purposes is constant. However, NH_4OH also dissociates:

$$NH_4OH \rightleftarrows NH_4^+ + OH^-$$

and K_b for this reaction $= \dfrac{[NH_4^+][OH^-]}{[NH_4OH]}$.

By making allowable approximations, a simple expression for the pH of this salt solution can be obtained.

(a) The salt concentration $[NH_4Cl]$ is equal to $[NH_4^+]$, which for dilute solutions is very nearly true.

(b) From the hydrolysis equation, assume $[H^+] = [NH_4OH]$.

(c) Designate $[NH_4^+]$ as $[S]$.

Then K_b may be written

$$K_b = \frac{[S][OH^-]}{[H^+]} .$$

Since both $[OH^-]$ and $[H^+]$ are unknown, it is necessary to remove one of these terms from the expression.

$$K_w = [H^+][OH^-]$$

where K_w includes $[H_2O]$ as part of the constant. Therefore

$$[OH^-] = \frac{K_w}{[H^+]}.$$

By substitution

$$K_b = \frac{[S]\dfrac{K_w}{[H^+]}}{[H^+]} = \frac{[S]K_w}{[H^+]^2}.$$

Solving for $[H^+]$:

$$[H^+]^2 K_b = [S]K_w$$

and

$$[H^+] = \sqrt{\frac{K_w[S]}{K_b}}.$$

Transposing this equation into pH form,

$$\log [H^+] = \log \sqrt{\frac{K_w[S]}{K_b}} = \tfrac{1}{2}(\log [S] + \log K_w - \log K_b)$$

Changing signs:

$$-\log [H^+] = \tfrac{1}{2}(\log K_b - \log [S] - \log K_w)$$

$$\boxed{pH = \frac{pK_w - pK_b - \log [S]}{2}}$$

Furthermore, since

$$K_w = [OH^-][H^+]$$

and

$$[H^+] = \frac{K_w}{[OH^-]},$$

therefore

$$K_h = \frac{\dfrac{[NH_4OH]K_w}{[OH^-]}}{[NH_4^+]}.$$

Since

$$[NH_4OH] = \frac{[NH_4^+][OH^-]}{K_b},$$

$$K_h = \frac{\dfrac{[NH_4^+][OH^-]}{K_b}\dfrac{K_w}{[OH^-]}}{[NH_4^+]} = \frac{K_w}{K_b}.$$

Prove in a similar fashion that

$$K_h = \frac{K_w}{K_a}.$$

The equation for *a salt of a weak acid and a strong base*, e.g., $NaCOOCH_3$, is

$$pH = \frac{pK_w + pK_a + \log [S]}{2}$$

(Attempt to derive: note the change of signs.)

The equation for *a salt of a weak acid and a weak base*, e.g., NH_4COOCH_3, is

$$pH = \frac{pK_w + pK_a - pK_b}{2}$$

(Attempt to derive.)

SUMMARY: The Henderson-Hasselbalch equation is used to determine the pH of buffer solutions.

1. $pH = pK_a + \log \dfrac{[salt]}{[acid]}$.

2. Two forms of equations can be used to calculate $[H^+]$ or pH of weak acids and bases, and their salts:

(a) For acids:

$$[H^+] = \sqrt{K_a[HA]} \qquad \text{or} \qquad pH = \frac{pK_a - \log [HA]}{2} .$$

(b) For bases:

$$[OH^-] = \sqrt{K_b[BOH]} \qquad \text{or} \qquad pOH = \frac{pK_b - \log [BOH]}{2} .$$

3. For salts of a strong acid and a weak base:

$$(a) \quad [H^+] = \sqrt{\frac{K_w}{K_b} [S]}$$

$$(b) \quad pH = \frac{pK_w - pK_b - \log [S]}{2}$$

$$(c) \quad pH = \frac{14 - pK_a - \log [BOH]}{2} .$$

4. For salts of a strong base and a weak acid:

$$(a) \quad [OH^-] = \sqrt{\frac{K_w}{K_a} [S]}$$

$$(b) \quad [H^+] = \sqrt{\frac{K_a K_w}{[S]}}$$

$$(c) \quad pH = \frac{pK_a + pK_w + \log [S]}{2} .$$

5. For salts of weak acids and bases:

(a) $[H^+] = \sqrt{\dfrac{K_w K_a}{K_b}}$ (b) $pH = \dfrac{pK_w + pK_a - pK_b}{2}$

6. (a) $pK_w = pH + pOH$ (b) $pK_b = 14 - pK_a$ (Why?)

(c) $pK_h = \dfrac{pK_w}{pK_a}$ or $= \dfrac{pK_w}{pK_b}$.

Table 2. Acid Dissociation Constants (Aqueous Solutions at 25° C)*

	K_{equil}	pK_b	pK_{a_1}	pK_{a_2}	pK_{a_3}
Acetic acid	1.75×10^{-5}		4.76(4.66, $\Gamma/2 = 0.1$)		
Ammonium hydroxide	1.80×10^{-5}	4.74	9.26		
Arginine			2.17	9.04	12.48
Aspartic acid			2.09	3.86	9.82
Benzoic acid	6.30×10^{-5}		4.20		
Carbonic acid	$K_1 = 4.31 \times 10^{-7}$ $K_2 = 5.60 \times 10^{-11}$		6.37	10.25	
Cysteine			1.71	8.37	10.78
Formic acid	1.77×10^{-4}		3.75		
Fumaric acid	$K_1 = 9.3 \times 10^{-4}$ $K_2 = 3.4 \times 10^{-5}$		3.02	4.32	
Glucose			12.2		
Glutamic acid			2.19	4.25	9.67
Glycine			2.34	9.60	
Guanidinium			13.5		
Histidine			1.82	6.0	9.17
Lactic acid	1.39×10^{-4}		3.86		
Lysine			2.18	8.95	10.53
Malic acid	$K_1 = 4.0 \times 10^{-4}$ $K_2 = 9.0 \times 10^{-6}$		3.40	5.05	
Methanol			17		
Oxalic acid	$K_1 = 6.5 \times 10^{-2}$ $K_2 = 6.1 \times 10^{-5}$		1.19	4.21	
Phosphoric acid	$K_1 = 7.5 \times 10^{-3}$ $K_2 = 6.2 \times 10^{-8}$ $K_3 = 4.8 \times 10^{-13}$		2.12	7.21(6.86,$\Gamma/2=0.1$)	12.32
Propionic acid	1.34×10^{-5}		4.87		
Pyruvic acid			2.49		
Serine			2.21	9.15	
Succinic acid	$K_1 = 6.4 \times 10^{-5}$ $K_2 = 2.7 \times 10^{-6}$		4.19	5.57	

* I. H. Segel, *Biochemical Calculations*. John Wiley and Sons, Inc., N.Y.

PRACTICAL LABORATORY EXAMPLES

Preparation of Buffers. Buffers are mixtures of weak acids or bases and their salts. The choice of buffer depends on the pH range required, and its capacity is dependent on concentration. Some common buffer mixtures are acetic acid-sodium acetate, pH buffer range 3 to 5, mono and disodium phosphate, pH buffer range 6 to 8, and tris(hydroxymethyl)aminomethane, pH buffer range 7 to 9.

(1) Prepare 1 liter of 0.1 M acetate buffer, pH 4.0, from a stock solution of 1 M HAc and 5 M NaOH.

This means that the total concentration of dissociated acetate from the salt and the undissociated acid should be equal to 0.1 M. This is the accepted interpretation of buffer concentration.

Therefore, the milliequivalents of total acetate required for this buffer is $1000 \times 0.1 = 100$.

Let $X =$ acetate contribution from the salt and $100 - X =$ acetate contribution from the acid. Then, since

$$pH = PK_a + \log \frac{[salt]}{[acid]},$$

$$4.0 = 4.76 + \log \frac{X}{100 - X}$$

$$-0.76 = \log \frac{X}{100 - X}; \qquad \text{taking antilog,}$$

$$-1 + 0.24 = \log \frac{X}{100 - X} \qquad \text{(Why?)}$$

$$\frac{X}{100 - X} = 0.175$$

and therefore, $X = 14.9$ me of acetate from the salt. Since the NaOH concentration is 5 M, the volume needed is:

$$5V = 14.9 \text{ me} \quad \text{and} \quad V = 2.98 \text{ ml} \qquad \text{(where } V_1 \times N_1 = \text{milliequivalents).}$$

When the 2.98 ml of 5 M NaOH are added to 100 ml of 1 M HAc, 14.9 ml of HAc will be neutralized and 85.1 ml of HAc will remain as acid.

Finally, 2.98 ml of 5 M NaOH are added to 100 ml of 1 M HAc, and the entire contents are diluted to 1 liter.

(2) Prepare 1 liter of 0.1 M acetate buffer, pH 4.0, from a stock solution of 1 M HAc and 1 M NaAc.

$$4.0 = 4.76 + \log X.$$

Let X = the ratio of $\dfrac{\text{salt}}{\text{acid}}$, and as above

$$-0.76 = \log X; \text{ take antilog}$$

$$-1 + 0.24 = \log X$$

$$X = 0.175.$$

Because the concentrations of both acid and salt are the same it can be assumed that V is the volume of salt required; and since we want 1000 ml of buffer, then $1000 - V$ is the volume of acid. Therefore:

$$\frac{V}{1000 - V} = 0.175; \quad \text{solving,} \quad V_{\text{salt}} = 149; \quad \text{and} \quad V_{\text{acid}} = 851.$$

Since the desired concentration of buffer is 0.1 M, 14.9 ml of 1 M NaAc should be added to 85.1 ml of 1 M HAc and diluted to 1 liter. *As you can see we arrive at the same milliequivalent values for salt and acid as in the previous problem.*

(3) How can you prepare 0.1 M phosphate buffer, pH 7.4, from 1 M NaH_2PO_4 and 1 M Na_2HPO_4 ($pK_a = 7.21$)? If you had the use of a pH meter how would you go about doing this? Dilute each solution 1/10 and add NaH_2PO_4 to Na_2HPO_4 until the desired pH is achieved.

(4) Make up the buffer in Problem 1 to have pH 4.7 at 0° C. First cool the electrodes to 0° C and standardize with buffers whose pH at 0° is known (0.05 M KH phthalate at 0° has a pH of 4.01; the common Beckman pH 7 standard has a pH of 7.12 at 0°). Hold your solutions at 0° during the pH adjustment. Mix cold 0.1 molar acetic acid with 0.1 molar NaAc until pH is 4.7.

Note: A common practice is to adjust and report the pH at 25° C even though the buffer may be used at other temperatures. This practice gives a convenient, reproducible method but cannot be used for certain studies, such as determining heats of ionization or effect of temperature on velocity of reactions.

(5) Prepare sodium acetate buffer, pH 5, of 0.1 ionic strength. Ionic strength is defined as $\Gamma/2 = \frac{1}{2} \sum_i C_i Z_i^2$, where C_i is the concentration of ion species i and Z_i is the charge on that ion. Thus, for monovalent salts the ionic strength is equal to the molar concentration of the salt. (Cf. introduction to Experiment 4 on ionic strength.)

Method 1. Dissolve either 0.1 mole of NaOH or of NaAc in about 800 ml of water. Add acetic acid to pH 5 (using a pH meter). Dilute to one liter.

Method 2. $\mathrm{pH} - \mathrm{pK} = \log \dfrac{\text{salt}}{\text{acid}} = 0.3.$

Therefore, $\dfrac{\text{salt}}{\text{acid}} = 2$, and since salt $= 0.1$ molar then acid $= 0.05$ molar. Add 0.1 moles of sodium acetate to 0.05 moles of acetic acid and dilute to one liter.

(6) Make a phosphate buffer, pH 7.4, of 0.05 ionic strength. Ionic strength $= \frac{1}{2} \sum C_i Z_i^2$; therefore, for all ion species the ionic strength is

(1) $\qquad \frac{1}{2}[\mathrm{Na^+}] + \frac{1}{2}[\mathrm{H_2PO_4^-}] + \frac{4}{2}[\mathrm{HPO_4^{-2}}] = 0.05.$

Since the $\mathrm{pK_a}$ of phosphate is 6.8 at 0.05 ionic strength,

(2) $\qquad \mathrm{pH} - \mathrm{pK_a} = 0.6 = \log \dfrac{[\mathrm{HPO_4^{-2}}]}{[\mathrm{H_2PO_4^-}]}.$

Taking the antilog,

(3) $\qquad \dfrac{[\mathrm{HPO_4^{-2}}]}{[\mathrm{H_2PO_4^-}]} = 4$

and for electrical neutrality

(4) $\qquad [\mathrm{Na^+}] = [\mathrm{H_2PO_4^-}] + 2[\mathrm{HPO_4^{-2}}].$

Substituting Equations 3 and 4 into Equation 1,

(5) $\quad \frac{1}{2}[\mathrm{H_2PO_4^-}] + 4[\mathrm{H_2PO_4^-}] + \frac{1}{2}[\mathrm{H_2PO_4^-}] + 8[\mathrm{H_2PO_4^-}] = 0.05;$

$$13[\mathrm{H_2PO_4^-}] = 0.05.$$

Therefore, $[\mathrm{H_2PO_4^-}] = 0.00385$. Using Equation 3 solve for $[\mathrm{HPO_4^{-2}}]$.

In order to obtain a buffer of the designated ionic strength dilute 0.00385 moles of $\mathrm{NaH_2PO_4}$ and 0.0154 moles of $\mathrm{Na_2HPO_4}$ to one liter.

(7) What is the pH of a solution after the addition of 25 cc of 0.20 N NaOH to 50 cc of 0.14 N HAc?

Find the ratio of salt to acid in milliequivalents (me); e.g., the me of HAc at the start is $0.14 \times 50 = 7.0$; the me of NaOH added is $0.20 \times 25 = 5.0$.

The me of HAc after NaOH addition is 2.0. Since $\mathrm{pH} = \mathrm{pK_a} + \log \dfrac{\text{salt}}{\text{acid}}$, then by substitution:

$$\mathrm{pH} = 4.76 + \log \tfrac{5}{2} = 4.76 + \log 5 - \log 2$$

$$= 4.76 + 0.69 - 0.30$$

$$= 5.15.$$

(8) What is the pH and $[H^+]$ of 0.5 M HAc?

$$[H^+] = \sqrt{K_a[HA]} = \sqrt{1.75 \times 10^{-5} \times 0.5} = \sqrt{8.75 \times 10^{-6}}$$

$$\log [H^+] = \tfrac{1}{2} \log 8.75 + \tfrac{1}{2} \log 10^{-6} = \frac{0.9420}{2} + \bar{3} = \bar{3} + 0.4710$$

$$\text{antilog } \bar{3}.4710 = 0.00296 = [H^+].$$

$$pH = -\log 0.00296 = -(-3 + 0.4713) = 2.53 \qquad \text{(note signs)},$$

or,

$$pH = \frac{pK_a - \log [HA]}{2} = \frac{4.76 - \log 0.5}{2}$$

$$= \frac{4.76 - (-0.31)}{2} = \frac{4.76 + 0.31}{2} = 2.53$$

as a check, since

$$-\log [H^+] = 2.53$$

$$\log [H^+] = -2.53.$$

The characteristic is -2 and the mantissa is -0.53. The mantissa must now be corrected to a positive value.

$$-0.53 = 1 - 0.53 - 1 = + 0.47 - 1 .$$

Therefore

$$\log [H^+] = -2 + (-1 + 0.47) = -3 + 0.47$$

$$\text{antilog } -3 + 0.47 \qquad \text{or } [H^+] = 0.00295 \text{ M}.$$

Alternately

$$-\log [H^+] = \log \frac{1}{[H^+]} = 2.53$$

$$\frac{1}{[H^+]} = \text{antilog of } 2.53 = 3.39 \times 10^2$$

and therefore

$$[H^+] = \frac{1}{3.39 \times 10^2} = 0.295 \times 10^{-2} \text{ M} = 0.00295 \text{ M}.$$

PROBLEMS

1. What is the pH and $[H^+]$ of (a) 0.5 M NaAc (b) 0.5 M NH_4OH (c) 0.5 M NH_4Cl (d) 0.5 M NH_4Ac. *Ans.* (a) $[H^+] = 5.90 \times 10^{-10}$; (b) $[H^+] = 3.31 \times 10^{-12}$; (c) $[H^+] = 1.65 \times 10^{-5}$; (d) $[H^+] = 9.81 \times 10^{-8}$.

2. Calculate the pH of 0.1 M KH_2PO_4. *Ans.* 4.67.

3. Calculate the concentrations of all ionic species of 0.1 M succinate, pH 4.59. *Ans.* 0.0266 M H_2 succinate, 0.0664 M H succinate^{-1}, 0.00698 M succinate^{-2}.

4. At what concentration of a weak acid expressed in terms of K_a will the acid be (a) 10 per cent, (b) 25 per cent, (c) 90 per cent dissociated? *Ans.* (a) $90K_a$, (b) $12K_a$, (c) $0.123K_a$.

5. Calculate the pK_a and pK_b of weak acid with K_a of (a) 6.23×10^{-4}, (b) 7.2×10^{-6}. *Ans.* (a) 3.21, 10.79. (b) 5.14, 8.86.

6. The K_a of a weak acid is 3×10^{-4}. Calculate (a) OH ion concentration, (b) the degree of dissociation of the acid in 0.15 M solution. *Ans.* (a) 1.49×10^{-12}. (b) 4.46 per cent.

REFERENCES

1. Segel, Irwin H. *Biochemical Calculations.* John Wiley and Sons, Inc., New York, 1968.
2. Montgomery, Rex, and Swenson, Charles A. *Quantitative Problems in the Biochemical Sciences.* W. H. Freeman and Co., San Francisco, 1969.
3. Christensen, H. N. *pH and Dissociation.* W. B. Saunders Co., Philadelphia, 1964.
4. Daniels, F. *Mathematical Preparation for Physical Chemistry.* McGraw-Hill Book Co., New York, 1958.
5. Finlayson, J. S. *Basic Biochemical Calculations.* Addison-Wesley Publishing Co., Reading, Mass., 1969.
6. Clark, W. M. *Topics in Physical Chemistry.* 2nd Ed. Williams and Wilkins Co., Baltimore, 1952.

CHAPTER I

THE MEASUREMENT OF HYDROGEN ION ACTIVITY IN AQUEOUS SOLUTION

THE USE OF BUFFER STANDARD AND INDICATORS

This method of estimating pH is based on the fact that a given acid-base indicator exists as two species of different color in a pH region bracketing its pK'. Actually indicators are weak acids capable of existing as the undissociated form HInd and as the anion $(-)$ Ind, which differ in color. By accurately preparing a series of buffer solutions having known pH values grouped around the pK' of the indicator, and containing the same concentration of indicator, a series of color standards is obtained in which the color of the indicator varies with pH. The approximate pH of an unknown solution then can be determined by adding the same concentration of indicator to it as is present in the standards. The color of the unknown is then compared visually or colorimetrically with the colors of a series of standard buffers. In this way it is possible to determine the pH of an unknown to within about 0.2 pH units.

This procedure is subject to certain errors. Protein-containing solutions often show large deviations from true pH because the protein may bind one or the other molecular species of indicator; a shift in the indicator equilibrium is thus brought about. A second error is caused by the presence of another colored component. This can be largely avoided, if the interference is not too great, by carrying out the color comparison with a "blank" tube of the colored substance *without* added indicator.

To estimate pH colorimetrically, it is necessary to have several series of buffer systems of known pH values covering the greater part of the pH scale, i.e., 1 to 13, along with appropriate indicators. No one buffer or indicator can cover the whole scale. As a general rule a buffer system and indicator are useful in a range of one full pH unit on either side of the pK' of the buffer or indicator. For this reason it is necessary to have a series of easily prepared buffer solutions of graded and known pH as well as indicators which have distinctly different colors for the acid and base forms.

THE USE OF THE pH METER

pH is formally defined as $\log \dfrac{1}{[H_3O^+]\gamma_{H_3O^+}}$; or, pH *is the logarithm of the reciprocal of the hydrogen ion activity.* Something approaching the true hydrogen ion activity can be measured by means of the hydrogen electrode. This consists, in principle, of a "platinized" electrode immersed in the solution to be tested. Hydrogen gas is passed through the solution. The "half-cell" thus formed consists of H_2 gas in equilibrium with H_3O^+ to yield an electrical potential. The potential at a hydrogen electrode under one atmosphere pressure of hydrogen in a solution of unit hydrogen ion (i.e., H_3O^+) activity is considered to be zero at 25°C. This is the standard reference electrode. The pH determined with the hydrogen electrode is somewhat arbitrary and conventions have been adopted for the relation of pH to electrical potential. The pH values determined by other procedures and in different laboratories can thus be compared on the same basis, but in all cases the standard solution used relating pH to emf must be specified.

Because the hydrogen electrode is not suitable for estimations of pH in solutions containing oxidizing or reducing substances it is necessary to use another method for such solutions. A convenient method employs the "glass electrode."

The glass electrode is a thin membrane of special glass on one side of which is a silver-silver chloride half cell, and on the other side of which is a solution whose pH is to be determined. The other half of the cell is a calomel electrode which consists of solid Hg_2Cl_2 over metallic mercury. Inside the glass cell containing the Ag-AgCl electrode there is 0.1 M HCl, which establishes a constant and definitely known pH (activity of hydrogen ions) on one side of the glass membrane. A KCl "bridge" connects the solution of unknown pH to the half cell.

It has been found experimentally that the electrical potential of the glass electrode varies with changes in the pH of an unknown solution. Furthermore, the potential is a linear function of pH. That is, if the potential of the glass electrode is plotted along one coordinate axis and pH along the other axis, a straight line is obtained. The slope of the line, $\dfrac{\Delta E}{\Delta pH}$, has the value of 0.0591 volts per pH unit. The equation for the relationship is $pH = \dfrac{E_g - E_{ref}}{0.0591}$ at 25° C. One limitation is that outside the pH range of 1 to 11 the linear relationship does not hold unless the electrode is fabricated of special glass.

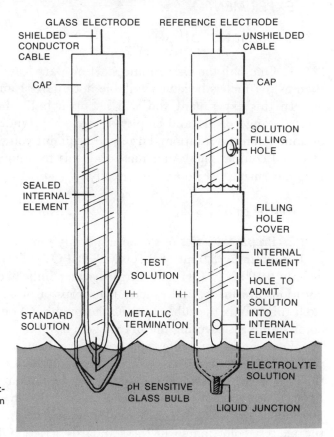

FIGURE I. pH measuring electrodes. (Courtesy of Beckman Instrument Co.)

This fact will serve to emphasize that only relative measurements of hydrogen ion concentration can be made with the glass electrode, since the absolute standard is the hydrogen electrode. However, when the glass electrode is employed two methods of standardizing it are available: by using buffers, the pH of which have been accurately determined by means of the hydrogen electrode, or by using chemicals of such purity that a set of standard buffers of known pH can be made in accordance with proportions stated in the literature, and whose pH has been verified by others.

The slope, $\dfrac{\Delta E}{\Delta pH}$, will have the value of 0.0591 only if the glass electrode and the potentiometer circuit are operating properly. Very often in routine use the glass electrode is standardized against only a single buffer with a pH near that of the unknown. It is obvious, however, that a single point cannot determine this ratio or a straight line. In addition, glass electrodes which have been allowed to dry or become coated with a film of grease may not give the theoretical ratio $\dfrac{\Delta E}{\Delta pH} = 0.06$. For accurate determination of hydrogen ion concentrations two or more standard buffers covering an appropriate range of pH therefore must be used.

EXPERIMENT 1

MEASUREMENT OF pH

1. One half the class should start on part 1 and the other half on part 2; then as pH meters become available both parts 1 and 2 should be completed.

In this experiment you will set up a buffer-indicator series covering a limited range of pH and employ these color standards to determine the pH of some unknown solutions, largely to acquaint you with the general technique of visual comparison. After mastering this technique, you will then compare it to measuring pH potentiometrically.

Work in pairs:

2. Prepare a series of eleven uniformly matched 15×180 mm test tubes, each containing 5.0 ml of 0.100 M KH_2PO_4. To these add the amounts of exactly 0.100 N NaOH and distilled water indicated in Table 3. After these additions are made, add exactly five drops of brom thymol blue indicator to each tube. Mix the tube contents thoroughly. You will now have a series of color standards covering the pH range 6.0 to 8.0.

Four unknown buffer solutions have been made up. Each student is required to determine the pH of these solutions and report his findings to the instructor. The students may practice by making unknowns for each other.

To test the pH of your unknown, add five drops of indicator to exactly 10.0 ml of the unknown in a similar matched test tube, mix well, and find the nearest color match in the standards series. Obviously uniform lighting is necessary. Usually daylight is far better than light from the tungsten lamp.

Work as individuals:

3. It is possible, by making use of the properties of a series of indicators of known pK′ and color change, to estimate pH *without* standard buffers. This is often a desirable preliminary step for approximating the pH range a given unknown falls into before attempting more exact determination by means of buffer standards. The indicators listed in Table 4 are useful for such pH estimations.

Estimate the pH of 0.1 M NaAc, 0.1 M HAc, 0.1 M NH_4OH, 0.1 M NH_4Cl, 0.1 M NH_4Ac, tap water, and distilled water by this means. Use one ml aliquots and a few drops of different indicators and attempt to observe which dye forms an intermediate color for a particular test solution. By careful selection or by a trial and error process, it will be possible to determine whether a given solution is either below or above a certain pH. The pH then can be pinpointed by additional trials with other indicators. This same principle is used with indicator paper, which is dipped into solutions and gives different colors depending upon the pH range of sensitivity of the dye-impregnated paper.

Record your estimate of the pH of these solutions.

Table 3. Phosphate Buffer Standards *

pH	0.1 M KH_2PO_4 ml	0.100 M NaOH ml	H_2O ml
6.0	5.0	0.56	4.44
6.2	5.0	0.86	4.14
6.4	5.0	1.26	3.74
6.6	5.0	1.77	3.23
6.8	5.0	2.36	2.64
7.0	5.0	2.95	2.05
7.2	5.0	3.49	1.51
7.4	5.0	3.93	1.07
7.6	5.0	4.27	0.73
7.8	5.0	4.52	0.48
8.0	5.0	4.69	0.31

* Based on a pK_2 of 6.86 for KH_2PO_4 at $\Gamma/2 = 0.1$

Work in pairs:

4. For this part of the experiment you will use the pH meter. As stated in the introduction, several buffer standards of known pH should be used to calibrate the pH meter. You will be supplied with two carefully prepared buffers which you will use for this purpose. Standardize the glass electrode with the buffer standard labeled pH 6.86. Then, after carefully rinsing the electrodes with distilled water, determine the pH of the other buffer standard. If the pH value you obtain is not within 0.05 pH units of the stated values, consult an instructor. The directions for standardization of the glass electrode and operation of the pH meter will be found on the instrument case.

Table 4. Indicators *

	pK_a	Useful pH Range	Color Change Acid	Color Change Transition	Color Change Basic
thymol blue (acid range)	1.5	0.5–2.5	red	orange	yellow
brom phenol blue	3.98	3.0–5.0	yellow	green	blue
brom cresol green	4.68	3.7–5.7	yellow	green	blue
chlorophenol red	5.98	5.0–7.0	yellow	orange	red
brom cresol purple	6.3	5.3–7.3	yellow	dark green	purple
brom thymol blue	7.0	6.0–8.0	yellow	green	blue
phenol red	7.9	6.9–8.9	yellow	orange	red
thymol blue (alkaline range)	8.9	7.9–9.9	yellow	green	blue
phenolphthalein	9.2	8.2–10.2	colorless	pink	red

* Clark, W. M. *Topics in Physical Chemistry.* 2nd Ed., p. 305. Williams and Wilkins Co., Baltimore, 1952.

After calibration is completed, return to the 0.1 M acid, base and salt solutions used above. This time accurately determine the pH of these solutions. Record your results.

Using the equations provided in the Introduction, calculate the theoretical pH of each of these solutions.

Compare these results by tabulation in a single table and explain the different pH values obtained by the three methods employed.

Verify by calculation whether one or two of the volumes given for NaOH in Table 3 for the given pH are correct. Use the value of pK_2 for phosphoric acid of 6.86. (Why?)

POTENTIOMETRIC TITRATION OF WEAK ACIDS

1. Potentiometric acidimetric titration consists of the stepwise titration of an acid or a base with small portions of a standardized base or strong acid delivered from a burette. The measurement of the electromotive force of the solution in a vessel which contains a glass electrode and a calomel electrode is achieved with a potentiometer. A potentiometer calibrated to read pH directly as well as emf is commonly called a pH meter.

A titration curve is obtained by plotting the pH readings on the ordinate against the milliequivalents of *standard acid or base added* on the abscissa. A smooth line is then drawn through these points. The end point of the titration of a particular species can be recognized by a sharp rise in the slope of the titration curve. In the titration curves of weak acids and weak bases the pK′ values may be estimated graphically, with fair accuracy, within the region of *minimum change of* pH *with addition of acid or base.*

From an inspection of the Henderson-Hasselbalch equation,

$$pH = pK' + \log \frac{[\text{proton acceptor}]}{[\text{proton donor}]} ,$$

it can be seen that pH = pK′ at the 50 per cent point of titration because [proton acceptor]/[proton donor] = 1. The weaker the acid the less distinct will be the end point, and for acids with pK′ > 10, the end point may be so poorly defined as to make an accurate interpretation of the titration curve impossible.

When titrating a mixture of acids or a polybasic acid, it has been found that *the titration curve for each species present is quite distinct provided that their* pK′ *values are more than two units apart* ($K_1 = 100K_2$). If their pK′ values differ by less than two units, the titration curves will be very poorly defined and may overlap to such an extent that there is no point of inflection between the two curves; that is, the titration of the second acid has begun before the titration of the first was complete so that no inflection in the curve indicating an end point for the titration of the first acid occurs. Such an overlap of titration curves would be observed in the titration of a mixture of lactic acid [pK′ = 3.9] and β-hydroxybutyric acid [pK′ = 4.4] and in the titration of succinic acid [pK$_1'$ = 4.18, pK$_2'$ = 5.55]. Protein solutions which have many titratable acid and base groups give a smooth curve, but individual amino acids in aqueous solution, which polymerize to form proteins, give distinctly polybasic curves.

Work in pairs:

2. Each pair of students is assigned one of the following metabolites found in tissue fluid; propionic, lactic, succinic, or malic acids provided as 0.1 M solutions. They will jointly proceed to determine the titration curve of their metabolite. Accurately pipette 50 ml of acid solution into a 150 ml beaker containing a small magnetic stirring bar. If the acid to be titrated is dibasic, use 25 ml of acid in the titration. Place the beaker on the magnetic stirrer and adjust the glass and calomel electrodes into the beaker with care so that there is no danger of breaking the glass electrode with the magnetic stirrer. Do a preliminary titration by adding 5.0 ml increments of the standard 0.1 M NaOH with stirring; record the initial pH and the pH found after each addition. The magnetic stirrer must be off during the actual measurement of the pH; otherwise it will interfere with the measurement. Note the regions of pH where the change of pH is *least* for each 5 ml addition of NaOH. Since one or more of the acids may have been prepared as the sodium or potassium salts in order to facilitate solution, this means that the sample may be either totally or partially neutralized. In order to determine the complete titration curve in these circumstances it will be necessary to add 5 ml amounts of 0.1 N HCl to a second 25 or 50 ml sample.

Proceed with the actual titration. Obtain a fresh sample and titrate it with standard base or standard acid. Add 1.0 ml increments of acid or base in those regions where the *least* change of pH was noted and also at the end points where the *greatest* change of pH took place with the addition of acid or base. Add 2 or 3 ml increments in the other portions of the curve. *It is important that the solutions be properly mixed before you take a reading of* pH *and to record the volume of titrant added each time you read the* pH. On graph paper, plot the pH values on the ordinate and the milliequivalents of standard acid or base added on the abscissa. Draw a continuous smooth line through the points. Now calculate from your data the change of pH per unit increment of acid or base, over a range of half a pH unit on each side of the visually estimated pK′ values. From the ratios of pH/milliequivalents acid or base and visual inspection, estimate as closely as possible the pK′ values.

There are alternate methods for determining the pK′ of weak acids. If the titration has been carried sufficiently far, the plateau of the curve reached at the end point represents the pH at which 100 per cent neutralization has been achieved. A point halfway down, therefore, would correspond to the pK′ of that acid. This becomes apparent from an examination of the Henderson-Hasselbach equation, since at half titration [HA] = [A⁻], and pH = pK′ at this point.

The Henderson-Hasselbalch equation can also be used to arrive at the pK$'$ on a statistically more accurate basis.

(1)
$$pH = pK' + \log \frac{[A^-]}{[HA]}$$

and

(2)
$$pK' = pH - \log \frac{[A^-]}{[HA]}$$

or,

(3)
$$pK' = pH + \log \frac{[HA]}{[A^-]} .$$

If, at each addition of titrant, the pH is measured, and the amounts of HAc and A$^-$ are calculated, as many values for pK$'$ can be obtained as there are increments of added base. These can be averaged and analyzed statistically to give a reliable value for pK$'$ under the conditions employed.

You should endeavor to do this. Express the pK$'$ of your acid in terms of the standard deviation (cf. p. 14).

Compare your pK$'$ values, obtained by each of the simple methods and by statistical analysis, with those provided in Table 2, page 25.

Can you offer some reasons for discrepancies?

From values of the pK$_a$ of the carboxylic acid groups in propionic and lactic acids and succinic and malic acids, what can you say about the influence of the hydroxyl group?

Do you see the inflection due to dissociation of the hydroxyl group? If not, why not?

Can you draw additional conclusions about variations of pK$_a$ of the COOH group from the data in Table 2, page 25?

REFERENCES

1. Clark, W. M. *Topics in Physical Chemistry*. 2nd Ed. Williams and Wilkins Co., Baltimore, 1952.
2. Clark, W. M. *The Determinations of Hydroxyl Ions*. Williams and Wilkins Co., Baltimore, 1928.
3. Bell, R. P. *Acids and Bases, Their Qualitative Behavior*. John Wiley and Sons, Inc., New York, 1952.
4. Bates, R. G. *Electrometric pH Determination, Theory and Practice*. John Wiley and Sons, Inc., New York, 1954.
5. Eisenman, G., Bates, R., Mattock, G., and Friedman, S. M. *The Glass Electrode*. John Wiley and Sons, Inc., New York, 1966.
6. *Laboratory Manual of Physiological Chemistry*. Department of Physiological Chemistry, Johns Hopkins University, Baltimore, 1957.
7. Willard, H. H., Merritt, L. L., Jr., and Dean, J. A. *Instrumental Methods of Analysis*. 4th Ed. Van Nostrand-Reinhold Books, New York, 1965.
8. Albert, A., and Serjeant, E. P. *Ionization Constants of Acids and Bases*. John Wiley and Sons, Inc., New York, 1962.

CHAPTER 2

AMPHOTERIC PROPERTIES OF AMINO ACIDS

ZWITTER ION CONCEPT OF AMINO ACIDS

Since amino acids have at least one amino group and at least one carboxyl group, they must be considered as *ampholytes* (substances which can act either as an acid or as a base). Sufficient evidence has accumulated indicating that amino acids actually exist in fully dissociated form as dipolar ions or "zwitterions" (i.e., "double ions"), $^+H_3N-CH_2-COO^-$ in aqueous solution, rather than in the undissociated form H_2N-CH_2-COOH. Such a molecule can react with acids

(1)　　$\underset{\textit{conjugate base}}{^+H_3N-CH_2-COO^-} + H^+ \overset{k_1'}{\rightleftarrows} \underset{\textit{conjugate acid}}{^+H_3N-CH_2-COOH}$

and with bases

(2)　　$\underset{\textit{conjugate acid}}{^+H_3N-CH_2-COO^-} + OH^- \overset{k_2'}{\rightleftarrows} \underset{\textit{conjugate base}}{H_2N-CH_2-COO^-} + H_2O.$

Titration curves of amino acids may be constructed by the same means you already used in Experiment 1. These curves indicate that an amino acid such as glycine has two dissociation steps, corresponding to the two dissociable groups. The pK′ may be determined readily for each step from a titration curve. Such a diagram also shows that there is a point in the curve where the amino acid behaves as a "neutral" salt. It is the pH at which the net charge on the amino acid is zero. This point is known as the *isoelectric point*. Both dissociable groups are fully but oppositely charged. For practical purposes this point may be taken as half-way between the two points of strongest buffering capacity. The isoelectric point (IEP) is given by:

$$\text{isoelectric pH} = \tfrac{1}{2}(pK_1' + pK_2')$$

where K_1' and K_2' are respectively the dissociation constants of the carboxyl and ammonium groups. pK_1' of glycine is 2.35; pK_2' is 9.78. The isoelectric point $= \tfrac{1}{2}(2.35 + 9.78) = 6.06$. At pH 6.06 glycine has no net charge and both groups are fully "ionized."

Thus by using the Henderson-Hasselbach equation, and knowing the pK' of each dissociable group, it is possible to calculate the ratio of the ion species concentrations at any given pH.

Certain amino acids have more than two dissociable groups, such as glutamic acid, which has two COOH groups and an NH_2 group; $pK_1' = 2.2$, $pK_2' = 4.3$, and $pK_3' = 9.7$. From an analysis of titration curves and correlation with the correct charged species at various pH's, it can be deduced that the IEP is determined by the two dissociable groups having the closest pK_a values. The IEP for glutamate, therefore, is 3.22. At the IEP the amino group exists almost completely as the positively charged ammonium ion. The α carboxyl group is essentially ionized and the γ carboxyl group exists undissociated. Hence at pH 3.22, glutamic acid has one positively charged group (NH_3^+), and one negatively charged γ carboxyl group. The amino acids also have a pH_m or point of maximum charge. In the simple cases pI (the pH at the IEP) $= pH_m$, but particularly for trifunctional amino acids and proteins, pH_m is at a different equivalence point than pI. It is the sum of pK_2 and pK_3, and $pH_m = 7.0$ for glutamic acid.

EXPERIMENT 2

Part A. Titration of Polybasic Amino Acids

Work in groups of four

In this experiment you will titrate an unknown amino acid having more than two dissociable groups. A different amino acid will be given to each student group, and titration will be carried out between pH 1 and 13. Standardize your pH meter, as previously, with standard buffers at pH 4.01 and 6.86. One pair of students titrates 10 ml of 0.1 M amino acid with 0.3 to 0.5 ml increments of standard acid. Record the initial pH and the pH after each addition. Record the volume of titrant added. Try adding some 1 M HCl to get below pH 1.0 and 1 M NaOH to get above pH 11.

The second pair of students titrates 10 ml of amino acid with standard base.

Plot milliequivalents of acid or base added versus pH.

Why can't you get to pH 1 with 0.1 M HCl?

To obtain a true titration curve of any substance, it is necessary to determine how much acid or base is consumed titrating the solvent (H_2O). Titrate a 10 ml sample of water very carefully, with droplets of acid or base (0.005 ml), and determine the pH after each addition until you have reached neutrality. Use this data to correct your curve when you make the plot.

What do you suspect that you might be titrating in the water?

Part B Formol Titration of Glycine

The dipolar character of amino acids makes it difficult to titrate the amino group quantitatively. However, in the presence of a large excess of formaldehyde it is possible to titrate an amino acid to a phenolphthalein end point.

The underlying principles may be illustrated as follows. Isoelectric glycine dissociates as a very weak acid (compare its pK_2' to that of NH_4^+).

$$(1) \quad \overset{\displaystyle R}{\underset{\displaystyle H}{^+H_3N-C-COO^-}} + H_2O \rightleftarrows \overset{\displaystyle R}{\underset{\displaystyle H}{H_2N-C-COO^-}} + H_3O^+.$$

Isoelectric form $\rightleftarrows$ Anionic form
    I   II

Among the components of this equilibrium, formaldehyde (formol) will react only with the amino group of the anionic species. It forms a series of ill-defined compounds which are unstable but are best illustrated as follows:

$$
(2) \quad HCHO + H_2N-\underset{\underset{H}{|}}{\overset{\overset{R}{|}}{C}}-COO^- \rightleftarrows HOCH_2-NH-\underset{\underset{H}{|}}{\overset{\overset{R}{|}}{C}}-COO^-
$$

and further,

(3)

$$
HCHO + HOCH_2-NH-\underset{\underset{H}{|}}{\overset{\overset{R}{|}}{C}}-COO^- \rightleftarrows (HOCH_2)_2-N-\underset{\underset{H}{|}}{\overset{\overset{R}{|}}{C}}-COO^-.
$$

The addition of a large excess (5 to 10 mol equivalents) of neutralized formol results essentially in the removal of the anionic form (species II) and the displacement of the equilibrium reaction from left to right. In effect, the dissociation of the isoelectric form (species I) is increased, i.e., the pK′ of the amino group is *lowered*, to a degree where the hydronium ion "released" in equation (1) can be titrated using phenolphthalein as an indicator.

It is important to note that what is actually *titrated* in the "formol" titration is the NH_3^+ *group, not a* COOH *group*. This is supported by the fact that NH_4Cl, an otherwise "neutral" substance, can be titrated in the presence of formaldehyde with NaOH to give an accurate measure of the NH_4^+ present, whereas formaldehyde added to acetic acid causes no change in its titration curve. It is also important to note that the formol titration begins either at neutral pH or the IEP of the amino acid so that no titration of COOH is possible, since at these pH's the COOH group is already fully deprotonated.

Work as individuals:

Pipette 10.0 ml of the 0.100 M glycine provided into a flask, add 2 drops of phenolphthalein, and titrate *cautiously* to the end point with 0.100 N NaOH. Record this volume. How does it compare with the volume you might expect from the titration of 10.0 ml of 0.1 N acetic acid? At the end point add to your flask 5.0 ml of the 2 per cent neutralized formaldehyde. Determine the volume of 0.1 N NaOH required to bring the partially formy-lated solution to a phenolphthalein end point.

Repeat this titration on a fresh 10 ml sample of glycine but this time using 5 ml of 4 per cent formaldehyde, and repeat with 8, 16, and 32 per cent formaldehyde, each time using a fresh sample of glycine. Proceed directly to the phenolphthalein end point after adding each formaldehyde solution. How do the values compare with the known concentration and volume of glycine you titrated? What concentration of formaldehyde provides the correct stoichiometry? Why should the formaldehyde be neutralized?

Compare the results and explain.

Predict the behavior of glutamic acid and lysine in the presence and absence of formaldehyde.

Note: This formol titration method has some advantages. In crude mixtures of amino acids, other groups dissociate in the pH 7 to 11 range, but by this method the contribution due to amines can be sorted out by differential titration in the presence and absence of formaldehyde.

Part C Paper Electrophoresis of Amino Acids

The net charge on polyelectrolytes, such as amino acids, can serve as a basis for separating one species from another. If a mixture of amino acids is chromatographed in an electrical field in the presence of a suitable support medium, individual amino acids can be separated from one another. These amino acids will have different charge configurations which can be predicted from a knowledge of the pK_a' of the dissociable groups and the pH at which the electrophoresis is carried out. A support medium which allows for rapid separation and convenient identification is cellulose acetate. Cellulose acetate can achieve suitable separations of amino acids in 15 to 20 minutes while paper requires upwards of 16 hours.

In this experiment the class will use formic-acetic acid buffer, pH 2.0, as a medium to separate amino acids electrophoretically. Five amino acids, lysine, phenylalanine, glutamic acid, serine, and alanine, in amounts of 1.0 μliter of 0.01 M solutions, will be spotted on cellulose acetate strips.

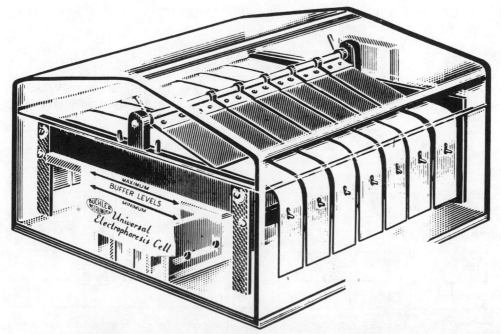

FIGURE 2. Electrophoresis cell. (Courtesy of Buchler Instruments Inc.)

Each student should gain some experience doing the following. Obtain oxoid cellulose acetate strips; handle only at the edges or ends. Use plastic gloves to put six strips on a hard piece of brown paper. Make a thin pencil line one inch from the center perpendicular to the edges. This allows for migration in the direction of the negative pole. (Why?) Be sure that the strip is properly oriented. The up side is that which is stamped with the manufacturers name. Put a code mark at one end of the strip to indicate the substance to be separated. Fill the buffer compartments with the appropriate buffer to the level indicated in the compartment. Very gently pick up the strips with forceps and float them on the required buffer solution. Do not submerge the paper until the buffer has soaked completely through. Once total wetting has been assured gently immerse the strip in the buffer for two minutes. *Uneven wetting will cause poor resolution.* Remove the wetted strips with forceps and place on a double thickness of filter paper. Blot gently and briefly. The paper should no longer appear wet (shiny) but only damp (dull). Set all of the strips in place on the bridge of the electrophoresis apparatus, and across the pan, thus joining the two buffer compartments. Apply tension to the strips and apply 1 μliter of sample to each appropriate strip, using a Pasteur pipette which has been drawn very fine. *One strip is reserved for multiple application of each of the five amino acids.* Be careful to apply the sample on the single straight pencil line; however, do not apply samples to the edge.

Close the apparatus, connect the leads, and set at 300 volts. Run for 15 minutes. Turn off the power supply; disconnect the leads, and remove the strips very carefully. *Make note of which end of the strips are at the cathode or anode.* Place the strips on filter paper and blot them with a second piece. While keeping the strips covered, dry with a heat gun. Spray lightly with ninhydrin reagent and continue heating. The full development of ninhydrin stain takes place at 125° in an oven in a few minutes.

Table 5

Amino Acid	pK$_1$	pK$_2$	pK$_3$	pI	Expected Charge at pH = 2	Electrode to Which Amino Acid Migrated*	
						Observed pH = 2	*Expected* pH = 2
lysine							
phenylalanine							
glutamic acid							
serine							
alanine							

* Indicate to which electrode the amino acid migrated by marking (+) for the anode and (−) for the cathode.

Can all of the amino acids used be identified on the basis of relative mobilities?

Are there color differences?

Do you think this method can be quantitated?

Complete Table 5 on page 44.

Discuss results in terms of the zwitterion concept.

PROBLEMS

1. Calculate how much solid NaOH should be added to 50 ml of 0.1 M glycine or 0.1 M lysine, to prepare a 0.1 M buffer, pH 9.1.

2. What are the charges of glycine, aspartic acid, lysine, and histidine solutions at pH 2, 5, 7, 9, 11? How would you expect these charge species to migrate in an electrical field at these different pH's? Give the order of migration.

3. You have two solutions of monosodium glutamate and glutamic acid at the same concentrations. In what proportions would you mix them to obtain a pH of 4.6? What would the proportions be if you had to start with mono and disodium glutamate?

4. An important amino acid having a third dissociating group is lysine. Instead of two carboxylic acid groups, it has an α and e-NH_2 group. What are the three principal charged species, and at what pH would you expect to find them? Write their structural configuration. Calculate pI and pH_m for lysine.

REFERENCES

1. Scherr, G. H. *Use of Cellulose Acetate Strips for Electrophoresis of Amino Acids.* Anal. Chem. **34** 777, 1962.
2. Block, R. J., Durrum, E. L., and Zweig, G. *A Manual of Paper Chromatography and Paper Electrophoresis.* Academic Press, New York, 1958.
3. *Laboratory Manual of Physiological Chemistry.* Department of Physiological Chemistry, Johns Hopkins University, Baltimore, 1957.
4. Zweig, G., and Whitaker, J. R. *Paper Chromatography and Electrophoresis.* Academic Press, New York, 1967.
5. Cohen, E. J., and Edsall, J. T. *Proteins, Amino Acids, and Peptides.* Reinhold Publishing Co., New York, 1943.

CHAPTER 3

ABSORPTION AND ANALYTICAL SPECTROPHOTOMETRY

PRINCIPLES OF SPECTROPHOTOMETRY

The spectrophotometer is an instrument which measures the intensity of transmitted light and is used extensively for bioanalytical determinations. It consists of two parts: a spectrometer and a photometer. By means of a radiant light source and a monochromator, the spectrometer is designed to emit discrete light frequencies, depending upon the limitations of the particular type of instrument used, but most commonly in the visible region from 400 to 700 nm. The photometer consists of a photoelectric cell sensitive to the range of emitted wavelengths of light, and a galvanometer which records the potentials induced in the photoelectric cell. These potentials are transcribed on a scale which reads out as per cent transmittance or absorbance.

Substances in solution exhibit color because they absorb at certain wavelengths of light and transmit others. Every substance absorbs radiant energy of one wavelength or another. This is a characteristic property of all substances which is as definitive as boiling points, melting points or refractive indices. For example, a solution of hemoglobin appears red to the eye because it is absorbing its complementary colors at the shorter wavelengths in the blue-green region. Certain other substances which may or may not appear colored to the eye can also absorb light in the ultraviolet range below 400 nm or in the infra-red range above 700 nm, each area of which exceeds the limits of detection of the eye. Many more materials are colored in the ultraviolet than in the visible region, and in the infra-red region almost all substances can absorb light. In each of these regions more refined instrumentation is required for measuring these additional properties.

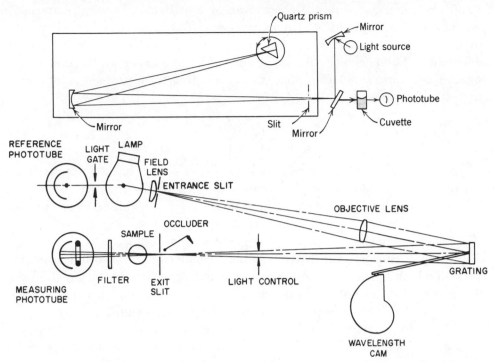

FIGURE 3. Top: Prism-type spectrophotometer. (Courtesy of Beckman Instrument Co.) Bottom: Diagram of light path through Spectronic 20 photometer. (Courtesy Analytical Systems Division, Bausch & Lomb, Inc., Rochester, N.Y.)

The ability of a solution to transmit light is known as the transmittance, T, of the solution. It is defined as the ratio of the intensity of light emerging from a solution, I, to the light entering the solution, I_0.

$$(1) \qquad T = \frac{I}{I_0}.$$

For analytical purposes, since most biochemicals are dissolved in aqueous solution, this fundamental relationship is modified to read: The transmittance

FIGURE 4. Transmittance of a solution.

T_c of a solution containing a concentration c of light–absorbing material is the ratio of I_e, the intensity of light emerging from a solution containing absorbant to I_r, the intensity of light emerging from a *reference solution* without absorbant. The reference solution is defined as containing all the material in the sample solution, including solutes but excluding the absorbant substance of interest. Therefore:

$$(2) \qquad T_c = \frac{I_e}{I_r}.$$

Since transmittance is a relative measure, it always has a value less than one in the presence of light absorbing material. If the intensity of entering light is reduced by half upon emergence, then $T_c = 0.5$. This may be expressed as per cent T_c when I_r is taken as 100%.

The Beer-Lambert law governing these relationships states that light absorption is proportional to the number of molecules of absorbing material through which light passes. Absorbance, therefore, changes with the thickness of the solution and with the concentration of absorbant in a manner characteristic for each absorbing material at a given wave length.

By assuming that the cell, which is used to measure absorbance, is of constant thickness it is possible to simplify the mathematical expression for this law in logarithmic form. The mathematical expression for the Beer-Lambert law at a given wave length is:

$$(3) \qquad -\frac{d(I_e)}{I_r} = \epsilon \cdot \ell \, dc$$

where ϵ is the proportionality constant, or as it is commonly known, the extinction coefficient. This differential equation states that the transmittance of a solution containing absorbant (i.e., the ratio of transmitted light, I_e, to incident light, I_r) is proportional to the concentration of absorbant in a cell of finite thickness ℓ.

Integrating between the limits I_r and I_e, and c_0 and c:

$$(4) \qquad -\int_{I_r}^{I_e} \frac{dI_e}{I_r} = \epsilon \cdot \ell \int_{c_0}^{c} dc.$$

Then

$$(5) \qquad -[\ln I]_{I_r}^{I_e} = \epsilon \cdot \ell [c]_{c_0}^{c}$$

and

$$(6) \qquad -[\ln I_e - \ln I_r] = \epsilon \cdot \ell [c - c_0].$$

Simplifying, since $c_0 = 0$,

$$(7) \qquad -\left[\ln \frac{I_e}{I_r}\right] = \epsilon \cdot \ell \cdot c$$

or to eliminate the negative sign:

$$(8) \qquad \ln \frac{I_r}{I_e} = \epsilon \cdot \ell \cdot c$$

and to the base ten:

$$(9) \qquad -\log T_c = \epsilon \cdot \ell \cdot c = \text{Absorbance}$$

where A (the absorbance) is the negative logarithm of T_c; ϵ is the molar extinction coefficient expressed as the logarithm to the base 10 at a given wavelength; ℓ is the length of the path in cm through which light passes, and is a constant (usually $= 1$ cm); and c is the concentration of absorbing substance in moles per liter.

When A is plotted versus c, a straight line should be obtained, because absorbance is directly proportional to concentration. The slope of the line is equal to $\epsilon\ell$ and the y intercept is zero. Since in general practice the length of the light path is constant ($\ell = 1$ cm) when calibrated spectrophotometric cuvettes are used, the value of ϵ can readily be ascertained.

Deviations from the Beer-Lambert law result from one or both of the following: (1) the light source is not completely monochromatic; (2) the wavelengths at which measurements are taken are not points of maximum absorbance, or the absorbant changes properties on dilution, i.e., it ionizes, dissociates, or combines with the solvent.

The absorbance spectrum of a particular substance is determined by its chemical and physical properties. The wavelengths of maximum and minimum absorbance can be ascertained, and this spectrum is characteristic of that particular substance for the conditions under which the spectrum was obtained. Variation of these conditions, i.e., changes of pH, oxidation-reduction states, temperature, absorbance properties of the curvettes themselves, as well as solvent employed, and other factors can affect the difference spectrum. These difference spectra, or the spectra of modified versus unmodified absorbant, are of particular significance for interpreting the properties of the absorbant. All ions, dyes, organic substances and biochemicals have characteristic absorption spectra. A simple ketone absorbs between 300 and 400 nm and in the infra-red region at 5.8 μ (or 5800 nanometers); chlorophyl absorbs at 440 to 480, 610, 640, and 680 nm. Each of these spectra is subject to shifts depending upon conditions.

The determination of the molar extinction coefficient at a given wavelength of maximum absorbancy is also of importance. This constant can be used as a means of identification as well as for measuring the concentration of the absorbent material. For instance, nicotinamide adenine dinucleotide (NAD^+) is an important coenzyme in metabolic oxidation-reduction reactions.. It is colorless in solution so far as the eye is concerned. When this compound is in the reduced state (NADH), its absorbance at 340 nm is frequently used as a means of assaying dehydrogenase enzyme activity.

The molar extinction coefficient at 340 nm reported for NAD^+ is 6.22×10^3. Therefore, when a 1 molar solution of this substance is put in a 1.00 cm light path of 340 nm wavelength light, its absorbance will be 6220. However, no instrument can read absorbances in this very extreme range, and dilutions have to be made in order to obtain an accurate reading.

As is frequently the case in dealing with biochemicals, when the molecular weight is not known, the extinction coefficient is expressed as that for a 1 per cent solution.

Example: A 1/1000 dilution of NADH has an absorbancy of 0.124 at 340 nm. What is the concentration of the original solution expressed in μmoles NADH per ml?

$$\frac{1 \text{ M}}{6.22 \times 10^3} = \frac{X}{0.124 \times 10^3}$$

$$X = \frac{0.124 \times 10^3}{6.22 \times 10^3}$$

$$X = 0.02 \text{ M.}$$

Since a 1 molar solution contains one mole of substance per liter or one millimole per milliliter, this solution contains 20 μmoles per ml. (To convert a molar solution to μmoles/ml, multiply by 1000.)

Example: Cytochrome *c*, a respiratory chain protein, has a molecular weight of approximately 13,000; a 1×10^{-5} M solution has been observed to have an absorbance maximum of 0.197 at 450 nm. What is the molar extinction coefficient at this wavelength? What is the concentration of cytochrome *c* in mg per cent?

$$A = \epsilon \cdot \ell \cdot c; \quad \epsilon = \frac{A}{\ell \cdot c}; \quad \epsilon = \frac{0.197}{1 \times 10^{-5}} = 1.97 \times 10^4.$$

A 1×10^{-5} solution will contain 0.130 g dissolved in one liter of water or 13 mg in 100 ml; therefore, the solution of cytochrome *c* is 13 mg per cent.

What are the units for the extinction coefficient?

For the determination of an unknown concentration of a substance the value for the unknown is read from a standard curve of A at known concentrations of absorbant.

Frequently, however, in many routine analytical procedures a single solution of known concentration is run along with the unknown after the standard curve has been determined. Since concentration is directly proportional to A,

$$\frac{C_u}{A_u} = \frac{C_s}{A_s}$$

where C_u = concentration of unknown; $A_u = A$ of unknown; C_s = concentration of standard; and $A_s = A$ of standard. Solving for C_u:

$$C_u = \frac{C_s A_u}{A_s}.$$

EXPERIMENT 3

Part A The Absorption Spectrum of Phenol Red

Work in groups of four:

In this experiment you will measure light absorption by the indicator dye, phenol red, at various wavelengths. From your experimental values for absorbance you will construct a graph of the absorption spectra of both the proton acceptor and donor forms of this weakly acidic dye.

An acid-base indicator such as phenol red can exist in two forms, as a proton acceptor and as a proton donor. Actually phenol red can exist in four different forms, depending upon pH. We are interested here in only the red and yellow species which undergo color change in the pH range of its pK_a'.

$$HA \rightleftharpoons H^+ + A^-.$$

yellow   *red*

Prepare three solutions as indicated in Table 6, using distilled water and clean, dry test tubes. Pour enough of each solution into a calibrated cuvette to fill it to within a few millimeters of the top. *Do not touch the cuvettes lower than a centimeter from the top, and do not drop them. Scratches and fingerprints will interfere with the measurements.* When the three cuvettes have been filled, wipe the outsides with a *lens tissue* to remove any solution which may have spilled.

Note that there is a mark etched on the side of the cuvette near the top, and a similar mark on the cuvette holder in the spectrophotometer. These two marks must be aligned each time the cuvette is placed in the holder. In this way, errors caused by differences in path length (because of non-uniform diameter of the cuvettes) is eliminated; also, any scratches made by the holder will not be placed in the light path.

Place tube 1 in the spectrophotometer, set the wavelength scale to 410 nm, and adjust the instrument to read zero absorbance. Determine and record the absorbance of tubes 2 and 3. Repeat this procedure at 10 nm intervals,

Table 6

	Reagents	Tube Number 1	Tube Number 2	Tube Number 3
A.	H_2O, ml	10	1.0	1.0
B.	1×10^{-4} M phenol red, ml		1.0	1.0
C.	0.1 M acetate buffer, pH 5.8, ml		8.0	
D.	0.1 M borate buffer, pH 9.8, ml			8.0

using tube 1 at each new wavelength to set the instrument at zero. Record the data.

On the graph paper plot λ in nm along the horizontal axis (abscissa) and absorbance along the vertical axis (ordinate) for tube 2.

Draw a smooth curve through the points obtained. Do the same for tube 3 on the same piece of graph paper. The two curves represent the absorption spectra for the proton donor and proton acceptor forms of phenol red. Note that the proton donor form has an *absorption maximum* between 400 and 450 nm, while the acceptor form has an absorption maximum between 500 nm and 600 nm. Note also that at the absorption maximum for the acceptor form, the donor form absorbs light very weakly. The quantitative determination of a colored material by spectrophotometric methods is usually conducted at the absorption maximum for the compound under study. You will make use of this principle in the next experiment in the determination of the pK_a' of phenol red.

Calculate the molar extinction coefficient of phenol red at λ_{max}.

Part B   Spectrophotometric Estimation of the pK_a' of Phenol Red

INTRODUCTION. As has been pointed out in the previous section, phenol red is a weak acid and the proton donor form dissociates reversibly in solution into a proton and a proton acceptor form:

$$HA \rightleftharpoons H^+ + A^-$$
$$\underset{yellow}{} \qquad \underset{red}{}$$

This dissociation can be expressed in terms of the usual Henderson-Hasselbalch equation:

$$(1) \qquad\qquad pH = pK_a' + \log \frac{[A^-]}{[HA]}$$

Let $[T] = [A^-] + [HA] =$ sum of the acceptor and donor forms of the acid, *or the total quantity of the acid present.* Then $[HA] = [T] - [A^-]$.

Substituting in Equation (1) we obtain

$$(2) \qquad\qquad pH = pK_a' + \log \frac{[A^-]}{[T] - [A^-]}$$

This is a linear equation of the form $y = ax + b$, where $b = pK_a'$; $a = 1$; $y = pH$; and $x = \log \dfrac{[A^-]}{[T] - [A^-]}$. As a consequence, if pH were plotted as ordinates and the corresponding values for $\log \dfrac{[A^-]}{[T] - [A^-]}$ were plotted as abscissas, a straight line intersecting the vertical axis at pK_a' should be obtained. (Why?)

In the preceding section, you have determined the absorption spectra of the acceptor and donor forms of phenol red, and have found that the acceptor form has an absorption maximum at about 550 nm. At this wavelength, the donor form absorbed very little light, and the absorption at 550 nm can therefore be assumed to be almost entirely due to the acceptor form. At 550 nm, the absorption of light by phenol red follows the Beer-Lambert law. This is:

(3) $$\text{Absorbancy} = \epsilon[A^-]$$

where ϵ includes the constant for the light path.

Therefore, if a series of measurements of the absorbance at 550 nm for various known concentrations of $[A^-]$ is plotted as ordinate versus the corresponding values for concentration as abscissa, a straight line is obtained. From this *standard curve* it is possible by measuring the absorbance to determine the concentration of A^- in a solution of unknown concentration, or to measure the proportion of A^- to HA present in a solution in which the total concentration of indicator is known. The determination of the concentration of A^- at several known pH values in a solution of *known total dye concentration* would yield a curve from which the pK_a' of the indicator may be obtained.

PROCEDURE: Prepare in your matched colorimeter tubes solutions containing the components indicated in Table 7.

Set the wavelength scale of the spectrophotometer to 550 nm. With tube 1 in place, set the instrument to read zero. Then determine and record the absorbance of the remaining tubes. Plot on graph paper the values obtained for the absorbance (ordinate) versus the molar concentration of phenol red (abscissa). Show your results to an instructor before proceeding.

What type of curve would be obtained if per cent T_c were plotted against concentration?

What dilutions have been made in this series?

Table 7

Reagents	Tube Number						
	1	**2**	**3**	**4**	**5**	**6**	**7**
0.1 M borate buffer, pH 9.8, ml	8.0	8.0	8.0	8.0	8.0	8.0	8.0
1×10^{-4} M phenol red, ml		0.1	0.2	0.4	0.6	0.8	1.0
H_2O, ml	2.0	1.9	1.8	1.6	1.4	1.2	1.0
molar concentration of phenol red		1×10^{-6}	2×10^{-6}	4×10^{-6}	6×10^{-6}	8×10^{-6}	10×10^{-6}
Absorbance at 550 nm							

Table 8

	Reagents	Tube Number						
		1	2	3	4	5	6	7
A.	0.1 M phosphate buffer, pH 7.2, ml	8.0	8.0					
B.	0.1 M phosphate buffer, pH 7.6, ml			8.0				
C.	0.1 M phosphate buffer, pH 7.8, ml				8.0			
D.	0.1 M borate buffer, pH 8.0, ml					8.0		
E.	0.1 M borate buffer, pH 8.4, ml						8.0	
F.	0.1 M borate buffer, pH 8.8, ml							8.0
G.	H_2O, ml	2.0	1.0	1.0	1.0	1.0	1.0	1.0
H.	1×10^{-4} M phenol red, ml		1.0	1.0	1.0	1.0	1.0	1.0
I.	absorbance at 550 nm							
J.	concentration of A^-, molar							
K.	$\dfrac{[A^-]}{[T] - [A^-]}$							
L.	$\log \dfrac{[A^-]}{[T] - [A^-]}$							

Have one of the other groups provide you with one of their solutions as an unknown and determine the concentration. Verify your answer. What is the reason for discrepancy if any?

The pK_a' of phenol red is in the vicinity of 7.8. Prepare a series of matched colorimeter tubes containing the components indicated in Table 8. Set the wavelength scale of the spectrophotometer to 550 nm. With tube 1 in place, set the instrument to read zero absorbance.

Measure and record the absorbance values of the remaining tubes. From your standard curve for the acceptor form of phenol red, convert the absorbance values to concentrations of A^-. Calculate the values of the quantities in rows K and L. On graph paper plot pH as ordinate versus the corresponding value from row L as abscissa. Draw the best straight line through the points obtained and extrapolate the line to the vertical axis. The point of intersection of the curve and the pH axis is the pK_a' of the indicator. Note that extrapolation to the vertical axis implies that at the intersection $\log \dfrac{[A^-]}{[T] - [A^-]}$ has the value zero. Consequently $\dfrac{[A^-]}{[T] - [A]}$ must be equal to unity at this point. As you already know, $pH = pK_a'$ when the proton donor and acceptor forms of a weak acid are present in equal concentration.

Report your results; provide an example of your calculations.

Part C   Cytochrome c Reduction

You have observed a shift in the absorption spectrum of a dye as a function of pH change. You will now observe changes in the absorption spectrum of cytochrome *c*, a respiratory pigment hemoprotein, resulting from changes in the oxidation-reduction state of this molecule. Cytochrome *c* is an enzyme of the respiratory chain involved in electron transport during the oxidation of carbon compounds by tissue. It is the only one of the cytochromes which has been isolated from the cell in very highly pure form. In this process it receives electrons from a carrier and is thereby reduced. It is subsequently re-oxidized by another respiratory component to which it passes its electrons. This process of reduction and re-oxidation will be examined with the spectrophotometer. From the absorption spectrum, the maxima of various peaks and the isosbestic points will be indicated.

Place the following in a 1 cm length cuvette: 2.3 ml of H_2O, 0.1 ml of 0.01 M $K_3Fe(CN)_6$, 0.5 ml of 0.2 M phosphate buffer, pH 7.4, and 0.1 ml of 1.5×10^{-4} M cytochrome *c*. An absorption spectrum between 380 and 630 mμ is taken against a blank containing the above reagents *without* cytochrome *c*. A few crystals of $Na_2S_2O_4$ are added to each cuvette with thorough mixing; readings are taken at 550 nm until there is no increase in absorbance. This is to avoid excesses of $Na_2S_2O_4$, which absorbs below 390 nm. When this is accomplished, a spectrum of reduced cytochrome *c* should be recorded.

A different spectrum of the oxidized and reduced forms may be obtained by using a reference cuvette containing the oxidized form of cytochrome *c* measured versus the reduced form. Success in this attempt will depend on the quality of your spectrophotometer.

Prepare a third cuvette containing oxidized cytochrome *c* as above. Use this tube as a blank and determine the difference spectrum against the tube containing reduced cytochrome *c*.

In all cases record the maxima and minima and check these against literature values.

If time and facilities permit, a demonstration of the use of a recording spectrophotometer should be attempted with these reagents.

PROBLEMS

1. A solution of 1×10^{-5} M Na_2ATP, the coenzyme adenosine triphosphate, has $T_c = 70.2$ per cent at 260 nm in a 1 cm cuvette. What would be the T_c at 0.5×10^{-5} M, 0.75×10^{-5} M, 1.5×10^{-5} M, and 2.0×10^{-5} M? *Ans.* 83.8 per cent, 76.8 per cent, 58.8 per cent, 49.2 per cent.

2. Plot T_c versus concentration for Na_2ATP. Is the curve linear? If not, why not? Calculate the absorbance and plot it versus concentration. Is the curve linear?

3. Determine ϵ for $\ell = 1$ cm for 1 M Na_2ATP from the slope of the curve obtained in problem 2. Verify your determination from literature values.

4. A uniform suspension of 50 μg/ml of *Micrococcus lysodeikticus* bacterial cell walls in phosphate buffer has an absorbance of 0.4 in a 1 cm cuvette

measured at 450 nm. Since low concentrations of this cell wall suspension obey the Beer-Lambert law, what is the concentration of a suspension having an absorbance of 0.28 at 450 nm? *Ans.* 35 μg/ml.

REFERENCES

1. Penzer, G. R. *Applications of Absorption Spectroscopy in Biochemistry.* J. Chem. Ed. **45** 693, 1968.
2. Segel, Irwin H. *Biochemical Calculations.* John Wiley and Sons, Inc., New York, 1968.
3. *Laboratory Manual of Physiological Chemistry.* Department of Physiological Chemistry, Johns Hopkins University, Baltimore, 1957.
4. Lemberg, R., and Legge, J. W. *Hematin Compounds and Bile Pigments.* Interscience Publications, John Wiley and Sons, Inc., New York, 1949.
5. Keilin, D., and Slater, E. C. *Aspects of Enzyme Research: Cytochrome.* Brit. Med. Bull. **9** 89, 1953.

CHAPTER 4

THE PREPARATION, CHEMICAL AND PHYSICAL PROPERTIES, AND QUANTITATIVE DETERMINATION OF PROTEINS

ELEMENTAL STRUCTURES, FUNCTION, AND NOMENCLATURE OF PROTEINS

Proteins are high molecular weight biopolymers of amino acids. There are 20 different naturally occurring amino acids found in proteins. Except for glycine, each has an *L*-configuration of the amino group on the alpha carbon atom. The proteins range in size from a molecular weight minimum of 6000 to a maximum of several million. Proteins vary in shape. They can be fibrous as silk fibroin or globular in shape as albumin. The basic shape is determined by peptide bonds formed between each pair of amino acids in a specific sequence. Other intermolecular forces, particularly hydrogen bonding, between amino acid residues within a single chain or between overlapping, interwinding, and parallel chains, determine secondary and tertiary structure. Changes in physical, chemical, and biological properties, i.e., the changes due to the denaturation of proteins, are associated with changes of non-covalent binding forces which ultimately determine the shape or geometry of the protein molecule.

When three amino acids are covalently linked as peptides, as for example, alanine, glycine, and tyrosine, the number of possible structural isomers is 3! or 6. The six isomers may be indicated by the sequences AGT, ATG, GAT,

GTA, TAG, TGA. A pentapeptide can exist in 120 arrangements and a decapeptide can have 3,628,800 structural isomers. Since there are at least 20 amino acids found in proteins, then the possible number of different proteins, each containing all 20 amino acids, is 20! or approximately 2.4×10^{18}. If we assume an unlimited molecular weight for a polypeptide chain, or if the chain can contain any number of repeated amino acids, then the number of structural isomers is infinite. The known number of proteins isolated and uniquely identified from natural sources is in the order of 4000. Consequently, there are an infinite number of possible combinations of amino acids in proteins, and those few already isolated are found to have varied physical, chemical, and physiological properties. They can be found in pure form as relatively inert and insoluble materials, such as hair and finger nails, or they can be crystallized as globular proteins, either as enzymes or other functional types of proteins.

All known enzymes, certain hormones, toxins, transport agents and many building blocks of subcellular structures are proteins. The typical cell may contain several thousand different types of proteins. A single cell has been found to contain as few as one or as many as a million molecules of a particular protein. Yet there is surprisingly little variation in the amino acid composition of many proteins; most have all 20 present. The folding of the polypeptide chains, which is mainly determined by the amino acid sequence, allows a wide variety of unique configurations believed to be largely responsible for a multiplicity of functional, chemical and physical properties. The latter, in turn, permit the purification or even crystallization of many proteins, an achievement which is of considerable importance in research.

Attempts to classify proteins on the basis of properties has led to ambiguity. A classification that has received recognition was suggested by a national committee in 1907. On the basis of the committee recommendation, proteins were divided into two general classes.

1. The *simple proteins* yield only amino acids on hydrolysis. Subgroups include albumins, globulins, prolamines, glutelins, certain enzymes and scleroproteins, keratins, collagens, myosins, silk fibroins, fibrinogens, and elastins.

2. The *conjugated proteins* contain a protein molecule bound to a non-protein prosthetic group. Sub-groups include nucleoproteins, lipoproteins, mucoproteins, chromoproteins, metaloproteins, flavoproteins, hemoproteins, glycoproteins, and others.

As technology advances, however, it is being increasingly shown that more and more simple proteins have other molecules bound to their structure, though not necessarily covalently or ionically bound to them. In fact, the nature of some of these associations are not clearly understood. Attempts at this juncture, therefore, to adhere to a rigid classification scheme, although of some value, seem unnecessary and premature.

SOLUBILITY PROPERTIES OF PROTEINS

The solubility of proteins is determined not only by the variety of residual groups of amino acids and the manner of folding of the peptide chain, but also by the properties of the solvent system in which the protein is contained. The four major factors affecting solubility all influence the secondary and tertiary structure of proteins.

These are (a) ionic strength, (b) pH, (c) temperature, and (d) the dielectric constant of the solvent. Peptide bond hydrolysis at the pH extremes, of course, will alter primary structure as well.

"Salting-out" and "Salting-in": Effects of Ionic Strength. The solubility of all ionic solutes, whether protein in nature or simple inorganic salts, is greatly affected by the presence of other ions. Each ion in solution is surrounded by an "ion atmosphere" of oppositely charged ions, and this ion atmosphere may cause great changes in the solubility of an ionic substance of interest.

Although the effect of neutral salts on protein solubility has often been measured in terms of "per cent saturation" of a solution with a salt such as ammonium sulfate, actually the effects of neutral salts on solubility of proteins have nothing to do with how saturated a solution is with respect to the neutral salt but rather are more closely related to the *ionic strength* of the solution.

Ionic strength is given by the equation

$$\text{Ionic strength} = \Gamma/2 = \tfrac{1}{2} \sum C_i z_i{}^2$$

where

$$C = \text{molar concentration of the ion}$$

$$z = \text{number of charges on the ion}$$

and the sign $\sum$ designates the *sum* of the products $C_i z_i{}^2$ for each ion present in the solution.

Examples: What are the ionic strengths of 0.2 M NaCl and 0.2 M Na$_2$SO$_4$?

For 0.2 M NaCl,

$$\begin{aligned}
\Gamma/2 &= \tfrac{1}{2}(C z_{\text{Na}}{}^2 + C z_{\text{Cl}}{}^2) \\
&= \tfrac{1}{2}(0.2 \cdot 1^2 + 0.2 \cdot 1^2) \\
&= \tfrac{1}{2}(0.2 + 0.2) \\
&= 0.2.
\end{aligned}$$

For 0.2 M Na$_2$SO$_4$,

$$\begin{aligned}
\Gamma/2 &= \tfrac{1}{2}(C z_{\text{Na}}{}^2 + C z_{\text{SO}_4}{}^2) \\
&= \tfrac{1}{2}(0.4 \cdot 1^2 + 0.2 \cdot 2^2) \\
&= \tfrac{1}{2}(0.4 + 0.8) \\
&= 0.6.
\end{aligned}$$

"Ionic strength," therefore, takes account not only of the concentration of the ion but also of its valence. This is very important because the "ion atmosphere" effects of di- and tri-valent ions on other ions are in proportion to the *square* of the charges. The interionic effects of polyvalent ions are therefore greater, on a *molar* basis, than the effects of monovalent ions.

The effect of a salt on the solubility of a relatively insoluble electrolyte not having a "common" ion depends on a number of factors. In general, the solubility of protein is *increased* by the presence of relatively *low* ionic strengths of neutral salts, whereas neutral salts present in very high ionic strengths will *depress* solubility of a salt or a protein. The *promoting* effect of a neutral salt on solubility is known as the "*salting-in*" effect, and the depression of solubility by high ionic strengths as the "*salting-out*" effect. Salting in stabilizes the charged groups of proteins; salting out ultimately involves a competition between protein and salt for water. As the salt concentration increases the salt successfully competes for water as water of hydration surrounding the charged salt ions. This causes the proteins to aggregate and precipitate out.

Proteins soluble in water or dilute salt solutions may be separated into two main classes, albumins and globulins, depending on the effect of neutral salts on their solubility.

Albumins are readily soluble in aqueous solutions of low ionic strength, but are salted out (i.e., precipitated from solution) by *very* high ionic strengths of neutral salts. Globulins are insoluble in pure water but are brought into solution (salted in) by low ionic strengths of neutral salts ($\Gamma/2 = <0.2$). The globulin may again be "salted out" at much higher ionic strengths. A distinction employed in the past for albumins and globulins was that albumins are precipitated out by fully saturated $(NH_4)_2SO_4$ while globulins are precipitated out at $\frac{1}{2}$ (50 per cent) saturation. The classification of soluble proteins into "albumins" and "globulins" is ambiguous since many proteins fall between these extremes in their solubility behavior.

The Isoelectric Point and Solubility: The Influence of pH. Experimentally or empirically, the isoelectric point of a protein is the pH at which a protein fails to migrate toward either the anode or cathode in a polar electrical field. Theoretically, it is the pH at which the net charge of all the dissociable groups, i.e., carboxyl, amino, imidazol, guanidinium, and so forth, is equal to zero. A protein surface will always have an electrical charge which is influenced by pH, electrolytes, and solvent, particularly water, and by hydrogen bonding; but there is one set of conditions involving the interrelationship of all of these factors at which the sum of the positive charges equals the sum of the negative charges. At this point the total charge, as distinguished from the net charge, will be at a maximum. The number of protein molecules having a net negative or net positive charge, however, is at a minimum or zero. Since like charges repel one another, in this circumstance certain molecules will be able to aggregate rather than to repel each other; but at either side of the IEP, all molecules will have either a net negative or positive charge. These molecules will repel each other and hence minimize the tendency toward aggregation and precipitation. In general, therefore, proteins are found to be least soluble at the IEP, and casein in particular is actually very insoluble.

Protein molecules are large and have a wide range of secondary and tertiary structural arrangements. For some proteins it is possible that there can be accumulations of like charges which contribute to localized molecular repulsion even at the IEP.

The isoelectric point or a particular charge configuration on a protein surface also depends to some extent on the nature of other ions in the solution. According to modern theories of electrolytes, each ion in solution is surrounded by an "atmosphere" of other ions and of water, which act as dipolar molecules. Therefore, the positive and negative charges of the protein molecule attract other ions of opposite charge, forming an "ion atmosphere" and causing a certain amount of "shielding" or "masking" of the charges on the protein. This leads to repulsion of other protein molecules at low salt concentration, and hence to solubilization, and to aggregation at high salt concentration. At some charged sites on the protein surface an oppositely charged "foreign" ion may be very strongly attracted or "bound" because of electrostatic and steric factors. Although such "ion binding" by proteins is quite reversible, it can alter the charge distribution and the net charge. For this reason the isoelectric point of a protein, defined as the pH at which there is no movement in an electrical field, will vary somewhat in value depending on the nature of other ions in solution. For many proteins the effects of other ions are not great, but for some, such as serum albumin, the isoelectric pH will vary greatly with changes in ionic composition of the solution. Serum albumin has the ability to "bind" ions, such as chloride, phosphate, and thiocyanate, very strongly, causing a shift in isoelectric pH. As an example of an important function of this property, a number of drugs and antibiotics are known to be bound to serum albumin, and consequently are believed to be transported by this means in blood. The true IEP of a protein, however, is that pH at which the effects of all other ions are at a minimum or zero. It is in reality the *isoionic point*.

Temperature Effects on Solubility. Some proteins exhibit no change in solubility below 35° C; others exhibit an increase in solubility while still others are less soluble below 35° C. On the other hand, all proteins begin a denaturation process above 35° which is complete at 80° C. In this range solubility decreases to the point of complete insolubility, or coagulation as in the case of egg albumin.

The Effects of the Dielectric Constant on Solubility. Organic solvents such as alcohols, ethers, and ketones, which lower the dielectric constants of aqueous solutions of proteins, also lower their solvating capacity, thus rendering proteins less soluble. Organic solvents such as dimethyl sulfoxide and formamide, which increase the dielectric constants of aqueous solutions, also increase protein solubility.

The Denaturation of Proteins.   The secondary structure of proteins arises from hydrogen bonding within the polypeptide chain between the carbonyl oxygen of one peptide linkage and the hydrogen of the amide group of another peptide linkage along the chain.

The tertiary structure of proteins results from interchain hydrogen bonds, polar and apolar attractions, and the covalent bonds of sulphydryl bridges, all of which play a significant role in determining the three-dimensional configuration of proteins.   When the intra- and inter-chain bonds are broken, proteins undergo a variety of structural modifications, both subtle and profound, which depend upon the extent and the type of disruption encountered. This phenomenon is referred to as a denaturation process, and it may be induced by a number of different chemical and physical means.   It may be induced by heat, extremes of pH, ultraviolet radiation, or pressure; by detergents or other surface agents; by urea or guanidine, which disrupt hydrogen bonds; by organic solvents and by heavy metal ions.   The experimental criteria used to measure the extent of denaturation are: (1) decrease in solubility, (2) increase in reactivity of residual groups of the peptide chain, (3) loss of biological activity, (4) changes of shape and size leading to changes in viscosity and sedimentation coefficients, (5) changes in absorption and fluorescence spectra, and (6) changes in optical rotary dispersion.

EXPERIMENT 4

This experiment is divided into four parts. In part A, groups of four students will begin the preparation of five different proteins. At a point designated in the instructions, the groups will pair off to complete their particular protein separation independently. Group 1 will prepare myosin from rabbit skeletal muscle; Group 2 will prepare casein from milk; Group 3 will prepare crystalline egg albumin from egg white; Group 5 will prepare crystalline globulin from squash or pumpkin seed meal. The proteins isolated by each group will be made up as 1 per cent solutions and used by the class in Part B. This part is concerned with the solubility properties of these and other proteins. Part C is merely an exercise on the color reactions of proteins. However, in Part D students will continue to work in pairs while performing a quantitative analysis of proteins by the biuret method. During waiting periods students are expected to complete Parts B, C, and D.

Part A. Separation of Proteins from Various Tissues

PREPARATION OF MYOSIN FROM RABBIT MUSCLE

Myosin makes up part of the contractile protein system of the myofibriles of skeletal muscle. During contraction, hydrolysis of the "energy rich" phosphate of adenosine triphosphate (ATP), by the enzyme ATPase in myosin, occurs. Myosin is a typical globulin protein, rod-like in shape, having a molecular weight of about 470,000. It is insoluble in distilled water and soluble in dilute neutral salt solutions; therefore, it acts like a globulin which can be "salted in" at low ionic strength and precipitated out at very low ionic strength or in water.

Procedure. These directions are for the preparation of myosin from one rabbit. Gently but firmly restrain the rabbit. The animal is injected intraperitoneally with 5.0 ml of 25 per cent $MgSO_4$. Wait until the animal is completely anesthetized (flaccid muscles and no eye reflex). It may be necessary to inject more $MgSO_4$ if the animal is not completely anesthetized after ten minutes. If so, another 5.0 ml should be injected.

A quicker way to sacrifice a rabbit is by an intracardial injection of air. Restrain the animal. Stretch out the fore and hind paws. Locate the sternum with your fingers and proceed posteriorly to the xiphoid process at the end of the sternum. Puncture through this extended cartilage at a 45° angle with a 2 inch 18 gauge needle attached to a 10 ml syringe. Go in approximately 1 to 1.5 inches or until the heart beat can be felt against the needle; then penetrate another $\frac{1}{4}$ inch into the ventricle. Withdraw $\frac{1}{2}$ ml blood as an assurance that the ventricle has been penetrated; then inject 3 to 4 ml of air. Continue to restrain the animal, especially as it goes into convulsions. It will be dead in a matter of seconds.

When anesthesia is complete, bleed the animal over the sink after cutting a neck artery. When the bleeding has stopped, remove the animal to the cold room, and carry out all succeeding operations in the cold. Skin the entire animal as quickly as possible and remove as much of the musculature as possible from all limbs, neck and back. Remove a total of 300 g of skeletal muscle. Pass the muscle through a chilled meat grinder.

At this point work in pairs

Divide ground muscle in half, and each pair of students is to prepare myosin separately. For every 150 grams of ground muscle obtained, add 0.5 liters of prechilled salt extraction solution (0.6 M KCl). Suspend the tissue in the extraction solution in a large jar and stir gently with a rod for 20 minutes. For each liter of extraction fluid add three liters of cold distilled water. Mix and pass the entire suspension through layers of *muslin* to get rid of large tissue material. Dilute the strained extract to 5 liters (final volume

SS-34 SUPERSPEED—
400 ml/20,000 rmp/48,200 × g

SM-24 SUPERSPEED—
360 ml/20,000 rpm/49,500 × g

GSA LARGE CAPACITY—
1,890 ml/13,000 rpm/27,500 × g

GS-3 HIGH SPEED—
3 liter/9,000 rpm/14,700 × g

SE-12 SUPERSPEED—
180 ml/20,000 rpm/41,435 × g

FIGURE 5. Refrigerated centrifuge and some available rotors for holding sample tubes. (Courtesy of Ivan Sorvall Inc.)

for 150 grams of original muscle tissue) and allow to stand overnight in the cold room.

Myosin will precipitate out on standing because the ionic strength is now so low it cannot remain in solution. This process of "salting in" and precipitation of myosin is reversible and could be repeated several times to free the myosin of other types of protein.

By the next laboratory period most of the myosin will have settled and much of the supernatant liquid can be removed by decantation or siphoning. The myosin gel can now be recovered by centrifugation. This is carried out in a refrigerated centrifuge with 250 ml plastic centrifuge containers. Centrifuge for five minutes at 13,000 rpm. After centrifugation, the myosin gel must be kept cold (0 to 5° C) *at all times*. As prepared in this fashion it is probably about 80 to 90 per cent pure. Resuspend the myosin in 0.6 M KCl as a one half per cent solution.

The preparation of this solution is very critical for the success of a subsequent experiment with myosin. Try to make a saturated solution, then filter or centrifuge off the undissolved protein. Save the clear supernatant. It may be useful, if there is time, to do a biuret determination, which is explained in Part D, page 77. You should set a reading of 0.100 absorbance for 1.0 ml of this solution. Label and store this myosin solution in the cold

Question: What operations in the isolation procedure have caused the removal of other soluble proteins found in the original muscle extract?

SEPARATION OF CASEIN

The lactating mammary gland ranks second to the photosynthesizing cell in the chain for sustaining life on this planet. Casein is the chief protein in milk. It is a phosphoprotein which exists in the four forms of alpha, beta, gamma and kappa casein, each having a different composition. Casein, therefore, is a heterogenous mixture of protein components. Like many other proteins derived from animal sources, e.g., meat (myosin) and eggs (albumin), it is a nutritionally adequate protein. Such proteins contain all of the essential amino acids required for normal growth and development.

Work in pairs

Each pair of students in this group will dilute 0.5 liters of skim milk with 1.5 liters of tap water. Over a 30 minute period, about 50 ml of 2 per cent HCl is added to bring the pH to 4.8 (use the pH meter). With care, the point at which precipitation of casein begins is readily observed. Stir for 10 minutes. Allow it to settle for $\frac{1}{2}$ hour and decant. Wash the precipitate twice by resuspending in distilled water to a volume of about 1 liter and decant.

Filter the casein on a large Büchner funnel using three thicknesses of paper. Suck dry. Transfer the moist residue to a liter beaker. Make a final suspension of the casein in 150 ml of distilled water. Filter and repeat the process until the filtrate is free of chloride ions. (Test with 1 per cent $AgNO_3$)

Agitate vigorously with the aid of a stirring motor. Decant water each time, or use the centrifuge with 250 ml cups to recover the precipitate.

Suspend the residue in 75 ml of 95 per cent ethanol in the same beaker, and stir vigorously for 5 minutes. Filter. Repeat the alcohol extraction once, then extract twice with 75 ml portions of ether in the same fashion. Filter the suspension and dry on a porous plate. The product should be a white and free-flowing powder.

What might you suspect to be in the supernatant?

Make up a 1 per cent solution of your preparation in 0.01 N NaOH. Label it and store in the cold.

PREPARATION OF VITELLIN

Vitellin is also a phosphoprotein. It is a globulin found in egg yolks, as distinguished from the ovalbumin of egg white. Both casein and vitellin have high serine content, and it has been shown that most of the phosphorus is esterified to the hydroxyl group of serine. Another phosphoprotein, phosvitin, can also be isolated from the vitellin fraction you are about to prepare, and like casein it too is not a homogeneous protein.

In collaboration with the group of students preparing crystalline ovalbumin, carefully separate the yolks of fresh eggs (less than 24 hours old). Include the chalaza in the egg white by trying to remove it from attachment to the yolk membrane. (Snip it with a scissors.)

Work in pairs

Each pair of students will transfer three yolks to a 200 ml graduated cylinder, and determine the volume. Dilute the yolks with an equal volume of 10 per cent NaCl. Break up the yolks and thoroughly mix. Transfer the mixture to a 500 ml separatory funnel. Extract *cautiously* with three volumes of ether, making sure that no pressure build-up takes place in the separatory funnel. Rotate the funnel from an upright to an upside down position, but do not shake. *Vent after inversion.* Repeat the ether extraction three times. If shaking has been too vigorous, separation of solvent layers can only be achieved by centrifugation.

Discard the ether fraction. Separate the aqueous layer, and *divide it in half.* Transfer one portion to a dialysis membrane. Add a few drops of toluene; knot the membrane, and dialyze against running tap water for two days. Filter the precipitate formed with a Büchner funnel. Suspend the protein with 50 ml of 95 per cent ethanol; extract with stirring for five minutes, and filter. Treat in the same way with absolute ethanol and then ether. Dry in the air.

Pour the second portion of the ether-extracted aqueous layer into a very large volume of tap water and stir well. Allow the precipitate formed to settle, and test an aliquot for complete precipitation by further dilution. When precipitation is complete, centrifuge or filter in a large Büchner funnel and dehydrate as above.

for complete precipitation by further dilution. When precipitation is complete, centrifuge or filter in a large Büchner funnel and dehydrate as above.

What is achieved both by the dialysis process and by the dilution of the salt extract?

Resuspend the vitellin as a 1 per cent solution in 10 per cent NaCl. Centrifuge to clear if necessary. Label and store in the cold room for use in the next experiment.

CRYSTALLIZATION OF EGG ALBUMIN

Ovalbumin is the major component of egg white (64 per cent). It was isolated in crystalline form by Hofmeister in 1890. It is a compact, spherical molecule of molecular weight 45,000.

Work in pairs

The egg whites from three eggs are used by each pair of students. Determine the volume. Stir the white, add 0.1 volume of 1 N acetic acid, and filter through cheese cloth to break up the membranes. Vigorously stir with a glass rod to speed up passage through the cheese cloth. Add an equal volume of *buffered* saturated ammonium sulfate solution and centrifuge off the precipitate of ovoglobulin after 30 to 60 minutes. Ovomucin, glycoprotein responsible for most of the viscosity of egg white, will also be removed in these two steps.

To the clear (yellow) supernatant, add buffered saturated ammonium sulfate slowly with stirring until the precipitate (turbidity) definitely ceases to dissolve. Stir the resulting opalescent solution occasionally. Allow the mixture to stand in the cold room for two to five days. Centrifuge off the albumin crystals. Add more salt if the yield appears to be too low. Recrystallization can be achieved by repeating $(NH_4)_2SO_4$ precipitation and washing but for purposes of this experiment you may stop at this stage. Resuspend the albumin in 0.9 per cent NaCl as a 1 per cent solution.

PREPARATION OF A CRYSTALLINE GLOBULIN FROM VEGETABLE SEEDS

Work in pairs

Fifty grams of freshly ground pumpkin or squash seed are extracted with 200 ml of 10 per cent NaCl in a 250 ml beaker at about 50° C for about one hour, with frequent stirring. The beaker is then heated in a water bath to 70° to 75° C (but not higher), cooled, and strained through two layers of gauze. It is then squeezed out and the fluid warmed to 60° C in a water bath. Now add 4 g of Filter Cel and mix thoroughly. Moisten a large folded filter paper in a funnel with 10 per cent NaCl solution. Filter the warm solution, returning the filtrate to the paper until the filtrate comes through fairly clear. If necessary, rewarm the solution which is being filtered. When you have obtained 100 ml of filtrate, transfer it to a liter flask, warm it in a water bath to 60° C, and add 400 ml of water heated to 60° C. Place this flask in your water bath containing water at 60° C for 30 minutes and then put away in your locker until the next period.

Now decant most of the supernatant fluid into a beaker and examine a drop of the heavy sediment under the microscope. Observe the octahedral crystals. The rest of the sediment may be filtered or centrifuged off. Place 1 or 2 ml of the supernatant in a test tube. Fill the tube with water. Note and explain the turbidity. Do a biuret test on the crystalline globulin dissolved at 1 per cent concentration in 10 per cent NaCl solution (see Part D). Save this solution for Part B.

Part B.   Solubility Properties of Proteins

Work in pairs throughout Part B

Complete the following table with yes or no answers after finishing parts a, b, and c.

Table 9

Type Protein	Heat Denaturation	Precipitation at 1/2 $(NH_4)_2SO_4$	Precipitation at full $(NH_4)_2SO_4$	Isoelectric Precipitation
casein				
egg albumin				
vitellin				
myosin				
vegetable seed globulin				

(a) To 5 ml of solutions of each of the above proteins add an equal volume of saturated $(NH_4)_2SO_4$. Is there a precipitate? If there is, remove it by centrifugation and add solid $(NH_4)_2SO_4$ until you achieve full saturation. Is there a precipitate? Record your results. What is your interpretation? Should there be precipitation in both cases? Follow this procedure with the other proteins in Table 9.

(b) Heat 5 ml of each of the protein solutions in Table 9 and record precipitations, if they occur.

(c) Dilute one per cent albumin with distilled water 1:2.5 and add 2 ml aliquots to six test tubes. To the first add a drop of 0.05 N NaOH; to the second, one drop of 0.05 N HCl; to the third, three drops of 0.05 N HCl; to the fourth, one drop of 0.05 N HAc; to the fifth, one drop of 0.05 N HAc and three drops of 10 per cent NaCl. The sixth tube is to be used as a control. Place all tubes in a beaker of water and heat slowly. Note the order of coagulation. What are the optimum conditions for heat denaturation?

ISOELECTRIC PRECIPITATION OF CASEIN

To 5 ml of an alkaline casein solution, add 0.10 N HCl drop by drop until maximum precipitation occurs. Continue the addition of HCl, shaking the tube after each addition, and note that the casein redissolves.

At this point add 0.10 N NaOH drop by drop. Note that the casein may be reprecipitated and then redissolved in an excess of the alkali. The protein precipitates at the point at which it is electrically neutral. To show that this point is not necessarily at pH 7.0 (neutrality), add 5 drops of bromcresol green indicator, and add 0.05 N HCl until the isoelectric point is reached. The solution at this point is acid (pH 4.7) according to the color change of the indicator at this pH.

Attempt this experiment with the other proteins.

Explain the formation of curd when milk sours.

What are other properties of protein solutions which are at a minimum at the isoelectric point?

Distinguish between electrical neutrality and hydrogen ion neutrality.

THE EFFECT OF pH ON MYOSIN SOLUBILITY

Outline of Method: A series of tubes, each containing a standard amount of myosin dissolved in 0.5 M KCl, is treated with varying but known concentrations of sodium acetate and acetic acid. The pH of each tube can therefore be calculated by the use of the Henderson-Hasselbalch equation. The relative solubility of myosin at the different pH values is determined by measuring the amount of protein remaining in solution after equilibration.

Procedure: Make up the solutions listed in Table 10 in 12.0 ml conical centrifuge tubes. Make additions in the order listed.

Add the myosin last to all tubes and mix well. Allow the tubes to stand

Table 10

| | Reagents | Tube Number | | | | | | | | |
		1	2	3	4	5	6	7	8	9
A.	1.0 M acetic acid, ml	2.0	1.9	1.7	1.4	1.0	0.6	0.3	0.1	
B.	1.0 M sodium acetate, ml	0.0	0.1	0.3	0.6	1.0	1.4	1.7	1.9	
C.	distilled water, ml	2.0	2.0	2.0	2.0	2.0	2.0	2.0	2.0	4.0
D.	stock KCl-myosin solution (0.5%), ml	2.0	2.0	2.0	2.0	2.0	2.0	2.0	2.0	2.0
E.	pH (Calculate)									

FIGURE 6.   General laboratory centrifuge. (Courtesy of Ivan Sorvall, Inc.)

at least two minutes. Then centrifuge all the tubes for one minute at $\frac{3}{4}$ of top speed in the table model clinical centrifuge (Fig. 6).

Remove 1.5 ml aliquots of the supernatant fluid from each tube and place in separate, clean spectrophotometric tubes correspondingly numbered. Add 1.5 ml of biuret reagent to each. The blank will consist of 1.5 ml of 0.9 per cent NaCl and 1.5 ml of biuret reagent.

Allow 10 minutes for the color to form. Determine the absorbance for each tube at 540 mμ in a spectrophotometer. Tube 9 will give the absorbance corresponding to the protein concentration of the stock KCl-myosin solution and from this can be calculated the amount of protein in each tube in Table 10.

From your standard curve determine the amount of protein remaining in solution in each of the experimental tubes. From this the amount of protein precipitated at each pH value can be determined. Knowing that the solubility of a protein is minimal at the isoelectric point, in what pH range is the isoelectric point of myosin?

Plot myosin solubility versus pH, after calculating the pH in each tube.

SEPARATION OF THE PLASMA PROTEINS

Normal blood plasma contains 6.5 to 7.5 grams of protein per 100 ml. Plasma proteins may be separated into three groups: fibrinogen, globulins, and albumins. Each of these protein types may be precipitated on the basis of their solubility in salt solutions, and a simple or crude fractionation is achieved by salting out these proteins at different ion strengths.

Salting out with ammonium sulfate (AS) is a useful early step on many procedures for isolating proteins, particularly enzymes. There are two ways by which this purification step can be achieved, by adding solid AS or by adding a neutralized saturated (100 per cent) solution of AS. Adding the solid AS has the advantage of minimizing volume increases, and the saturated solution has the advantage of convenience. In this experiment you will isolate and reconstitute in 2 per cent NaCl your fibrinogen and your albumin fraction which will be used in a subsequent experiment.

To 10 ml of plasma in a small beaker add enough saturated ammonium sulfate solution so that the plasma is 25 per cent saturated. Centrifuge off the precipitate of fibrinogen. Then raise the ammonium sulfate concentration to 33 per cent of saturation by adding more ammonium sulfate solution. Spin down the precipitate of euglobulins with the table model centrifuge. Bring the supernatant from the euglobulins to 46 per cent of saturation by adding more ammonium sulfate solution. Centrifuge off the pseudoglobulins. To the supernatant of the pseudoglobulins add ammonium sulfate solution until the mixture is 64 per cent saturated. Centrifuge off the albumin. Precipitate the remainder of the albumin by adding solid ammonium sulfate to the final albumin filtrate until it is fully saturated. This type of procedure has been used to purify enzymes as well as proteins. However, it is important to remember that simple salt fractionations of this type do not yield pure component proteins, since there is considerable overlap between fractions, and more than 30 different proteins have been shown by electrophoresis to be present in plasma. Reconstitute the fibrinogen and albumin precipitates in 5 ml of 2 per cent NaCl.

The following equation can be used for calculating the amount of solid AS required to bring a solution to a specific saturation at zero degrees Centigrade, which is the usual temperature used in enzyme or protein purification.

$$X = \frac{50.6[S_2 - S_1]}{1 - 0.3S_2}$$

where X = grams of solid AS to be added to 100 ml of solution of S_1 saturation to change it to S_2 saturation

S_1 = saturation of initial solution

S_2 = saturation of final solution

Note: S_1 and S_2 are fractions of saturation at $0°$. To change them to per cent multiply by 100.

Derivation of the formula is based on these facts. (1) The solubility of AS at $0°$ is 70.6 g in 100 g of H_2O. (2) The specific volume of solid AS is 0.565 ml/g. (3) The concentration of AS in a saturated solution at $0°$ is 50.6 g per 100 ml.

For example: On adding 70.6 g of AS to 100 g of H_2O there is a substantial increase in volume which may be determined by calculation. 70.6 g of AS has a volume of $70.6 \times 0.565 = 39.8$ ml. Therefore, a saturated solution made up by adding 70.6 g AS to 100 g of H_2O will have a total volume of 139.8 ml. Its final concentration will be 70.6/139.8 or 50.6 g/100 ml of solution.

At 23° the solubility of AS is 76.2 g/100 g H_2O, and the factor in the equation becomes 53.3.

A second formula involves changes of saturations of solutions by adding saturated AS:

$$Y = \frac{V_1[S_2 - S_1]}{1 - S_2}$$

where

$Y =$ ml of saturated AS required to be added to V_1
$V_1 =$ initial volume of solution at S_1
$S_2 =$ saturation desired

Note: If the solution of AS is less than saturation, then the factor 1 has to be changed correspondingly. Left as it is it stands for 100 per cent.

Proteins may be salted out of solution by a number of salts at high concentration. At very low concentrations, less than 0.5 M, salts ($NaCl$, $MgSO_4$, K_2HPO_4, and others) show differences in their ability to precipitate proteins, but these differences are small enough that it is believed that specific protein salt complexes are not involved in reducing solubility. Salts become hydrated in solution and thus decrease water availability for interaction with polar groups.

Include a calculation for your plasma fractionation in your report.

THE CLOTTING OF FIBRINOGEN

When blood clots, fibrinogen is converted to fibrin by the proteolytic action of a third protein, thrombin. Thrombin is not normally present in blood as thrombin, but as prothrombin, its precursor. Upon injury, prothrombin is enzymatically converted to thrombin by the action of thromboplastin in the presence of calcium. By this sequence of reactions, clotting takes place. Clotting can be prevented either by removing calcium from the blood (by adding oxalate or citrate to the serum) or by removing thrombin.

Add three to five drops of prepared thrombin to your reconstituted fibrinogen and to the albumin precipitated with either 64 per cent or 100 per cent saturated ammonium sulfate. Observe both solutions for clot formation.

SOLUBILITY OF SERUM ALBUMIN AND EFFECT OF ALCOHOL CONCENTRATION

Place tubes containing reagents as shown in Table 11 in an ice bath to perform this experiment. Record the tubes in which precipitation occurs.

Table 11

	Reagents	Tube Number				
		1	2	3	4	5
A.	1 per cent serum albumin at pH 4.8, ml	5	5	5	5	5
B.	H_2O (cold), ml	5	3.95	2.9	1.85	0.8
C.	95 per cent ethanol (cold), ml		1.05	2.1	3.15	4.2
D.	percent alcohol present (calculate)					

What are the conclusions from these data?

Allow all tubes to come to room temperature by placing them into a 40° C bath. Cool again to 0°. *Explain results.*

PRECIPITATION OF PROTEIN BY METALS

Proteins form insoluble salts with certain metallic ions. This reaction is useful in deproteinizing a solution. It is also the basis for the administration of milk or egg white to one who has swallowed a metallic poison such as mercuric bichloride. When a protein solution is made basic, the net charge on the protein will be negative (Why?). Therefore, in the presence of heavy metal cations, metal proteinate salts will form, which may be soluble or insoluble depending on the nature of the metal used. For the most part, Na, K, and Mg salts of proteins are soluble, but salts of Hg, Pb, Cu, and Zn are not.

(a) To 3 ml of clear 1 per cent albumin solution made slightly basic, add 1 ml of 0.005 M $HgCl_2$.

(b) Add a few drops of glacial acetic acid and note whether the precipitate dissolves.

(c) Repeat (a) using 0.005 M PbAc. Allow the solution to stand a few minutes and then add 1 ml of neutral 0.05 M ethylene diamine tetraacetic acid, a potent metal chelating agent. What are your observations?

PRECIPITATION OF PROTEIN BY ACIDS

When protein solutions are made acidic the net charge is positive. The proteins act as bases and can form salts with added acids or anions. Certain of these salts are insoluble. Acids such as tungstic acid are used both for the isolation of protein and for the deproteinization of solutions. Protein-free blood filtrates are prepared in this way. Most deproteinization steps in analytical procedures take advantage of this property, and agents such as trichloroacetic and perchloric acids are frequently used.

(a) To 3 ml of 1 per cent albumin solution add a few drops of sodium tungstate solution. Why doesn't the insoluble albumin tungstate form? What is the effect of acidifying this solution with dilute acetic acid? Show that the precipitate is not an isoelectric precipitate.

(b) Omit the tungstate and repeat (a) using picric acid, trichloracetic acid, or sulfosalicylic acid, and try some of the available proteins with these reagents. In these cases it is not necessary to add acid. Try adding dilute base. Do any protein precipitates redissolve?

Part C. Qualitative Color Reactions (Exercise)

A number of amino acids give characteristic color tests with certain reagents. These tests are actually tests for various functional groups. They are thus specific for these functional groups and not for the amino acids themselves. The tests have nevertheless been widely used because they are quite simple to perform, and because very often, the amino acids are the only

Table 12

Test*	Reagent Used	Positive Test (Color)	Specificity
ninhydrin	Ninhydrin in water-saturated butanol		
biuret	Alkaline $CuSO_4$ in sodium potassium tartrate solution		
Millon's	Mercuric and mercurous nitrates in $HNO_3 + HNO_2$		
Hopkins-Cole	Glyoxylic acid in H_2SO_4		
xanthoproteic (Mulder's)	Concentrated HNO_3, then NaOH		
Sakaguchi	Alkaline sodium hypochlorite α-naphthol		
Erlich diazo reaction	Diazobenzene sulfonic acid		
nitroprusside	Sodium nitro- prusside, ammonia		
Folin–Ciocalteau phenol reagent	Na_2WO_4 Na_2MoO_4 H_3PO_4		

* See F. Koch and M. F. Hanke, *Practical Methods in Biochemistry*, Williams and Wilkins Co., Baltimore, 1948. Also B. L. Oser, ed., *Hawk's Physiological Chemistry*, McGraw-Hill, New York, 1965.

chemical components of body fluid or tissue containing the particular functional group in any quantity. This is particularly true for proteins which give positive color reactions to reagents specific for a given amino acid.

These tests are rarely used in biochemical research or clinical chemistry. They are, however, used in conjunction with ion exchange separation of amino acids. A quantitative adaptation of the ninhydrin reaction is almost exclusively used for this purpose, but in rare instances one of the special color tests such as the Sakaguchi or Pauli test may be used to confirm the identity of a specific amino acid. A quantitative adaptation of the biuret test is used in most laboratories for the determination of protein in biological fluids. However, because of the historical interest, and possible occasional use of such tests, Table 12 should be completed after conducting a literature search of these methods.

Part D.   Quantitative Determination of Protein by the Biuret Reaction

COMPARISON OF METHODS

There are many methods for the determination of the amount of protein in a natural product. A particular method is chosen in accordance with the materials to be analyzed and the type of information needed. Some of the common methods are listed here.

(a) Kjeldahl analysis for total nitrogen after appropriate extraction and precipitation or with a correction for non-protein nitrogen. A conversion factor for grams protein per gram of nitrogen may be used. See A. Hiller, J. Plazin, and D. D. Van Slyke, *J. Biol. Chem.* **176** 1401 (1948). See also P. L. Kirk, *Anal. Chem.* **22** 354 (1950).

(b) Biuret assay using a known protein to establish a standard curve. A. G. Gornall, C. J. Bardawill, and M. M. David, *J. Biol. Chem.* **177** 757 (1949).

(c) Lowry method. A modification of the Folin-Ciocalteau Method based on the presence of tyrosine and tryptophan in proteins. O. H. Lowry et al., *J. Biol. Chem.* **193** 265 (1951).

(d) Turbidity measurements after precipitation in a controlled manner using a known protein as a standard. P. L. Kirk, *Advances in Protein Chemistry* **3** 139 (1947).

(e) Ultraviolet absorption analysis at $\epsilon 280/260$ nm after removal of non-protein materials by dialysis or fractionation. Appropriate extinction coefficients for the particular proteins studied must be used ($1.55 \times \epsilon 280 - 0.775 \times \epsilon 260 \times$ dilution factor = mg protein per ml). O. Warburg and W. Christian, *Biochim. Z.* **310** 384 (1941).

(f) $\Delta 215$–225 method based on peptide bond absorption in the 195–225 nm region. ($\Delta 215$–225 $\times$ 154 = μg protein per ml). J. B. Murphy and M. W. Kies, *Biochem. Biophys. Acta* **45** 382 (1960).

(g) Measurement of the specific gravity of a solution after appropriate purification. E. A. Kabat and N. Mayer, *Experimental Immunochemistry*, C. C. Thomas Co., Springfield, Ill., 1961.

(h) Measurement of the refractive index of a protein solution. G. E. Perlman and L. G. Longsworth, *J. Am. Chem. Soc.* **70** 2719 (1948).

(i) Direct weighing of the purified and dried sample.

(j) Radioactive amino acids which are incorporated into material which precipitates in hot 5 per cent trichloroacetic acid are considered to measure protein synthesis in metabolic studies.

Methods, a, b, c, d, e, and f are used in the analysis of foods and tissues, and to follow the purification of tissues. Methods g, h, and i are applicable only after fairly high purity has been attained, or if the presence of salts and other non-protein materials can be taken into account by a correction factor.

The two most widely used spectrophotometer methods are the Lowry method (sensitivity 1 to 200 μg) and the Biuret method (sensitivity 0.25 to 200 mg). The other two spectrophotometric methods (280/260 sensitivity 0.05 to 2.0 μg, and Δ215–225 sensitivity 10 to 100 μg) require a much more sophisticated spectrophotometer. The major shortcoming of these methods relates to the fact that different proteins have different absorption spectra, and no one protein standard can adequately serve for the protein of interest. The best arrangement is to make a standard reference curve with the protein under study, but this is not always possible, since a pure sample of the protein being measured may not be available.

THE CHEMICAL BASIS FOR BIURET DETERMINATION

Ammonia and substituted ammonias, including the amino acids, form complex ions with Cu^{+2} and other metallic ions. The cuprammonium ion, $Cu(NH_3)_4^{+2}$, is the simplest example of such a complex; simple amino acids like glycine form complexes of similar structure.

In addition to such *simple* Cu complexes with amino acids there are more intricate complex ions formed between Cu^{+2} and peptides or proteins, in which the Cu^{+2} is linked to more than one group of the peptide.

These are formed especially in alkaline solutions, and such complexes have an intense pink or violet color quite different from the blue complexes of the simple cuprammonium type. Substances having two or more peptide linkages, such as peptides or proteins, can be detected by making a solution of the substance alkaline and adding small amounts of Cu^{+2}. A pink or violet color is considered positive; a blue color, such as may be given by a simple amino acid, is not to be confused with the positive "biuret" test.

The test is called the "biuret reaction" because the substance biuret gives a pink color under the same circumstances. Biuret has the structure:

$$NH_2—\underset{\underset{O}{\|}}{C}—NH—\underset{\underset{O}{\|}}{C}—NH_2$$

It should be emphasized that biuret does not itself exist in tissues, body fluids, or proteins, but it gives a color similar to that given by peptides because it contains what may be regarded as two peptide linkages.

The absorption spectra of the copper complexes formed by different proteins are similar although not identical; therefore, while serum albumin may be used as a color standard, for any protein it may not be ideal for the particular protein under investigation.

PREPARATION OF STANDARD CURVE FOR BIURET REACTION

A standard albumin solution in 0.9 per cent NaCl will be furnished. This is a primary standard and will have been previously analyzed by an independent method and will contain 1.00 mg of serum albumin per ml.

Table 13

	Reagents	Tube Number					
		1	2	3	4	5	6
A.	standard serum albumin 1.00 mg/ml, ml		0.1	0.2	0.5	1.0	1.4
B.	10 per cent Na deoxycholate, ml	0.1	0.1	0.1	0.1	0.1	0.1
C.	0.9 per cent NaCl, ml	1.4	1.3	1.2	0.9	0.4	
D.	biuret reagent, ml	1.5	1.5	1.5	1.5	1.5	1.5

Prepare a series of matched colorimeter tubes containing increasing amounts of a standard solution of serum albumin as in Table 13 together with a "blank" tube containing all reagents except the protein. Add reagents in the order given, adding the biuret reagent last, mix thoroughly and read tubes 2 through 5 against the "blank" (tube 1), which is set at zero absorbance. The readings are taken at 540 nm after 30 minutes. The colors developed are stable over a 60 minute period. Check 1.0 ml of your myosin solution for protein concentration. Record your results.

What is the limit of sensitivity of this method?

REFERENCES

1. Cohen, E. J., and Edsall, J. T. *Proteins, Amino Acids and Peptides as Ions and Dipolar Ions.* Reinhold Publishing Corp., New York, 1943.
2. Edsall, J. T., and Wyman, J. *Biophysical Chemistry.* Academic Press Inc., New York, 1958.
3. Colowick, S. P., and Kaplan, N. O. (Eds.) *Methods in Enzymology.* Academic Press Inc., New York, 1955.
4. Sherage, H. *Protein Structure.* Academic Press Inc., New York, 1961.
5. *Laboratory Manual of Physiological Chemistry.* Department of Physiological Chemistry, Johns Hopkins University, Baltimore, 1957.
6. Kleiner, I. S., and Dotti, L. B. *Laboratory Instructions in Biochemistry.* C. V. Mosby Co., St. Louis, 1966.

7. Carter, H. E., et al. (Eds.) *Biochemical Preparations* (in 12 volumes). John Wiley and Sons, Inc., New York, 1949–68.
8. Daniel, L. J., and Leslie, N. A. *Laboratory Experiments in Biochemistry.* Academic Press Inc., New York, 1966.
9. Neurath, H., and Bailey, K. *The Proteins* (in two volumes). Academic Press Inc., 1953–54.
10. Alexander, P., and Block, R. J. *Laboratory Manual of Analytical Methods in Protein Chemistry.* Pergamon Press, Long Island City, N.Y., 1960.

CHAPTER 5

THE ELECTROPHORETIC SEPARATION OF SERUM PROTEINS

INTRODUCTION

There are two main methods of electrophoresis: (a) moving boundary electrophoresis, and (b) zone electrophoresis. In moving boundary electrophoresis the proteins travel through a buffer medium; in zone electrophoresis, the support medium is not solely a buffer and the mobile ions or proteins move on paper, cellulose acetate, starch gel, silica gel, polyurethane foam, or on acrylamide polymers. The apparatus and methodology of the former method is both expensive and elaborate. In addition, large samples are required and resolution of proteins is limited. Zone electrophoresis is much less expensive, much simpler to operate, requires μliter quantities of sample, and is much more sensitive. In this method the charged species are separated into discrete zones, hence the name; however, in the case of free boundary electrophoresis the separation of boundaries is observed by special schlieren optical systems. Both methods may be quantitated, but again, zone electrophoresis is more convenient and less expensive. The separated zones of proteins can be made visible with dyes which bind to proteins, and scanned photometrically with densitometers which have the capacity to integrate the curves of the absorption densitograph. By this means, a simultaneously quantitative and qualitative analysis of the mobile charged species is obtained upon separation.

The rate of migration (mobility) of an ion in an electrical field is the sum of two forces, the driving force and the resisting force. The driving force depends on the number of charges per molecule, the sign of the charge on the ion, and the degree of dissociation, which is a function of pH. The driving force is also dependent upon the magnitude of the electrical field potential, the time of exposure to the field force and the changes in temperature during separation. The factors which offer resistance to electrophoretic mobility require considerable mathematical treatment. Among the considerations are the size and shape of the ion, the viscosity of the medium, the concentration of the ion, its solubility and the absorptive properties of the support medium. These factors will be considered one at a time.

The charge on a molecule may result from ionization of the dissociable groups or it may be induced on a neutral molecule by association with an electrolyte. Ionization depends upon solution pH and the pK' of the dissociable groups, and mobility is directly related to the degree of dissociation. Since amino acids are amphoteric, they can occur as more than one charged species. Glycine, for example, has a pK' of 2.35 for the COOH group, a pK' of 9.78 for the NH_2 group, and a pI of 6.1 overall. At 6.1, which is the isoelectric point for glycine, the net charge is zero. At pH 2.35 glycine is protonated, and it will have a net positive charge; at pH 9.78 the COOH group will be dissociated and glycine will have a net negative charge. The degree of dissociation and therefore the percentage of all of the species charged at any given pH can be calculated from the Henderson–Hasselbalch equation upon selection of the appropriate pK'. Differential migration of ions, i.e., their separation in an electrical field, therefore results from the different amounts of charged ions at a given pH. Consequently, the greater the number of a particular ion species, the higher its mobility will be.

The force F exerted on a charged molecule is the product of the electrical field strength S and the net charge Q on the mobile ion.

$$F = SQ = \frac{V}{d} Q$$

where V = voltage and d = the distance between electrodes. The field strength S is the electrical potential in volts divided by d in centimeters. The rate of migration of an ion, therefore, can be altered by changing the distance between electrodes as well as by changing the electrical potential.

In the absence of a resisting force, ions would accelerate in an electrical field. They assume a constant velocity because of this resistance. The resisting force F' is a function of size, shape, and viscosity in accordance with the Stokes equation:

$$F' = 6\pi r \eta v$$

where r = ion radius of a spherical molecule; η = viscosity; and v = ion velocity, the change in ion distance per unit time.

When the resisting force exceeds the driving force, no ion migration takes place; when the driving force exceeds the resisting force, there is an accelerated migration; but when a constant velocity is achieved the driving force F is equal to the resisting force F':

$$SQ = 6\pi r \eta v.$$

Since μ, the electrical mobility, is the rate of migration of an ion in a field of unit strength, S becomes a constant equal to one and the velocity becomes the mobility; then

$$\mu = \frac{Q}{6\pi r \eta}$$

Thus electrophoretic mobility is directly proportional to net charge and inversely proportional to the size of the molecule and to solution viscosity.

The distance traveled by the ion is also directly proportional to the time of exposure to the electrical potential under carefully controlled conditions. However, when there is a variation of temperature, mobility increases as temperature increases, and

$$\log \mu = a\,\frac{1}{T} + b.$$

A second effect of temperature is produced through evaporation of the solvent from the support medium. At low voltages, convection cooling may be adequate, but at high voltages more efficient cooling is required.

Since it is impossible to carry out electrophoresis in the absence of other electrolytes, particularly buffers used to maintain a constant pH, mobility may be expressed as:

$$\mu = \frac{4\pi Ce}{0.327 \times 10^8 D}\,\sqrt{\Gamma}$$

where e is the electric charge (not the net charge), D is the dielectric constant, Γ is the ionic strength, and C is a constant.

An ideal support medium should not adsorb the ion. In extreme cases adsorption can prevent migration altogether. A number of methods may be used to determine the extent of adsorption, but essentially any tailing or changing of the shape of the original spot is an indication of adsorption effects.

In order to achieve reproducible results, a direct current power supply should be capable of maintaining either a constant voltage or a constant current without traces of alternating current. During electrophoresis the support medium, which in this case is cellulose acetate, will offer resistance to current flow which inevitably produces heat. Maintenance of a constant temperature will compensate for heat effects resulting from internal resistance. At the same time both voltage and current will vary as resistance fluctuates, because: $I = \dfrac{E}{I}$, and $R = \dfrac{E}{I}$. Since resistance changes do take place, the power supply should be able to compensate for them and provide a constant E or I.

EXPERIMENT 5

Part A. Procedure

Depending on the equipment available, it may be possible to compare different electrophoretic techniques for sensitivity and resolution. One group of students may demonstrate the separation of serum proteins using cellulose acetate, while another group can use acrylamide gel discs for the same purpose. It may also be desirable to conduct only one type of separation, and to demonstrate the other.

SEPARATION OF SERUM PROTEINS USING CELLULOSE ACETATE STRIPS

Work in groups of four:

Several different samples of human serum have been made available including samples of Versatol, a commercial preparation of serum from purified protein components. These samples are reconstituted to simulate normal serum and various types of abnormal serum which arise from different types of disorders. Each student should get some experience applying samples to the strips.

(a) Remove several celluose acetate strips from the package with the aid of forceps, and place them upon a clean sheet of paper. Number the strips and draw a light pencil application line across them 7 cm from the end.

(b) Using forceps, float the strips on the surface of 0.1 M barbital buffer, pH 8.6, and allow them to soak in the buffer for 20 minutes by submerging each strip below the liquid surface.

(c) Remove the strips from the buffer solution and place them on sheets of filter paper. Gently, but completely, blot out the excess buffer. Leave the strips on the paper pad to provide a damp resilient backing.

(d) Load the sample applicator, with the aid of a capillary tube, with no more than 2 μliters of serum, and gently apply the sample along the application line at right angles to the long edge. Hold the applicator vertically, and press the wires gently and evenly against the strip for 2 to 3 seconds.

(e) At this point the strips can be handled with clean hands or with forceps and loaded onto the bridge, which has been removed from the electrophoresis chamber. Position the strips uniformly and tautly. While handling, try to keep the strips in a horizontal position. Secure the strips with the tension bars, and place the bridge in the chamber containing buffer.

(f) After closing the chamber, carry out the electrophoretic separation at a constant current of 3.0 milliampers *per strip* for a period of 30 to 50 minutes. It will be necessary to readjust the current one or two times at the beginning of the run.

(g) At the conclusion of the run turn off the current. Remove the bridge with the strips, and carefully place each strip separately and consecutively in the following solutions, provided in plastic trays. Agitate the strips at each step.

(1) Two minutes in 0.5 per cent Ponceau-S stain in 5 per cent trichloro-acetic acid.

(2) 20 seconds successively in three rinse baths of 5 per cent acetic acid.

(3) 45 seconds in a 100 per cent methanol dehydration bath.

(4) 45 seconds in a clearing bath of 10 per cent acetic acid in methanol.

(h) Lay the strips out on a clean glass plate, making sure not to trap air bubbles underneath. Press the strips flat with the end of a second glass plate.

(i) Dry the strips uniformly, using mild heat carefully applied with a hot air gun or an oven. Be sure that the edges do not melt. Since cleaning often obscures the marking of the samples it is best to relabel them with a felt pen.

Mount the developed strips on sheets of paper and fasten them with adhesive tape.

(j) Count the number of bands and compare to results on the same sample using acrylamide gel separation. If a densitometer is available, obtain a densitogram from the samples.

SEPARATION OF SERUM PROTEINS USING ACRYLAMIDE GELS

Work in groups of four:

In this separation the support medium is a polymer of acrylamide cross-linked with N,N dimethyl bis-acrylamide.

$$CH_2{=}CH{-}\overset{\overset{\text{O}}{\|}}{C}{-}NH_2 \xrightarrow[\substack{Na_2S_2O_4 \\ riboflavin}]{h\nu} {-}\left[{-}CH_2{-}\underset{\underset{NH_2}{\overset{|}{C}{=}O}}{\overset{|}{CH}}{-}CH_2{-}\underset{\underset{NH_2}{\overset{|}{C}{=}O}}{\overset{|}{CH}}{-}\right]{-}{}_x$$

In moving boundary electrophoresis the sharpness or resolution of the zones occupied by the proteins diminishes with time because the zone spreads as a result of diffusion. In zone electrophoresis, using cellulose strips, stack gels, or other macromolecular gels, this effect is minimized to some extent and sharper zones can be produced. When acrylamide cross-linked gels are employed, the gel presents different pore sizes, depending upon the concentration of gel employed. The average pore size for a 7 per cent gel is 50 Å. If a protein is large and globular, as for example, serum β lipoprotein with dimensions of

185 × 185 Å, this protein will experience extreme frictional resistance and will not be able to pass through the gel.

On the other hand, serum albumin, which is cigar-shaped (150 × 40 Å), will also experience resistance, but it can orient itself and readily pass through the gel. This is referred to as the sieving effect of the gel. In the stack gel procedure the sample of protein is suspended in a small volume of gel which is allowed to polymerize on top of two other columns of gels. The first is a 2.5 per cent solution of stacking gel at pH 6.7 which allows all of the proteins to pass through, but, as the name implies, stacks them in a very thin line prior to passage through the 7 per cent separating gel, which resolves the proteins into disc shaped zones.

Standard uniform bore straight glass tubing, 70 mm × 7 mm, has been prepared containing a 45 mm length of 7 per cent separating gel. The separating gel was placed in these tubes in the same manner as indicated below for placing stacking gel. All that remains now is to add stacking and sample gel in accordance with the instructions below.

(a) Make sure that your tubes are perpendicular to the plane of the table top; add 0.2 ml of stacking gel to the top of the separating gel. Layer this gel immediately with distilled water, using a special water layering syringe to release the water very slowly and gently without disturbing the gel. This allows the formation of a distinct, sharp boundary between the water and the gel. Turn on the fluorescent lights and allow light catalyzed polymerization to take place over a period of 20 to 30 minutes.

Polymerization is detected by noting a faint opaqueness in the gel.

(b) Turn off the light and remove the water carefully with absorbent tissue. Be sure not to disturb the gel surface during this operation. *Transfer operations to the cold room.*

(c) Rinse the gel surface with a drop or two of stacking gel solution, and quickly but cautiously remove it. This operation gets rid of the water.

(d) Make a mixture of 0.01 ml of serum sample and 0.5 ml of stacking gel and transfer a 0.15 ml aliquot to the top of the polymerized stacking gel.

(e) Allow the sample gel to polymerize for 10 to 20 minutes.

(f) Remove the red cap at the bottom of each tube by a gentle twisting action, and transfer the tubes to the upper electrophoretic chamber. Position the tubes even with the upper edge of the bath surface. From this point on, the order of the sample tubes must be kept; otherwise, mixup of samples will occur.

(g) Pour the lower chamber buffer, which does not have tracking dye, into the lower chamber. Pour the upper chamber buffer, containing tracking dye, into the upper chamber. With the aid of a syringe and needle, hang a drop of buffer on the bottom of each tube in order to make sure that no air bubbles are trapped at the lower end of the tube during electrophoresis. Take the same precautions at the upper end by forcing air bubbles out with the syringe.

(h) Raise the lower chamber to make contact with the tubes from below.

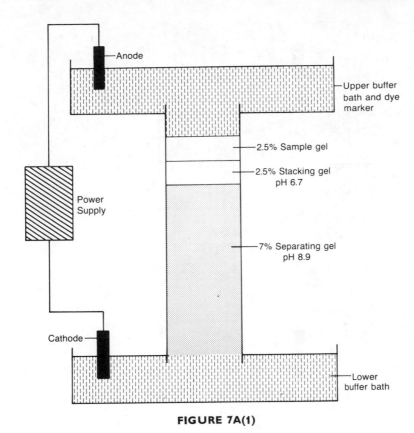

FIGURE 7A(1)

If necessary, close off the buffer chamber. Be sure at this point that the lower chamber is connected to the anode lead and the upper chamber to the cathode lead. (Why?) Turn on the power and adjust the current initially to 2 milliamperes per tube. Within a few minutes, a thin band of tracking dye from the upper chamber will travel through to the separating gel.

(i) At this point increase the current to 5 milliamperes per tube. The run is terminated when the tracking dye is 3 mm from the bottom (in about 30 to 40 minutes).

(j) Remove the tubes from the upper bath with a twisting action and immerse them in the proper order in an ice water bath.

(k) Using a 10 ml syringe with a 22 gauge needle, dislodge the gel from the glass tubing by forcing ice water between the gel and the glass surface. Fit a rubber dropper bulb filled with cold water to the tube, and gently rock the gel until it breaks free and slides easily up and down the tube. Squeeze the bulb and allow the gel to fall into the palm of your hand.

(l) Place the gel into a test tube containing 0.5 per cent amidoschwarz stain in 7 per cent acetic acid.

(m) Stain the gels 30 to 60 minutes.

(n) Set up the destaining tubes, which have been plugged with a drop of separating gel, in the electrophoresis apparatus.

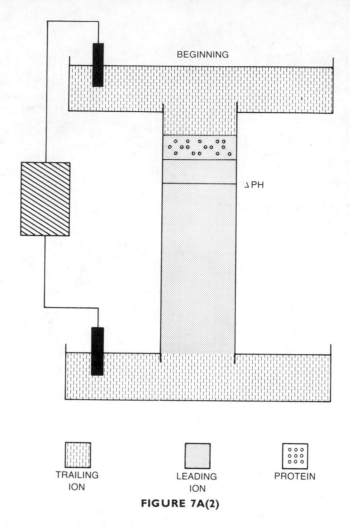

TRAILING ION

LEADING ION

PROTEIN

FIGURE 7A(2)

(o) Put a plastic disposable glove on one hand; pour the gel column from the tube into the palm of your hand and rinse off the excess stain with distilled water, then with 7 per cent acetic acid.

(p) Transfer the gels to the destaining tubes and fill them with 7 per cent acetic acid. Use glass plugs in the place of destaining tubes if there are not enough to fill all the holes in the electrophoresis chamber.

(q) Fill the lower and upper chambers with 7 per cent acetic acid. As in the electrophoretic run, have a drop of 7 per cent acetic acid at the bottom and at the top of each tube to eliminate air bubble entrapment.

(r) Connect the power as before, but reverse the polarity switch and turn on the power. Do this in the cold room. Adjust the current to 5 ma per tube and destain until the lower portion of the gel is absolutely clear.

(s) After destaining, turn the power off. Remove the destaining tubes, and transfer the gels to test tubes containing 7 per cent acetic acid.

(t) Compare your results with these obtained by cellulose acetate separation. Obtain a densitograph from the gel plug if a densitometer is available.

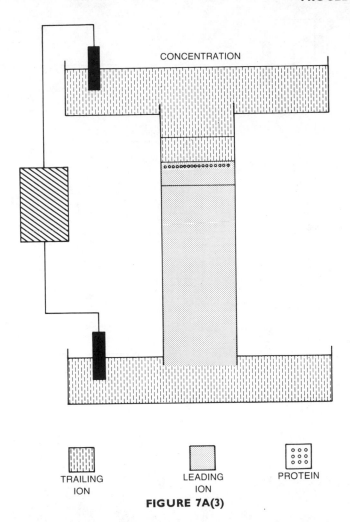

CONCENTRATION

TRAILING
ION

LEADING
ION

PROTEIN

FIGURE 7A(3)

SIMPLIFIED ALTERNATE ACRYLAMIDE GEL PROCEDURES

Preparing sample gel, stacking gel and separating gel can be lengthy and tedious. The formation of the former two gels prevent convection during electrophoresis but take no part in separation. These steps can be avoided in two ways: (a) by suspending the sample in a high concentration of sucrose or urea, or (b) by using a gel filtration suspension in place of spacer and sample gels. These are procedures developed by Clark and Broome (cf. references). Another simplification involves the use of the dye coomassie blue. In this case no destaining is necessary. What follows is a description of a simplified procedure employing two of these modifications.

(a) Standard uniform bore straight glass tubing (70 mm × 7 mm) containing previously prepared separating gel, and layered with a buffer solution during polymerization in order to insure the producing of a flat gel surface, are inserted in place in the upper gel chamber.

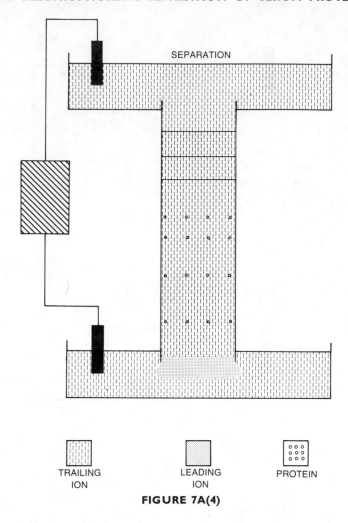

SEPARATION

TRAILING
ION

LEADING
ION

PROTEIN

FIGURE 7A(4)

(b) The sample is prepared by thoroughly mixing 0.02 ml of human serum or other protein samples, such as brain microsomes or mitochondria previously solubilized with 10 μliters of Triton X 100, with 0.8 ml of 5 per cent sucrose.

(c) A 0.2 ml aliquot of this mixture is carefully placed on the separating gel surface by allowing the dense sucrose-sample solution to drop through the buffer solution which is still over the gel.

(d) Now follow instructions in the previous section from g through l, but be sure when adding the upper chamber buffer to fill it very cautiously so as not to flush out the sample solution. It is essential that no bubbles be trapped either on the top or on the bottom of the gel columns. The run should take about 30 minutes.

(e) While electrophoretic separation is taking place, the fixing, storing, and staining solutions are prepared. The staining solution consists of 2 ml of coomassie blue stain per tube in 1 × 7.5 cm test tubes, using one for each gel. The fixing solution consists of 2 ml of 12 per cent trichloroacetic acid

FIGURE 7B

(Courtesy of Buchler Instruments, Inc.)

in similar tubes. The destaining solution is 2 ml 7 per cent acetic acid, which has been placed in a third set of tubes.

(f) When electrophoresis is completed, the gels are removed as directed in k and l, and placed in the tubes containing 12 per cent TCA for 30 minutes.

(g) Remove the gels in serial fashion so as not to get them mixed up. Rinse the gels several times with distilled water. Transfer them to the tubes containing coomassie stain, and agitate for 10 sec. Leave them in the stain for 1 hour.

Put a plastic disposable glove on one hand

(h) Remove the gels from the stain and rinse with 7 per cent HAc. Store the stained and washed gels in tubes containing 7 per cent HAc. The background stain will diffuse out of the gel in about 12 hours. This solution can be decanted and replaced with fresh 7 per cent HAc.

(i) If a densitometer is available, a densitographic profile of the protein bands in the gel can be obtained.

REFERENCES

1. Block, R. J., Duyran, E. L., and Zweig, G. *A Manual of Paper Chromatography and Paper Electrophoresis*. Academic Press, New York, 1958.
2. Ornstein, L., and Davis, B. J. *Disc Electrophoresis*. Preprint by Distillation Industries, Eastman Kodak Co., 1962.
3. Davis, B. J. *Disc Electrophoresis, Methods and Applications to Human Serum Proteins*. Ann. N.Y. Acad. Sci. **121** 321 (1965).
4. Ornstein, L. *Disc Electrophoresis, Background and Theory*. Ann. N.Y. Acad. Sci. **121** 404 (1965).
5. Williams, D. E., and Reisfeld, R. A. *Disc Electrophoresis in Polyacrylamide Gels: Extension to New Conditions of pH and Buffer*. Ann. N.Y. Acad. Sci. **121** 373 (1965).
6. Clarke, T. J. *Simplified "Disc" (Polyacrylamide Gel) Electrophoresis*. Ann. N.Y. Acad. Sci. **121** 428 (1965).
7. Broome, J. *A Rapid Method of Disc Electrophoresis*. Nature **199** 179 (1963).

CHAPTER 6

THE IDENTIFICATION OF *N*-TERMINAL GROUPS OF PROTEINS BY SANGER'S METHOD

INTRODUCTION

As recently as 1945 the problem of determining the exact sequence of amino acids in a protein was so formidable that few scientists believed such details of protein structures would be known within their lifetime. Today the complete sequence of insulin, a number of pituitary peptide hormones, many enzymes, myoglobin, and hemoglobin are known.

The technical advance which showed the way to this new knowledge was accomplished by Frederick Sanger. He identified the *N*-terminal amino acid on peptide chains by labeling the free amino end with a dinitrophenyl group (DNP). After hydrolyzing the protein, he identified the DNP amino acid derivative by paper chromatography. The usefulness of this reaction for determining the *N*-terminal amino acid in a polypeptide was dependent upon the stability of the DNP group toward acid hydrolysis. Fortunately, in most DNP amino acid derivatives it is stable. The reagent employed by Sanger for this purpose was 1-fluoro, 2,4-dinitrobenzene (FDNB).

Since the introduction of FDNB as a tool for sequence analysis, many more reagents have been tried. Among the most promising and most useful techniques, even superseding FDNB, is the method introduced by Edman in 1950. In this procedure phenylisothiocyanate reacts with the *N*-terminal amino acid. Treatment of the adduct, a phenylthiocarbamate, with H^+ causes cyclization and the formation of a phenylthiohydantoin derivative of the *N*-terminal amino acid. The phenylthiohydantoin can be hydrolyzed and identified. The great advantage of this technique over that of Sanger's method is that the process is repeatable, hence allowing the sequence analysis of five to eight amino acids at the *N*-terminal end of a polypeptide.

Enzymatic and chemical methods for identifying the *C*-terminal amino acid sequences are also used. Presently, however, because of advances in technology, sequence determination is almost a routine matter. The Edman procedure, the chromatographic separation of amino acids and peptide, and the quantitative analysis of amino acids and peptides have all been automated. Each year, more and more proteins are being identified sequentially because of these rapid technological innovations.

From a practical standpoint, however, the original Sanger method takes less time and requires less skill, has educational utility, and can be satisfactorily performed by a conscientious novice, and for this reason remains the method of choice for students.

The DNP labeling of an *N*-terminal amino acid may be formulated as follows:

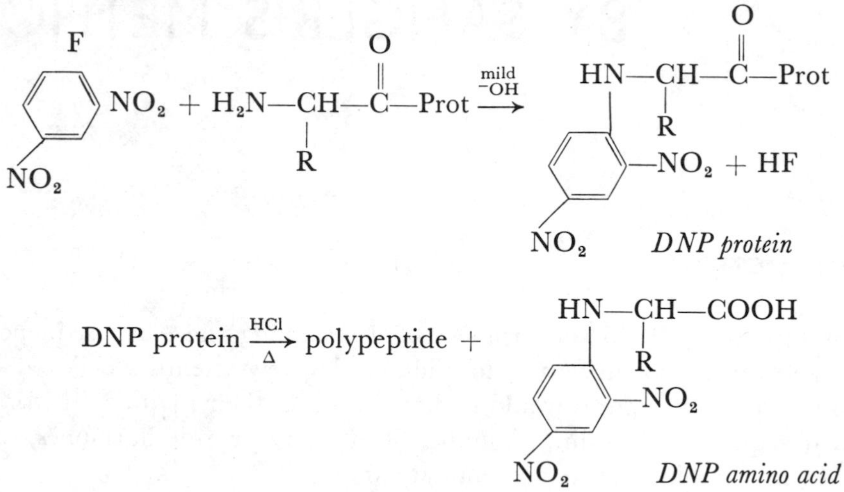

During the acid hydrolysis of DNP protein, DNP proline, DNP glycine, and DNP cystine are destroyed. Varying conditions of hydrolysis may compensate for this in the case of glycine. If cystine is suspected to be at the end, oxidation to cysteic acid is the best course for detection. A DNP proline end terminal group may be recognized without hydrolysis because DNP proline peptide has a specific absorption peak at 375 mμ in 1 per cent $NaHCO_3$.

EXPERIMENT 6

Work in pairs:

Each pair of students will analyze a different crystalline protein, either insulin, lysozyme, or ribonuclease. The experiment is divided so that one partner prepares the DNP amino acids from the proteins, and the other partner prepares three reference DNP amino acids. After each student has accomplished his task, they should work together, spotting the chromatograms and interpreting the results.

Part A. Preparation of the DNP Protein

The proteins will be provided as liquid suspensions or as dry solids. Whichever is the case, either pipette an appropriate volume or weigh out 4 mg and transfer the sample to an 18×150 mm test tube. Add 1.0 ml of water, 20 mg of $NaHCO_3$, and 1.0 ml of 2 per cent FDNB in 95 per cent ethanol. Add a glass bead and shake the mixture for 2 hours at room temperature.

Acidify the solution cautiously with 3 to 5 drops of concentrated HCl. Extract the acidified solution with peroxide-free ether (stored over $FeSO_4$). Add 5 ml aliquots of ether, shake, and remove the ether by decantation or by means of a disposable Pasteur pipette. Repeat this several times until the ether no longer becomes colored; this removes ethanol, excess FDNB, and dinitrophenol, and leaves an aqueous suspension of the insoluble DNP protein.

Part B. Hydrolysis of the DNP Protein and Extraction of DNP Amino Acids

(a) Add 4 ml of 8.5 N HCl to the DNP protein solution or paste. (HCl used for hydrolysis should be redistilled in glass to remove traces of heavy metal ions.) Cap the test tube with a glass marble and heat at 105° to 108° C in an autoclave (4 to 5 psi) for four hours in the case of insulin, sixteen hours for the other proteins. Allow the autoclave to cool without exhausting. This step should hydrolyze the peptide bonds and leave a mixture of amino acids and the DNP amino acids you are interested in identifying.

(b) Add 10 ml distilled water. Extract the DNP amino acids with four 5 ml aliquots of peroxide-free ether.

Extraction is done by adding ether to the tube, stoppering, shaking, carefully releasing the pressure, and allowing the two layers to separate. Collect the ether layer into a 50 ml Erlenmeyer flask with a capillary medicine dropper or by careful aspiration with a disposable Pasteur pipette. Evaporate off the ether using a warm water bath and a stream of nitrogen. Redissolve the DNP derivatives in a few drops of acetone. Store in the dark since DNP amino acids are sensitive to light.

(c) After extraction of the DNP protein hydrolysate with ether, the aqueous solution contains mono ϵ-DNP-lysine, O-DNP tyrosine, and DNP histidine. DNP arginine-Di DNP-histidine partitions between the two phases. In order to detect these, the water phase would have to be chromatographed as well. This is not necessary in this experiment.

Part C. Preparation of Reference DNP Amino Acids

(a) Check with your instructor in order to make sure that at least one of the amino acids you select is known to be the *N*-terminal amino acid of the protein you are investigating.

(b) First, 3 to 4 mg of each amino acid and 10 mg of $NaHCO_3$ are dissolved in 1.0 ml of water in separate test tubes. To these mixtures are added 1.0 ml of a 2.0 per cent FDNB solution in ethanol. The mixtures are shaken for two hours at room temperature and then diluted with 2 ml of water.

(c) Extract the mixture several times with 2 ml of ether to remove all excess FDNB. The final ether extraction should be clear. Discard the ether. Acidify the solution with five drops of concentrated HCl and extract the DNP amino acids three times with 5 ml of ether or until the ether solution is clear. Collect this ether extract in a 50 ml Erlenmeyer flask. Evaporate it just to dryness with a stream of nitrogen and a hot plate, and redissolve in 0.5 ml of ethanol. Store the solution or the dried residue in the dark until needed for chromatography.

Part D. Chromatographic Identification of DNP Amino Acids

Chromatography is a method of fractionation based on the selective adsorption of substances as they pass through a column of an adsorbent. Materials which have been used to chromatograph amino acids include charcoal, starch, cellulose powders, ion exchange resins and filter paper, either as strips or in compressed piles.

Paper partition chromatography, first described by Martin and Synge in 1944, has virtually revolutionized the art of detecting and identifying small amounts of organic and inorganic substances. It permits separation of mixtures on a very small scale that no other simple method affords. The technique has found widespread application in many branches of biochemistry. In addition to amino acids, it has been used to separate and identify carbohydrates, proteins, and peptides, sterols, and steroid hormones, fatty acids and other organic acids. For detection of the compounds on the paper, specific color

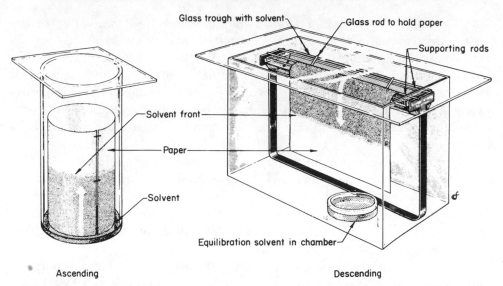

FIGURE 8. Paper chromatography. (From *Experimental Biochemistry*, John M. Clark, Jr., ed. W. H. Freeman and Company. Copyright © 1964.)

tests, u.v. absorption, fluorescence, radioautography, and enzymatic response have been used.

Several factors are of importance in the separation of substances by paper chromatography. These involve partition, a combination of partition and adsorption, and in some cases ion exchange. The predominant factor in the separation of solutes on paper is the partition between two immiscible phases. Theoretically, the paper functions as an inert support for the aqueous solvent (by hydrogen bonding and apolar attractions), although other functions of the paper, such as capillary movement of the solutes (adsorption) and polar attractions due to artifacts in the paper (ion exchange), may also occur. Partition takes place between the stationary aqueous phase and the moving solvent phase. The solute moves in the direction of solvent flow at a velocity which is governed by the differential attraction between the stationary aqueous phase and the moving non-polar organic phase. A high velocity of migration in the direction of flow indicates greater attraction of the solute for the solvent, whereas low migration velocity indicates greater attraction for the stationary aqueous phase. Factors other than charge are important for determining the rate of migration of a solute. These are molecular weight, type of paper used, size of the sampling chamber, temperature, and so forth. The migration rate of the solute in the direction of solvent flow is characterized by the term R_f, which is the distance the solute migrates from the starting point divided by the distance traveled by the solvent front:

$$R_f = \frac{\text{Movement of the solute band}}{\text{Movement of solvent front}}$$

The R_f of any solute in a system cannot be more than 1.0, and in fact, is usually less than 1.

The DNP amino acid reference solutions and DNP derivative of the protein hydrolysate are spotted on 40 × 40 cm square sheets of chromatography paper *in duplicate*. The paper used is Whatman #2 filter paper which has been previously soaked in 0.1 molar phthalate buffer, pH 6.0, and dried. Pencil mark an origin line 4 cm above the factory cut edge of the chromatography paper. Practice on a scrap of filter paper applying acetone spots with a capillary until you can always keep the spot at 5 mm diameter or about 5 μl volume.

Use ascending chromatography with benzyl alcohol-ethanol (90:10), saturated with phthalate buffer, pH 6.0, as solvent. When you are ready to spot the experimental samples, start 4 cm from the lead edge and apply eight spots 3 cm apart. Care should be exercised to keep the paper clean. Make two applications of each amino acid sample, and six of the protein sample. Drying the spots may be facilitated with a hot air gun.

The sheet is curled into a cylindrical shape and the overlapping portions (half an inch) are pinned together with capillary tubing. Finally, a 1 × 1 inch rectangle of paper is cut off from one of the lower corners to prevent the overlap from reaching the bottom of the cylinder, and thus eliminating subsequent deformations of the chromatographic front due to an excess rise of solvent up the joint.

Twelve to sixteen hours will be required for development, and it should be done in the dark. Upon completion of the run, the papers are *air* dried. Inspect for degree of separation. Mark off the spots and determine the distance traveled by the DNP amino acids by roughly estimating the center of the spot and measuring the distance from the origin. This distance divided by the total distance that the solvent has traveled is the R_f value, and is specific for each derivative under a given set of chromatographic conditions.

Table 14

Protein	R_f of DNP Amino Acids from Proteins	R_f of Known Amino Acids		
horse heart myoglobin		phe	leu	ala
papain		lys	try	
lysozyme		val	ser	
insulin				
ribonuclease				

Complete Table 14.

Make a statement of your conclusions.

What is the function of these proteins?

Examine the sample under visible and UV light (2537 Å). Some sample, particularly the unknown, may be contaminated with free dinitrophenol. This spot is readily recognized since its yellow color disappears upon exposure to HCl vapors.

REFERENCES

1. Sanger, F. *The Free Amino Groups of Insulin*. Biochem. J. **39** 507 (1945).
2. Fraenkel-Conrat, H., Harris, J. I., and Levy, A. L. "Recent Developments in Techniques for Terminal and Sequence Studies in Peptides and Proteins," in *Methods of Biochemical Analysis* (D. Glick, ed.). John Wiley and Sons, Inc., New York, 1955.
3. Edsall, J. T., and Wyman, J. "The Sequence of Amino Acids and Residues in Peptides," in *Biophysical Chemistry* (Vol. 1). Academic Press, New York, 1958.
4. Alexander, P., and Block, R. J. *Laboratory Manual of Analytical Methods in Protein Chemistry*. Pergamon Press, Long Island City, N.Y., 1960.
5. Canfield, R. F. *Peptide Sequence Determination by Edman Degradation*. J. Biol. Chem. **238** 2698 (1963).
6. *Experimental Biochemistry Laboratory Manual*. Department of Biochemistry, University of Wisconsin, 1967.

CHAPTER 7

THE CATALYTIC PROPERTIES OF AMYLASE, CATALASE, AND UREASE

INTRODUCTION

In 1926 Sumner crystallized the enzyme urease from jack bean meal and identified it unequivocally as a protein. Until Sumner's achievement, no association had been made between proteins and enzymes, and his discovery was vigorously disputed and even ridiculed for many years. Many investigators believed that enzymes were some kind of vital life force which could not be attributed to something so common as a protein, much less be crystallized like an ordinary substance. Many investigators refused to accept this important discovery. More than 100 crystalline enzymes have been isolated since 1926; all are proteins, and now the protein nature of enzymes is taken as a matter of course. Sumner ultimately was awarded a Nobel prize for his efforts and perseverance.

At present, enzymology occupies a very important position in medical, biological, and industrial research. Quantitative determinations of certain enzymes have become important diagnostic tools in the practice of medicine. For instance, increased levels of acid phosphatase in the blood are associated with metastatic carcinoma of the prostate; alkaline phosphatase levels reflect changes in bone formation and in bone dissolution processes; increased levels of oxalacetate-glutamate transaminase and lactic dehydrogenase are indicative of myocardial infarction and liver damage. Glutamate transaminase is associated with acute viral hepatitis, α-amylase with pancreatitis, and alkaline phosphatase with a variety of diseases of bone, jaundice, liver neoplasm, and multiple myeloma.

It also appears certain that many chemotherapeutic agents exert their characteristic effect by inhibiting particular enzymes. In industrial research there is considerable effort to utilize enzymes in the mass production of important biochemicals, in the fixation of atmospheric nitrogen, and in the cracking and synthesis of petroleum products, to cite a few examples. Lastly, but not least of all, the biological significance of enzymes as universal agents of all metabolic activity in the cell is continually being unfolded.

Enzymes as catalysts are extremely efficient. An appreciation of this efficiency may be gained from the fact that under optimal conditions rates are 10^8 to 10^{11} times more rapid than corresponding non-enzymatic reactions. The number of molecules of substrate (reactant) converted to product per molecule of enzyme per minute, that is to say the turnover number, ranges from 10^3 to 5×10^6. Most enzymes are very highly specific toward the substrate utilized and products formed, but are responsible, nevertheless, for catalyzing a host of complex reactions, including hydrolytic reactions, polymerizations, oxidation-reduction reactions, dehydrogenations, aldol condensations, acyl transfers, and free radical reactions.

As contrasted to laboratory conditions for carrying out similar reactions, enzymes make it possible to perform the same reactions under very moderate conditions, usually at neutral pH and at ambient temperatures.

In addition, there are many delicate control mechanisms governing enzyme reaction rates. For example, the rate of enzyme synthesis and concentration within the cell is genetically controlled by complex feedback mechanisms involving both substrates and products. Enzyme or protein biosynthesis, which in itself is an exceedingly complex metabolic process, also is enzyme catalyzed. In a more restricted sense, a particular reaction rate can also be controlled by other types of substrate and product inhibition mechanisms. Monitors of enzyme catalysis include monovalent and divalent cations, such as Mg^{+2} and Ca^{+2}, or Na^+ and K^+, which can act either antagonistically or synergistically with respect to some enzyme reaction. Some enzymes are activated by K^+ and inhibited by Na^+, or vice versa, and certain enzymes can be stimulated by either ion alone, though there may be a differential effect; but others also show an even greater stimulation when both are present simultaneously. Enzymes also exist as inactive forms that have to be modified chemically or enzymatically in order to be converted into a viable state. In these cases, a physiological event occurs prior to enzyme action which triggers a sequence of reactions involving enzyme activation. Coenzymes and prosthetic groups are complex organic molecules, as for example nicotinamide adenine dinucleotide and flavin adenine dinucleotide, both of which are essential for certain catalytic processes to take place. In general, prosthetic groups are firmly bound to the enzyme as metal complexes of Mg^{+2}, Mn^{+2}, or Fe^{+2}, and porphyrin. These can be found in enzymes such as catalase, the respiratory cytochromes, and the magnesium-chlorophyll complexes of chloroplast proteins of plants. Coenzymes, on the other hand, are loosely bound to enzymes or bind to form a complex with the enzyme only during catalysis. In either case, however, prosthetic groups or coenzymes associated with certain enzymes are an essential part of the catalytic process.

A systematic classification and nomenclature for enzyme-catalyzed reactions was established by the Commission of Enzymes of the International Union of Biochemistry in 1964. An enzyme unit was defined as that amount which will catalyze the transformation of 1 μmole of substrate per minute under standard conditions at 30° C. There are two special cases: (a) When the substrate is a bio-polymer in which more than one bond is attacked, then 1 μequivalent of the group attacked should be substituted. For example, the number of peptide bonds or glycosidic bonds broken in a protein or a polysaccharide is taken as the measure rather than the number of complete molecules hydrolyzed. (b) In the case of bimolecular reaction, 1 μmole of substrate A or B is taken as the basis, but if $A = B$, as when the reaction is between two identical molecules, the basis should be 2 μmoles of A. Thus in all cases one cycle of the reaction is taken as the measure of the rate.

There are several ways to follow enzyme reaction: (1) by measuring substrate disappearance, (2) by determining the quantity of product formed, or (3) by measuring physical changes of prosthetic or coenzyme groups during catalysis.

Each enzyme exhibits optimum reaction rates under conditions involving activator, coenzyme, substrate, and enzyme concentrations, as well as pH, temperature, extent of product accumulation, and on denaturation of the enzyme during assay.

EXPERIMENT 7

Part A.   Amylase

Introduction.   α-Amylase (trivial name); International Union of Biochemistry number 3.2.1.1. Systematic name: 1,4-glucan 4-glucanohydrolase. Molecular weight 45,000.   α-Amylase has been crystallized from saliva, human pancreas, and the microorganisms Pseudomonas and Aspergillus. Among the amylases there are two broad groups, α- and β-amylases. The β-amylases rapidly hydrolyze the amylose portion of starch to maltose. They hydrolyze α-1,4-glucan links in polysaccharides so as to remove successive maltose units from the non-reducing ends of the chains. The α-amylases, in contrast to the β-amylases, cause a rapid loss of the capacity of amylose to give a blue color with iodine; also, the rate of appearance of maltose is much slower in the α-amylase catalyzed reaction than in the β-amylase catalyzed one. They hydrolyze α-1,4-glucan links in polysaccharides containing three or more α-1,4-linked glucose units to a series of ill-defined products: soluble starch, erythrodextrin, achrodextrins, and maltotriose and maltose.

Saliva, produced by submaxillary, sublingual, and parotid glands, as well as by mucous membranes and the buccal glands of mouth, throat and esophagus, contains about 99.5 per cent water; the solid material consists of salivary amylase (ptyalin), several proteins (one of which is mucin, a glycoprotein), a number or inorganic ions, e.g., Ca^{+2}, Na^+, K^+, Mg^{+2}, Cl^-, HCO_3^-, and phosphates, and bio-organic compounds such as amino acids, urea, uric acid and cholesterol.

PROCEDURE

Work in pairs throughout this experiment.

Collection of Saliva.   After rinsing your mouth thoroughly with water, chew on a small piece of paraffin wax so as to stimulate the flow of saliva.

Transfer the accumulated saliva to a small beaker, and use it for the following experiments. An occasional individual will be found whose saliva shows a lack of, or weak, salivary amylase activity; do not be concerned.

Comparison of Activity of Dialyzed and Undialyzed Saliva.   Secure a five inch length of cellophane dialysis tubing and knot one end very tightly. With the aid of a funnel, add about 5 ml of saliva to the bag and tie off the upper end. Immerse the bag in about 400 ml of distilled water and allow dialysis to proceed for 60 minutes or longer; change the water after 20 and 40 minutes and occasionally agitate the dialysis bag.

While dialysis is proceeding, carry out preliminary experiments to estimate amylase activity in the original saliva. It is desirable to learn the dilution that is necessary to hydrolyze starch to the "achromic point." The "achromic point" is reached when the addition of the digestion mixture to iodine ceases to produce any change in color in approximately five minutes at 37° C.

Preincubate 10 ml of a 1 per cent solution of soluble starch in an 18 × 150 mm test tube for 2 minutes at 37°. Add 2 ml of diluted saliva (1:20 dilution of original saliva with distilled water), mix, and incubate in the 37° water bath. At 30 second intervals, transfer three drops of the incubation mixture to a depression in a spot plate into which you have previously placed two drops of 0.01 M iodine solution. Note which sample fails to give a positive starch-iodine test. If the incubation time for this is less than three minutes make an appropriate dilution of the original saliva and repeat the determinations. Continue until you find the dilution that, under these conditions, yields an achromic point between three and eight minutes. Go on to the last part of this section as time permits.

ACTIVATION OF SALIVARY AMYLASE. After dialysis is complete, carefully rinse the contents of the bag into a graduated cylinder and add sufficient distilled water to dilute the dialysate to the concentration found to produce the desired achromic point. Transfer 5 ml of the original saliva to a graduated cylinder and make the same dilution. Make certain that both solutions are well mixed.

Prepare five mixtures as in Table 15, without the saliva.

Temperature equilibrate all mixtures in the 37° water bath for five minutes. At zero time, add saliva as indicated in the table. At one-minute intervals, test three drop aliquots of each by the spot plate procedure given above and determine the time required for each to reach the "achromic point."

What can you conclude about the property of this enzyme from this experiment?

Table 15

	Additions	Test Tube Number				
		1	2	3	4	5
A.	1 per cent starch, ml	10	10	10	10	10
B.	0.1 M NaCl, ml	1				1
C.	distilled H_2O, ml			1	1	
D.	0.1 M Na_2SO_4, ml		1			
E.	diluted dialyzed saliva, ml	2	2	2		
F.	diluted original saliva, ml				2	2

How does dialysis of saliva affect the ability to hydrolyze starch?

Test for Reducing Groups in the Digestion Mixture. Again preincubate 10 ml of a 1 per cent solution of starch at 37° for two minutes. Divide a circle of filter paper into 16 numbered segments, corresponding to the intervals of 0, 0.5, 1, 2 and so on, to a final interval of 15 minutes. At zero time add 0.2 ml of undiluted saliva to the starch solution. At the prescribed time intervals, transfer three drops of the digestion mixture to the spot plate. In addition, dip a clean stirring rod into the digestion mixture and spot it once or twice at the appropriate position on the filter paper. Allow it to dry; test the sample in the spot plate for the achromic point with iodine as before. Repeat this at each time interval. Record the time at which the achromic point appears. After all the spots have been collected on the filter paper and dried in an oven, test for reducing sugars in the digestion mixture. Spray the paper in the hood with aniline phthalate reagent. Return the paper to the oven and heat for 15 minutes. A brown color will develop in the presence of reducing sugars.

What do you conclude?

Part B. Catalase

Introduction. Catalase (trivial name); International Union of Biochemistry number 1.11.1.6. Systematic name: hydrogen-peroxide:hydrogen-peroxide oxidoreductase. Molecular weight 250,000. Catalase is an enzyme present in all living cells (with a few exceptions, as among anaerobic microorganisms), but is especially abundant in blood and liver. It has been prepared in crystalline form from both sources. The enzyme is highly specific and will act upon no natural substrate other than H_2O_2, though it will utilize short-chained alkyl peroxides as electron donors. It catalyzes the following reaction: $H_2O_2 \rightleftarrows H_2O + \frac{1}{2}O_2$. Hydrogen peroxide is a product of a number of cellular oxidations and catalase is believed to be present in cells in order to prevent H_2O_2 accumulation. It is one of the most active enzymes known, having a turnover number of 5×10^6 at zero degrees.

THE CATALASE REACTION. The activity of catalase preparations has been determined in a number of ways, most of which are subject to inherent errors. A convenient though approximate method is to determine the undecomposed H_2O_2 by titration with permanganate after incubating the enzyme with an excess of H_2O_2. The enzyme is allowed to act on a dilute solution of peroxide for five minutes, and the reaction is stopped by the addition of strong sulphuric acid, which destroys the enzyme. The titration reaction is:

$$2MnO_4^= + 5H_2O_2 + 6H^+ \rightleftarrows 5O_2 + 2Mn^{+2} + 8H_2O$$

THE EFFECTS OF CYANIDE ON CATALASE. Cyanide ions form very stable complexes and inactivate enzymes containing ferric iron, but have little effect on enzymes containing ferrous iron. For example, cyanide does combine with methemoglobin but not with hemoglobin, which contain ferrous iron. The cytochromes, catalase, and peroxidases contain the ferric form and are therefore strongly inhibited by cyanide.

PROCEDURE

Work in pairs.

A 1:1000 dilution of fresh blood is used as the enzyme source. Place 10 ml of cold distilled water into a 50 ml flask surrounded by crushed ice. Obtain a 10 lambda (10 μliter) sample of fresh finger-tip blood by puncturing with a sterile lancet. Discharge the blood into the chilled water. Draw water into the pipette several times to insure complete transfer of blood. Swirl the flask for a minute in order to get a homogeneous mixture. Maintain the diluted blood at ice temperature throughout the experiment to minimize heat inactivation.

Prepare 18 × 150 mm tubes as indicated in Table 16. All reagents should be brought to ice temperature and the reaction run at zero degrees. Be sure to mix contents by swirling after each addition.

Table 16

	Reagents	Test Tube Number				
		1	2	3	4	5
A.	0.02 M phosphate buffer, pH 7.0, ml	10	10	10	10	10
B.	0.05 M H_2O_2 in 0.02 M phosphate buffer, pH 7.0, ml	2	2	2	2	2
C.	water, ml	1.0	1.0	1.0	0.5	
D.	5 × 10^{-5} M KCN, ml				0.5	1.00
E.	catalase, ml	0.5	0.5		0.5	0.5
F.	heated catalase ml			0.5		
G.	ml 0.005 M $KMnO_4$ used					
H.	H_2O_2 remaining					
I.	H_2O_2 decomposed					
J.	specific activity					

Tube 1 is a reagent blank and represents the total amount of H_2O_2 added to each tube. In this case the addition of 2 ml of 6 N H_2SO_4 precedes the addition of enzyme; *it should be added to this tube prior to initiation of the reactions in the other tubes.* Calculate the amount of 0.005 M $KMnO_4$ needed to react with 2.0 ml of 0.05 M H_2O_2. If tube 1 does not approach this value, your peroxide solution has decomposed and you must get a fresh one. Terminate the reaction in tubes 2 to 5 after exactly five minutes by adding 2 ml of 6N-H_2SO_4 to each tube. After terminating the reaction at the appropriate time, transfer the inactivated enzyme mixture to a 50 ml beaker.

The H_2O_2 which has not been decomposed by the catalase will be stable in the acid solution for at least half an hour. During this period, complete the titration of the peroxide with 0.005 M $KMnO_4$. The first drop of permanganate in excess over that required for the oxidation of peroxide turns the solution pink. This is the end point.

CALCULATION OF ACTIVITY: The activity of catalase may be expressed as the μmoles of H_2O_2 decomposed by enzyme action in five minutes at zero degrees per ml of blood.

To begin with, since a 0.05 M solution of H_2O_2 was used at the start of the experiment, 100 μmoles of H_2O_2 was present in each tube at the start. In order to do your calculations, the number of μmoles of H_2O_2 found in tube 1 is the zero time control, and if enzyme action did take place, then some amount of H_2O_2 less than the 100 μmoles at the start will be found. The stoichiometry of the decomposition of H_2O_2 by permanganate indicates that for every two moles of MnO_4^- added, 5 moles of H_2O_2 decomposes. The factor is 2.5. In order to calculate the μmoles of H_2O_2 *remaining,* the μmoles of MnO_4^- *added* must be multiplied by 2.5. By subtracting the amount of H_2O_2 undecomposed from the amount of H_2O_2 present at the start (Tube 1), the μmoles of H_2O_2 decomposed is obtained. Since a 1:1000 dilution of blood was made, and a 0.5 ml aliquot was taken for assay, then the μmoles of H_2O_2 decomposed by catalase must be multiplied by 2000 in order to obtain the specific activity of the catalase according to the definition given above.

What are your observations and conclusions? Calculate the per cent inhibitions in tubes 3 to 5 and report them.

Part C. Urease

Introduction. Urease (trivial name); International Union of Biochemistry number 3.5.1.5. Systematic name: urea amidohydrolase. Molecular weight 483,000. Urease has six equal structural subunits. It is a highly specific enzyme present in several plant tissues such as soy bean and jack bean. The enzyme catalyzes the reaction:

$$H_2N-\overset{\overset{\displaystyle O}{\|}}{C}-NH_2 + 3H_2O \rightarrow CO_2 + 2NH_4OH$$

Urease is very widely used in clinical laboratories as a catalyst for the quantitative determination of urea. The ammonia formed may be titrated to give a measure of the amount of urea in blood or urine.

Although urease is not found extensively in nature, through Sumner's work this enzyme has influenced the development of modern enzymology more than any other. An important principle of enzyme action was also deduced from work on urease, namely the function of —SH (sulfhydryl groups) in enzyme catalysis. Urease appears to have 3 or 4 sulfhydryl active sites. Urease is one of a large number of enzymes the activity of which is dependent upon the presence of intact —SH groups of cysteine linked as part of the peptide chain of the enzyme. Oxidation of these groups may induce a reversible inactivation of the enzyme. For illustration:

$$2 \text{ Enz—SH} \rightleftarrows \text{Enz—S—S—Enz} + 2(\text{H})$$

(active form)   (inactive form)

Such inactive —S—S— forms can presumably be reactivated by reaction with —SH compounds such as glutathione (Structure?) or cysteine.

$$\text{Enz—S—S—Enz} + 2\text{GSH} \rightleftarrows \text{GSSG} + \text{Enz—SH}$$

(inactive)   gluthathione   glutathione   (active)
                (reduced)   (oxidized)

A more stable change, particularly in the case of urease, is the reaction of the —SH groups with certain heavy metal ions such as Hg^{+2}, Ag^+, or Cu^{+2} to form a "mercaptide."

$$\text{Enz—SH} + \text{Ag}^+ \rightarrow \text{Enzyme—SAg} + \text{H}^+$$

(active form)   silver mercaptide
                (inactive)

Such inactive mercaptide derivatives of enzyme molecules may in some cases be reactivated. For instance, inactivation produced by the heavy metal derivative *phenylmercuric acetate* may be reversed by reduced glutathione and other —SH compounds. You will attempt to demonstrate such a reactivation.

(1)   Enz—S-$\boxed{\text{—H} + \text{AcO—}}$-Hg—$C_6H_5$

(active)   Phenylmercuric
                            acetate

$$\rightleftarrows \text{HOAc} + \text{Enz—S—Hg—}C_6H_5$$

(inactive)

(2)   Enz—S—Hg—C_6H_5 + HS—CH_2—$CHNH_2$—CO_2H

(inactive)   (Cysteine)

$$\rightarrow \text{Enz—SH} + C_6H_5\text{—Hg—S—}CH_2\text{—}CHNH_2\text{—}CO_2H$$

(active)

It is believed that one of the important means of biological control of enzymatic activity is exerted through reversible inactivation by reactions involving the

—SH group in those enzymes in which a free —SH group is required for activity.

PROCEDURE

For this experiment use 8×150 mm test tubes, following the instructions in Table 17. Use the urease enzyme already prepared for you.

Table 17

		Tube Number			
	Reagents	1	2	3	4
A.	0.05 M tris buffer, pH 7.2, ml	10	10	10	10
B.	1×10^{-3} M phenyl mercuric acetate, ml			1	1
C.	1×10^{-2} M cysteine, ml				1
D.	standard urease, ml	1	1	1	1
E.	H_2O, ml	2	2	1	
F.	0.3 M urea in 0.05 M tris buffer, pH 7.2, ml	10	10	10	10
G.	titration volumes				
H.	corrected titration values				
I.	total μmoles of urea hydrolyzed				

To prepare the "blank" (tube 1), add all reagents, but add four drops of 1 per cent $HgCl_2$ *before adding the enzyme*. Transfer a 10.0 ml aliquot to a 50 ml flask, add two drops of 0.04 per cent methyl red indicator and titrate with 0.05 HCl. Use a 10 ml burette. Titrate to the first pink color (pH at this point?). Record the volume of acid used. This value is to be subtracted from all other titration values you will determine; the blank represents the acid required to neutralize the components of the starting medium. Since base is produced in this reaction, the experimental tubes will require more acid to achieve neutralization.

Now reagents A through E are added to tubes 2, 3, and 4, temporarily omitting the urea-tris substrate. The phenyl mercuric acetate and cysteine are allowed to incubate with urease at 37° in a water bath for 10 minutes. The reaction is initiated by the addition of the urea-tris mixture, and each tube is incubated for 15 minutes at 37°. At the end of that time a few drops of $HgCl_2$ are added with mixing in order to stop catalysis.

Titrate 10 ml aliquots from each tube, using methyl red as above for the blank. If there is time repeat these titrations several times.

Calculations

Activity (total μmoles urea split per minute at 25°)

$$\frac{[\text{Vol unknown} - \text{Vol tube 1}] \times \text{molarity of HCl} \times 1000}{15 \times 2} \times \frac{23}{10}$$

(Why?)

Why should —SH groups be essential for activity of some enzymes?

Some enzyme molecules contain many —SH groups. Are *all* required for enzyme activity?

Calculate your results in terms of μmoles of urea split per minute at 25° C; record in Table 17.

Once again you must calculate the μmoles of substrate disappearing based on a known amount of urea at the beginning of the experiment. However, in this case, you must be aware that the reaction is measured by the amount of base produced. The stoichiometry in this case is that for every μmole of urea decomposed, two μmoles of base are produced. The titration volume in tube 1 represents the amount of base present in the absence of enzyme activity and has to be subtracted. The incubation time was 15 minutes and a 10 ml aliquot was removed from a total volume of 23 ml.

Report the per cent inhibition.

REFERENCES

1. *Experimental Biochemistry Laboratory Manual*, Department of Biochemistry, University of Wisconsin, 1967.
2. *Laboratory Manual of Physiological Chemistry*, Department of Physiological Chemistry, Johns Hopkins University, 1957.
3. *Laboratory Biochemistry*, Department of Biological Chemistry, University of Michigan, 1955.

CHAPTER 8

THE ISOLATION OF DNA AND RNA

DNA and RNA are carriers of genetic information in the cell, but in addition, they can have structural and metabolic functions. The isolation of intact native DNA and RNA is a major problem, and only in exceptional cases have intact nucleic acids been isolated. The isolation of DNA is such a formidable problem that it is reasonably claimed that few, if any, investigators have isolated DNA with over 90 per cent of the molecules in their native state. Most DNA is of such large size that chemical and enzymatic rupture at one nucleotide in a thousand is enough to destroy its function and macromolecular structure. Nevertheless, many problems of biochemical genetics, metabolic regulations, and cell differentiation are now studied with techniques which require that DNA be obtained with less than one cleavage per 10,000 nucleotides. Extreme care in removal of hydrolytic enzymes, prevention of bacterial contamination, and use of mild chemical and physical processes are essential for isolating native nucleic acid. Fortunately, the composition, identity of unusual bases, amount, rate of synthesis, subcellular location and many other important properties of DNA can be measured even if partial degradation has destroyed the biological function. Also, small nucleic acids such as amino acid transfer RNA (s-RNA) are less labile because of their small "target size," and can be isolated and purified by processes considered too harsh for isolating viral or messenger RNA. A typical cell contains 50 or more different kinds of s-RNA, and isolation and structure or sequence determination of these nucleic acids is a major problem of modern biochemistry. The detailed structures of a number of this type of nucleic acid have recently become known.

DEGRADATION OF NUCLEIC ACIDS

Degradation of a nucleic acid to its component nucleotides, nucleosides bases, sugars, and phosphates is useful for analysis and identification. Exonucleases degrade the chain specifically from one end or the other. Some endonucleases cleave only at pyrimidine bases, others cleave only at guanylic acid and still other enzymes cleave at the four common bases. In addition to the four common bases of nucleic acids, over a dozen other purines and pyrimidines have been shown to occur in natural nucleic acids. The end products of enzymic degradation can be oligonucleotides, 5′-mononucleotides or 3′-mononucleotides, depending on the enzyme chosen. Some ribonucleases cleave the 2′,3′ cyclic phosphate mononucleotide as a first product, and slowly cleave this to the 3′-mononucleotide. Nucleosides can be obtained by phosphatase action on the 3′ or 5′-mononucleotides, and eventually there can be cleavage of the βN-glycosidic linkage to yield free bases.

Chemical degradation of DNA does not lead to simple products. DNA is resistant to alkaline hydrolysis, but high pH completely degrades RNA to mononucleotides. The purine bases of DNA can be removed from the chain by mild acid (10 hours at 37° in 9.1 N HCl). One hour at 100° in 12 N perchloric acid hydrolyses DNA to free its constituent bases, and also produces phosphate and levulinic acid.

Aqueous acid or base catalyzed degradation of RNA is markedly different from that of DNA because of the 2′-hydroxyl group. In both acids and bases, the diester cleavage proceeds via the cyclic 2′,3′-phosphate formation. Base catalysis gives mixtures of the 2′ and 3′-mononucleotides as final products. Instability of the purine ribose glycosidic link in acid makes quantitative production of the 2′,3′-mononucleotides difficult. After one hour in 1 N HCl at 100° the products are mixed 2′,3′-nucleotides of the pyrimidines, purine bases, and phosphate and ribose.

ISOLATION OF NUCLEIC ACIDS

The major steps in nucleic acid isolation are cell rupture by osmotic shock; homogenization; enzymatic digestion or mild mechanical rupture, sometimes followed by isolation of nuclei, mitochondria, viruses or other subcellular structures. Since nucleic acids bind very strongly to cations and to cationic proteins such as histones and protamines, it is necessary to separate one from the other with a minimum breakage of the polynucleotide chain. Aqueous phenolic solutions containing detergents and chelating agents have been most successfully employed for this purpose. This solvating mixture produces a two phase system when added to broken tissue; the nucleic acids enter the aqueous layer as the denatured proteins dissolve in the phenolic layer or precipitate at the interface. Protein dissociation from the nucleic acids is facilitated by the action of added anionic detergents or concentrated salts. Chelating agents are also added to remove polyvalent metals which can form salts with the phosphate groups of the nucleic acids. Finally, the pH is made mildly alkaline so as to reduce electrostatic interactions of nucleic acids with other ions.

All of these purification steps should be conducted in a manner which minimizes degradation by DNAse or RNAse (remove protein quickly, keep solutions cool, and work rapidly in the first few steps), or by chemical hydrolysis (avoid pH below 2 or temperature above 90° C), and by hydrodynamic shear (avoid rapid stirring and do not pipette DNA through any capillary orifices). In addition, the steps must prevent denaturation (strand separation) by electrostatic repulsion, which is serious if DNA is exposed to salt of less than 0.01 molar, by heating to temperatures above 70° C, or by changing the ionic form of the bases (at pHs below 3 or above 10). Meta cresol and phenol are often used to extract nucleic acid because (a) a mixture of the two can be cooled to 5° without crystallizing phenol, and (b) the mixture is a good deproteinizing agent. Addition of 8-hydroxyquinoline to the mixture improves the yield by retarding ribonuclease action. rRNA can be isolated devoid of mRNA by extraction with naphthalene-1,5-disulfonate instead of 4-aminosalicylate. The rRNA with mRNA isolated in this manner will stimulate amino acid incorporation into polypeptides in a cell-free system; on the other hand rRNA extracted by disulfonate will not.

THE SEPARATION OF PURE RIBOSOMAL RNA AND NUCLEAR DNA FROM RAT LIVER

Work in Pairs.

One student is responsible for RNA and the other for DNA. Do not discard any fraction until the experiment is completed.

(a) One fresh rat liver frozen in liquid N_2 is broken down in a blender precooled to 5° C with a mixture of 6 per cent (w/v) sodium 4-aminosalicylate, 1 per cent NaCl, and phenol-cresol mixture using 5 ml of each solution for 1 g of liver. *Wear plastic gloves to prevent phenol-cresol burns.*

(b) The mixture is stirred for 20 minutes at ambient temperature and then centrifuged at 8000 × g for 10 minutes at 5°; if any emulsion remains in the top phase, this is removed and centrifuged again. The top liquid phases are combined.

(c) NaCl, 3 g per 100 ml of the top phase, is added and the mixture is extracted with 0.5 volume of phenol-cresol for 10 minutes at ambient temperature.

(d) After extraction, the mixture is centrifuged at 10,000 × g for 15 minutes at 2°. The mixture separates into three phases: a lower clear yellow phase, a solid intermediate phase, and a cloudy white aqueous upper phase. The upper aqueous phase is removed. Avoid the protein precipitate of the middle layer, recentrifuging if necessary at 20,000 × g for 5 minutes to clear it. Mix the aqueous phase with 2 volumes of ethanol-m-cresol (v/v 9:1), and allow it to stand for 30 to 60 minutes at 2°.

(e) The precipitate is centrifuged off and collected at 20,000 × g for 5 minutes at 5° and thoroughly extracted twice with 12 ml of cold 3 M NaAc, pH 6.0. This removes DNA, glycogen, and sRNA. The ribosomal RNA is centrifuged off each time at 20,000 × g for 5 minutes at 5°.

From this point the students are to work separately.

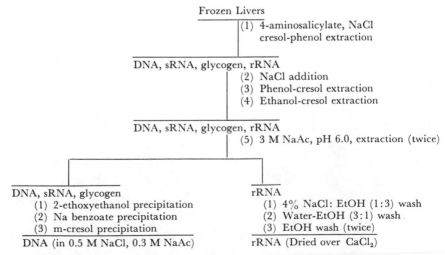

(f) The ribosomal RNA is washed once with a cold mixture of 12 ml of 4% NaCl:ethanol 1:3; once with ethanol-water 3:1 and twice with ethanol;

and dried in vacuum over $CaCl_2$. The yield should be about 40 mg from one 7 g liver. Dissolve approximately 10 mg in 0.01 M NaOH (1 mg/ml concentration) for future analysis.

(g) The DNA collected in NaAc extract is now separated from the glycogen and the sRNA. The 3 M NaAc solution is mixed with an equal volume of 2-ethoxyethanol; the precipitate is centrifuged off at 20,000 × g for 5 minutes, and is redissolved in 1 M NaCl. If no precipitation occurs, continue to add 2-ethoxyethanol until it does.

(h) Add Na benzoate to 20 per cent concentration. Centrifuge for 10 minutes at 20,000 × g to remove the glycogen.

(i) DNA is precipitated from the supernatant fraction by adding 0.2 volumes of m-cresol. DNA separates as a gel and is centrifuged off at 48,000 × g for 10 minutes. The sRNA is not precipitated by either m-cresol or Na benzoate.

(j) The gel sediment is redissolved in a minimum of 0.5 M NaCl–0.3 M NaAc mixture, and precipitated as fibres with an equal or greater volume of ethoxyethanol.

(k) These fibres are redissolved in 0.15 M NaCl-NaAc mixture and dialyzed at 2° against this same salt solution. The DNA prepared in this way is free from protein and has a sedimentation constant of 16 to 18 S. The DNA is best kept in solution.

SEPARATION OF DNA FROM BACTERIAL OR YEAST CELLS*

To 2 to 3 grams of wet packed E. coli (yeast) cells or 0.5 gram of lyophilized *M. lysodeikticus* cells, add 30 ml of 0.15 M NaCl in 0.1 M EDTA buffered to pH 8. Stir in 2 ml of 25 per cent SDS (sodium dodecyl sulphate); transfer the upper broken cell suspension to a 250 ml erlenmeyer flask and warm to 60° (in a water bath) for 10 minutes. Cool to room temperature, add 5 ml of 6 M $NaClO_4$, stir, and then shake for 15 minutes in a glass or Teflon stoppered flask with an equal volume of chloroform-isoamyl alcohol (24:1 v/v). Centrifuge the resulting emulsion for 3 minutes in a low speed centrifuge at 5000 rpm.

Carefully remove the upper (aqueous) phase with a disposable pipette; avoid the copious precipitate at the interface. Recentrifuge if necessary in water to clear; place in a 125 ml erlenmeyer flask, and gently layer 70 ml of ethanol over the solution. Mix gently with a stirring rod, collecting the fibrous precipitate by winding it on the rod and squeezing excess liquid out by turning the spooled mass against the side of the flask. Transfer the spooled mass to a glass-stoppered tube containing 10 ml of H_2O plus 1 ml of the saline-EDTA solution. As soon as the DNA dissolves, add 0.5 ml of 25 per cent SDS, then 2 ml of 6 M $NaClO_4$ and 10 to 15 ml of chloroform-isoamyl alcohol. Repeat the deproteinization step several times by shaking, centrifuging, and recovering the upper phase. Again spool out the DNA. After adding several volumes of ethanol and after pressing out the excess solvent, dissolve in 2 ml of water

---

* If yeast is used, grind the yeast cake vigorously for 10 minutes in a chilled mortar and pestle with 2 volumes of acid-washed sea sand; add saline–EDTA and stir in SDS.

containing 0.2 ml of saline-EDTA buffer. This stock solution may be frozen if tests of yield and DNA properties are to be performed.

See Experiment 9 for procedure for analysis of DNA and RNA. Calculate the yield in each case, expressed as mg nucleic acid per gram of tissue wet weight.

REFERENCES

1. Colowick, S. P., and Kaplan, N. O., eds. *Methods in Enzymology, Vols. III and XII*. Academic Press, New York, 1968.
2. *Experimental Biochemistry Laboratory Manual*. Department of Biochemistry, University of Wisconsin, 1967.

CHAPTER 9

THE DISTRIBUTION OF NUCLEIC ACIDS IN SUB-CELLULAR PARTICLES

The cytological distribution of a particular constituent often suggests a possible metabolic role for that substance. In recent years, cytochemical procedures have become increasingly sensitive and quantitative. A major breakthrough resulted when new techniques made it possible to isolate and purify sub-cellular particles for further *in vitro* studies by chemical, enzymatic, and electron microscopic procedures. Some structures that have been isolated by differential centrifugation are cell walls, cell membranes, nuclei, mitochondria, chloroplasts, microsomes, ribosomes, lysosomes, mitotic apparati, secretory granules, and nucleoli. Several of these structures have been broken into even smaller specialized sub-units by application of additional refined techniques of density gradient and zonal centrifugation, size exclusion chromatography, and gel electrophoresis.

Long chain polynucleotides are hydrophilic polyanions resembling proteins in some of their chemical and physical properties. They are soluble in dilute salt solutions; they bind polyvalent cations strongly, particularly basic proteins and cationic detergents; and they aggregate or precipitate in the presence of other polycations. Like proteins, polynucleotides can be precipitated by concentrated salt solutions (particularly ammonium sulfate), by alcohols in the presence of salts, and by trichloroacetic acid (TCA) or perchloric acid (PCA).

Ordinarily, nucleoprotein complexes are dissociated by 1 M NaCl or NaClO$_4$, and the separated nucleic acids are precipitated with ethanol. Another useful procedure takes advantage of the solubility properties of polynucleotides in TCA or PCA. At 0° C., nucleoproteins (and proteins in general) are precipitated out by these reagents; but at 90° C. the nucleo-proteins are degraded. The nucleic acids, on the other hand, remain in solution after cooling, while all proteins are rendered insoluble. This affords a means of separating proteins from nucleic acids. By procedures similar to the one described in this experiment, TCA extraction of nucleic acids can be employed successfully with a variety of tissues; some tissues, however, may require special consideration. In the case of brain tissue, for example, care in the selection of solvents for lipid extraction is necessary, both to insure the complete removal of lipids and to prevent the loss of nucleic acids. Ethanol and ether are unsuitable for this purpose; a better choice is chloroform-methanol (2:1). It is also desirable to extract the lipids before performing the TCA extraction of the proteins.

A second problem with brain tissue arises from interference with the diphenylamine reaction most commonly used for analysis of DNA. In this case, DNA and RNA must be separated from one another prior to analysis. This can be accomplished by the selective hydrolysis of RNA in alkaline solution. RNA is sensitive to alkaline hydrolysis, while DNA is not: the mechanism of hydrolytic cleavage of RNA involves phosphodiester cyclization at the C_2 and C_3 hydroxyl groups of ribose, and since there is no C_2 hydroxyl group in DNA, it is not hydrolyzed by alkali. The intact DNA can be separated from the degraded RNA by acidification with cold TCA or PCA (see references 8 and 9).

With minor modifications, the scheme described in this experiment can be employed to analyze either whole tissue or tissue organelles for total lipid and protein as well as for total DNA and RNA. When this is desired, lipid extraction should precede protein and nucleic acid separation.

EXPERIMENT 9

In this experiment you will examine the distribution of ribose nucleic acid (RNA) and deoxyribose nucleic acid (DNA) in sub-cellular components from liver. Students are to work in pairs; one member of each pair will prepare cell nuclei, and his partner will begin preparing mitochondria, microsomes, and the soluble cytoplasmic fraction. However, since the preparation of nuclei is faster, the student completing this procedure should assist his partner when feasible.

Each pair of students will excise one fresh rat liver. The liver is cut in half and the weight of each half is recorded. Each liver half is then washed once with the medium designated for each preparation and then minced and homogenized in the same medium. The procedure outlined for the preparation of the sub-cellular components is given in Flow Sheet I.

At this point, one ml of the 20 *per cent homogenate* is saved to determine the total nucleic DNA and RNA of the entire liver.

After these preparations, the nucleic, mitochondrial, microsomal, and soluble fractions are treated according to the procedure outlined in Flow Sheet II. This general procedure can be applied to a variety of tissues. As an example, for neuronal tissue a discontinuous centrifugation separation is required in order to obtain a pure preparation of mitochondria, and during the extraction procedure additional steps should be included, e.g., extraction with 5 ml of EtOH/ether/chloroform (2/2/1) and ether/acetone (2/1), in order to ensure complete lipid extraction.

After isolation, the sub-cellular fractions are freed of lipid, protein, and low molecular weight metabolites and then analyzed for pentose and deoxypentose content as a measure of RNA and DNA.

FLOW SHEET I. PREPARATION OF SUB-CELLULAR COMPONENTS

Record and make up all volumes accurately.

A. Nuclei

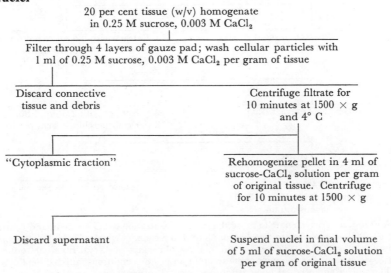

117

B. Mitochondria, Microsomes, and Supernatant

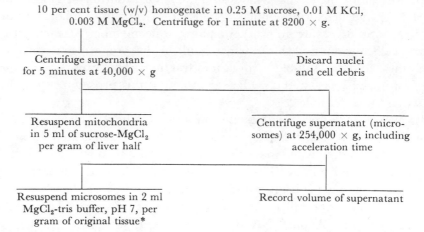

10 per cent tissue (w/v) homogenate in 0.25 M sucrose, 0.01 M KCl, 0.003 M MgCl$_2$. Centrifuge for 1 minute at 8200 × g.

Centrifuge supernatant for 5 minutes at 40,000 × g	Discard nuclei and cell debris
Resuspend mitochondria in 5 ml of sucrose-MgCl$_2$ per gram of liver half	Centrifuge supernatant (microsomes) at 254,000 × g, including acceleration time
Resuspend microsomes in 2 ml MgCl$_2$-tris buffer, pH 7, per gram of original tissue*	Record volume of supernatant

FLOW SHEET II. EXTRACTION PROCEDURE FOR TOTAL NUCLEIC ACID ANALYSIS OF TISSUES AND TISSUE FRACTIONS

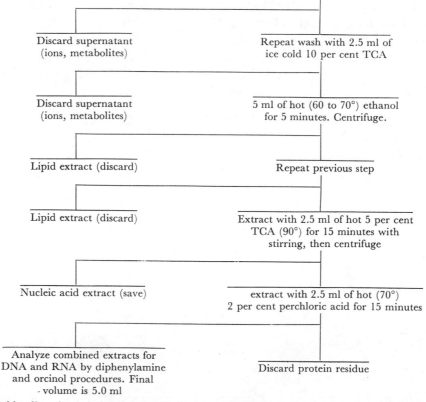

1 ml of the original 20 per cent homogenate and 1 ml of each of tissue fraction† are placed in heavy-walled glass conical centrifuge tubes with 2.5 ml of cold 10 per cent TCA in an ice bath, then stirred and centrifuged.‡

Discard supernatant (ions, metabolites)	Repeat wash with 2.5 ml of ice cold 10 per cent TCA
Discard supernatant (ions, metabolites)	5 ml of hot (60 to 70°) ethanol for 5 minutes. Centrifuge.
Lipid extract (discard)	Repeat previous step
Lipid extract (discard)	Extract with 2.5 ml of hot 5 per cent TCA (90°) for 15 minutes with stirring, then centrifuge
Nucleic acid extract (save)	extract with 2.5 ml of hot (70°) 2 per cent perchloric acid for 15 minutes
Analyze combined extracts for DNA and RNA by diphenylamine and orcinol procedures. Final volume is 5.0 ml	Discard protein residue

* Add sodium deoxycholate to bring the final concentration to 0.3 per cent, and centrifuge for one hour at 100,000 × g to prepare free ribosomes from microsomes which also include the membranes of the endoplasmic reticulum. Polysomes, protein assembly units of several ribosomes adhering to RNA messages, will be found in the microsome fraction under the conditions outlined here.

† The nuclei and original homogenate represent 1 gram of original tissue per 5 ml of solution The "mitochondria," "microsomes," and "supernatant" are 1 gram per 10 ml.

‡ All centrifugations should be 3 to 5 minutes at 2000 × g on the clinical table model centrifuge.

Table 18. DNA Determination

Reagents	Tube Number									
	1	2	3	4	5	6	7	8	9	10
A. DNA standard (50 mg/100 ml) in 1 N perchloric acid, ml		0.1 (0.05 mg)	0.2 (0.1 mg)	0.4 (0.2 mg)	0.8 (0.4 mg)					
B. extracts, ml						1.0	1.0	1.0	1.0	1.0
						Homogenate	Nuclei	Mitochondria	Microsomes	Supernatant
C. 1 N PCA	1.0	0.9	0.8	0.6	0.2					
D. diphenylamine reagent	2.0	2.0	2.0	2.0	2.0	2.0	2.0	2.0	2.0	2.0

Table 19. RNA Determination

Reagents	Tube Number									
	1	2	3	4	5	6	7	8	9	10
A. RNA standard (50 mg/liter) in 5 per cent TCA, ml		0.2 (0.01 mg)	0.5 (0.025 mg)	1.0 (0.05 mg)	2.0 (0.10 mg)					
B. 5 per cent TCA, ml	2.0	1.8	1.5	1.0	0.0	1.5	1.5	1.5	1.5	1.5
C. extracts, ml						0.5	0.5	0.5	0.5	0.5
						Homogenate	Nuclei	Mitochondria	Microsomes	Supernatant
D. orcinol reagent	2.0	2.0	2.0	2.0	2.0	2.0	2.0	2.0	2.0	2.0

Table 20

	Volume of Extract 1	Absorbance, RNA Determination 2	RNA in 0.5 ml Aliquot, mg 3	RNA in each Fraction, mg 4	Total RNA in Liver 5	Per Cent Recovery 6	Absorbance, DNA Determination 7	DNA in 1.0 ml Aliquot 8	DNA in each Fraction 9	Total DNA in Liver 10	Per Cent Recovery 11
homogenate	1.0					100					100
nuclei	5.0										
mitochondria	5.0										
microsomes	2.0										
soluble supernatant											

Store all fractions at 0°, or immediately proceed to Flow Sheet II. Otherwise nucleases will cleave the nucleic acids to fragments which no longer precipitate in TCA, and recovery of the nucleic acids will be diminished.

All extracts at this point can be stored and DNA and RNA analysis carried out during the next laboratory period. *Do not discard any of these fractions until all analyses are complete.*

THE DETERMINATION OF DNA & RNA IN EACH FRACTION

Place the reagents in test tubes as directed in Table 18. Cap with a marble and heat for 10 minutes in a boiling water bath. Cool and read absorbance at 600 mμ in the spectrophotometer, using tube 1 for a blank. Determine the DNA content of tubes 6 to 10 from a plot of your standard curve, which is made by plotting A versus the mg of DNA added to tubes 2 to 5.

Add the reagents to test tubes as directed in Table 19. Mix and heat for 20 minutes in a boiling water bath. Record the absorbancy at 640 mμ using Tube 1 for a blank. Determine RNA concentration from a plot of your standard curve, made by plotting A versus the mg of RNA added to tubes 2 to 5.

Record your data in Table 20.

SAMPLE CALCULATIONS

Assume that the original wet weight of liver is 4.8 grams. One member of the team received 2.2 grams of liver to isolate nuclei, and the other, therefore, started with 2.6 grams to isolate the mitochondria, microsomes, and solubles. In the former case the volume of the 20 per cent homogenate would be 11 ml, and in the latter case the 10 per cent homogenate volume would be 26 ml. One ml would have been removed from the 20 per cent homogenate for the analysis of total DNA and RNA in liver. In the experiment itself, the volume of the soluble fraction will have to be that found after the last centrifugation described in Flow Sheet I, but for purposes of calculation it is assumed to be 20 ml.

All of the nuclei from 2.2 g of liver should be contained in a volume of 5 ml if the directions in Flow Sheet IA are followed. However, a correction for the removal of the 1 ml aliquot, a factor of 11/10, will have to be applied later, as will be seen.

All of the fractions, including nuclei, mitochondrial, microsomal, and soluble, are subjected to the nucleic acid extraction procedure outlined in Flow Sheet IIB and in each case a 1 ml aliquot is used to extract the nucleic acids. The nucleic acids extracted are now contained in 5 ml of TCA, PCA combined.

Finally, for the RNA analysis of each fraction, a 0.5 ml aliquot was removed to determine RNA content. The RNA content of this 0.5 ml aliquot is determined from the standard curve, and the figure obtained is entered in column 3 of Table 20. In order to calculate the RNA content of each 5 ml TCA extract, this figure must be multiplied by 2 × 5 or 10. In order to arrive at the RNA content of the original tissue extract obtained from approximately one-half of the liver, an additional correction needs to be made. The volume of sucrose-MgCl used to suspend all the mitochondria was 5 ml, for microsomes it was 2 ml, and for the soluble fraction it was 20 ml. Therefore, the RNA content in the mitochondrial fraction for one-half the liver would be 10 × 5 or 50, for microsomes it would be 10 × 2 or 20, and for the soluble fraction it would be 10 × 20 or 200. Enter these figures in column 4.

The last step concerns a correction for the RNA content in the entire liver. Under the assumption that the liver weighed 4.8 g, and the portion used for Part B of Flow Sheet I weighed 2.6 g, in each of the cases this would be a factor 4.8/2.6.

$$\text{Total RNA(Mc)} = \frac{\text{mg RNA}}{0.5 \text{ ml TCA aliquot}} \times \frac{50 \times 4.8}{2.6} \qquad (\text{or} \times 92.3)$$

$$\text{Total RNA(Ms)} = \frac{\text{mg RNA}}{0.5 \text{ ml TCA aliquot}} \times \frac{20 \times 4.8}{2.6} \qquad (\text{or} \times 36.9)$$

$$\text{Total RNA(Su)} = \frac{\text{mg RNA}}{0.5 \text{ ml TCA aliquot}} \times \frac{200 \times 4.8}{2.6} \qquad (\text{or} \times 369)$$

In a similar manner, the figures for DNA in these three fractions may be obtained, but with one additional consideration: the combined TCA PCA aliquot used for chemical analysis was 1.0 ml rather than 0.5 ml.

$$\text{Total DNA(Mc)} = \frac{\text{mg DNA}}{1.0 \text{ ml aliquot}} \times \frac{25 \times 4.8}{2.6} \qquad (\text{or} \times 46.1)$$

$$\text{Total DNA(Ms)} = \frac{\text{mg DNA}}{1.0 \text{ ml aliquot}} \times \frac{10 \times 4.8}{2.6} \qquad (\text{or} \times 18.46)$$

$$\text{Total DNA(Su)} = \frac{\text{mg DNA}}{1.0 \text{ ml aliquot}} \times \frac{100 \times 4.8}{2.6} \qquad (\text{or} \times 184.6)$$

In the case of the nucleic fraction, the correction for the removal of the 1 ml aliquot prior to further isolation requires an additional correction and the use of 2.2 instead of 2.6 as the weight of liver used.

$$\text{Total RNA(Nu)} = \frac{\text{mg RNA}}{0.5 \text{ ml aliquot}} \times \frac{50 \times 4.8}{2.2} \times \frac{11}{10} \qquad (\text{or} \times 120)$$

$$\text{Total DNA(Nu)} = \frac{\text{mg DNA}}{1.0 \text{ ml aliquot}} \times \frac{25 \times 4.8}{2.2} \times \frac{11}{10} \qquad (\text{or} \times 60)$$

Lastly, for calculations applying to the original homogenate fraction which was saved to obtain the total RNA and DNA in liver,

$$\text{Total RNA(Hm)} = \frac{\text{mg RNA}}{0.5 \text{ ml aliquot}} \times 10 \times \frac{11}{1} \times \frac{4.8}{2.2} \quad (\text{or} \times 240)$$

$$\text{Total DNA(Hm)} = \frac{\text{mg DNA}}{0.5 \text{ ml aliquot}} \times 5 \times \frac{11}{1} \times \frac{4.8}{2.2} \quad (\text{or} \times 120)$$

In your calculations, be sure to remember to use your own experimental figures.

Present all of your data in a tabular, orderly form. Include your graphs and liver weight.

You should expect to find 95 per cent of the DNA in the nucleic fraction. The RNA is distributed throughout each fraction, 10 to 20 per cent in the nucleus microsomes, and 5 to 10 per cent in soluble fraction. Do your results indicate this type of distribution? Where did you find the remaining 5 per cent of DNA?

From our understanding of hereditary mechanisms it can be concluded that the DNA content of somatic cells is constant in each cell. Thus, by determining the total DNA content of the nuclear fraction of a tissue, it is possible to calculate the number of cells in that tissue. The total DNA per nucleus ranges from 6.4 to 7.2×10^{-12} grams in mammalian tissues. Assume a value of 7×10^{-12} grams per nucleus for rat liver, and calculate from your data the number of cells in the liver removed from your experimental animal. Turn in your calculations with your laboratory report.

REFERENCES

1. Colowick, S. P., and Kaplan, N. O., eds. *Methods in Enzymology, Vols. III and XII.* Academic Press, New York, 1968.
2. Dishe, Z. In *The Nucleic Acids, Vol. I*, Chargaff, E., and Davidson, J. N., eds. Academic Press, New York, 1955.
3. Burton, K. *The Conditions and Mechanism of Diphenylamine Reaction for Colorimetric Estimation of Deoxynucleic Acids.* Biochem. J. **62** 315 (1956).
4. *Experimental Biochemistry Laboratory Manual.* Department of Biochemistry, University of Wisconsin, 1967.
5. Schneider, W. C. *Phosphorus Compounds in Animal Tissues. I. Extraction and Estimation of Desoxypentose Nucleic Acids and Pentose Nucleic Acids.* J. Biol. Chem. **161** 293 (1945).
6. Ceriotti, G. *Determination of Nucleic Acids in Animal Tissues.* J. Biol. Chem. **214** 59 (1955).
7. Schmidt, G., and Tannhauser, S. J. *A Method for the Determination of Deoxyribonucleic Acid, Ribonucleic Acid and Phosphoproteins in Animal Tissues.* J. Biol. Chem. **161** 83 (1945).
8. Tsanev, R., and Markov, G. G. *Substances Interfering with Spectrophotometric Estimation of Nucleic Acids and Their Elimination by the Two Wave Method.* Biochim. Biophys. Acta **42** 442 (1960).
9. Santen, R. J., and Agranoff, B. W. *Studies in the Estimation of Deoxyribonucleic Acid and Ribonucleic Acid in Rat Brain.* Biochim. Biophys. Acta **72** 251 (1963).

CHAPTER 10

THE MEASUREMENT OF HYPERCHROMIC SHIFTS AND VISCOSITY CHANGES IN PURE DNA

THE MEASUREMENT OF HYPERCHROMIC SHIFTS

Hyperchromicity. Optical rotation and hyperchromic shifts observed upon melting of the Watson-Crick base pairs are useful quantitative measurements for determining the fraction of the DNA which is the double helix form. The high temperature at which the helix melts can be maintained in specially designed instruments, but one can also separate the strands by raising the pH, adding organic solvents, or chemically modifying the bases.

Hyperchromic shifts, or changes in the absorbance properties of nucleic acids in solution, provide a rapid and convenient method for analyzing helical content and denaturation processes. Hypochromism refers to *reduction* in ultra-violet absorption of a nucleic acid based on simple calculations of the sum of expected light absorption of its known constituent nucleotides. The more ordered a nucleic structure is, the greater is the hypochromic effect; i.e., DNA > RNA; double-stranded DNA > single-stranded DNA; polynucleotides > oligonucleotides > tri- and di-nucleotides. The reverse effect would be hyperchromism, which is the change or *increase* in absorbance in the ultra-violet region resulting from the disruption or denaturation of a molecule into a less ordered structure. A hyperchromic shift therefore can take place when double-stranded DNA is denatured to a single-stranded form, or when a polynucleotide is hydrolyzed to nucleotide sub-units.

EXPERIMENT 10

Part A. Hyperchromic Shifts Due to pH Change, Heat, and Detergent Action

Prepare a DNA stock solution in 0.15 M NaCl (15 ml), such that the absorbance at 260 mμ is 1.0 $\pm$ 0.03, from your own DNA (see Experiment 8) or a commercial preparation. Choose one of the denaturant solution sets from Table 21 and make up the solutions by adding 2.00 ml of stock DNA to 2.00 ml

Table 21

Set I pH Effect	Set II Heat Plus Formaldehyde*	Set III Heat and Quick Cool†	Set IV Dimethyl Formamide (DMF)‡
0.05 M borate, pH 8.5	0.05 M borate, pH 8.5	0.05 M borate, pH 8.5, to 50° and then chill	0.02 M borate, pH 8.5
0.05 M borate, pH 9.5	0.05 M borate + 2 per cent CH$_2$O	to 60° and then chill	0.02 M borate in 20 per cent DMF
0.05 M borate, pH 10.5	0.05 M borate + 2 per cent CH$_2$O (heat to 60° for 5, 15, and 30 min)	to 70° and then chill	0.02 M borate in 40 per cent DMF
0.05 M borate, pH 11.5	0.05 M borate + 2 per cent CH$_2$O + 1 M NaCl (heat to 60° for 5, 15, and 30 min)	to 80° and then chill	0.02 M borate in 80 per cent DMF
0.05 M borate, pH 12.5		to 90° and then chill	
		to 95° and then chill	

* When the bases are separated by heating they become available for reaction with formaldehyde, and are thus prevented from renaturing when the solution is cooled. In this way the spectra at room temperature can be used to determine the extent of strand separation.

† Heat test tube to temperature required, hold 5 minutes, plunge into ice bath and swirl for rapid cooling. Then warm to room temperature and measure absorbancy at 260 mμ. Try to "renature" one of the denatured samples by cooling slowly.

‡ Use spectral grade DMF, and measure absorbancies at 270 mμ.

of each denaturant solution in the set. In set IV use 1.0 ml of stock DNA, with $A = 2 \pm 0.5$ at 270 mμ in this case, and add 5ml of dimethyl formamide. Plot the absorbancy at 260 mμ versus the parameter varied in your series. Measure the change in absorbancy against appropriate buffer blanks.

QUESTIONS

How would you attempt to renature denatured DNA? What is meant by "annealing" a DNA solution? What kinds of biological questions are studied by annealing DNA and RNA mixtures?

HYPERCHROMIC SHIFTS ARISING FROM ENZYME ACTION

Deoxyribonuclease from beef pancreas (DNAase I) splits phosphodiester linkages, preferentially adjacent to a pyrimidine nucleotide, to yield 5' phosphate polynucleotides, with an average chain length of a limit digest of a tetranucleotide.

Place 2.5 ml of 0.03 M Tris buffer, pH 7.5, containing 0.15 M NaCl and 0.03 M MgCl$_2$, in each of two cuvettes. Add 0.4 ml of 0.15 M NaCl, containing 0.5 mg/ml of DNA. At zero time, add 0.1 ml of water to the blank and 0.1 ml of DNAase (40 μg/ml in 0.1 M MgCl$_2$) to the other tube. Adjust the absorbance to zero at 260 mμ with the blank, and read the absorbance in the experimental tube every minute for five minutes. Determine ΔA_{260}/min; continue to observe the absorbance until it reaches a plateau.

Plot the per cent increase in A_{260} as a function of time.

The specific activity of the enzyme may be expressed in these terms also:

$$\text{Units/mg protein} = \frac{\Delta A/\text{min (initial rate)}}{\text{mg enzyme used}}$$

Part B. Viscosity Properties of DNA

Drawings of chemical structures of large molecules tell us something about the relations of adjacent atoms, but they are often misleading with respect to the configuration of conglomerates of atoms in polymers and other large molecules. Some experimental approaches produce only general information about shape and are very simple to carry out. Viscosity measurements are among these, and can give very useful information about overall molecular configuration.

The basic equation for expressing viscosity is given as:

(1)
$$\eta = \frac{F}{A\left(\dfrac{v}{D}\right)}$$

The viscosity is expressed as a force F, per unit of area A, per v/D. The latter expression is the shear, and it has units of reciprocal seconds (sec^{-1}). It is a velocity gradient, which means it is an expression of the change in velocity along a line perpendicular to the parallel walls of two plates or capillary walls; D is the diameter.

The viscosity, therefore, is the force per unit area of these plates or capillary walls to move these plates relative to one another in a given fluid, or to move the fluid between the capillary walls. It is force (dynes) per unit of area (cm^{-2}) per unit of shear (sec^{-1}). The dyne is the force which gives a mass of 1 gram an acceleration of 1 cm/sec and has units of gram cm/sec². The units of viscosity, therefore, are

$$\frac{\text{gram cm/sec}^2}{\dfrac{\text{cm}^2}{\text{sec}}} = \text{gram cm}^{-1}\,\text{sec}^{-1}$$

One unit of viscosity in these units is a *poise*. Water has a viscosity of 0.01 poise (1 *centipoise*) at 20° C.

The most common method of measuring viscosity is by use of the Ostwald viscometer. It consists of a bulb and a large reservoir, connected by a capillary tube of small diameter. A diagram and instructions for use of the viscometer will be found on page 205. In viscosity measurement, the time necessary for a volume of fluid to drain through the capillary tube is directly proportional to the viscosity of the fluid. The viscometer must first be calibrated with solutions of known viscosity, such as glycerol-water mixtures, which yield a constant factor relating the time flow from a particular viscometer to the actual viscosity of the fluid. On the other hand, if the diameter of the capillary tube and the volume of the bulb were known, it would be possible to calculate the time necessary for a fluid of known viscosity to drain past the two reference marks in accordance with the definition of absolute viscosity given above.

At this point it will be useful to introduce several terms.

The absolute viscosity η_{soln} is proportional to the time required for a solution to flow through a capillary of known dimensions under standard conditions of temperature and pressure.

The relative viscosity is defined as the ratio of the apparent or measured viscosity η of a solvent containing solute and the apparent or measured viscosity of the solvent alone, η_0.

$$(2) \quad \eta_{\text{rel}} = \text{relative viscosity} = \frac{\eta}{\eta_0} \cong \frac{\text{time (sec) for the solution to drain}}{\text{time (sec) for the solvent to drain}}$$

Einstein derived an equation which measures the *increase* in viscosity of a solvent into which solid spheres are mixed.

$$(3) \quad \frac{\eta}{\eta_0} = 1 + \frac{Vf^4}{4} \quad \text{or} \quad \eta = \eta_0\left[1 + \frac{Vf^4}{4}\right]$$

where V is a constant for a particular solute and f is the ratio of the length to the radius of the solute molecules.

This equation can also be given in the form:

$$(4) \qquad \frac{\eta}{\eta_0} = 1 + 2.5\phi \qquad \text{or} \qquad \eta = \eta_0[1 + 2.5\phi]$$

where $1 + 2.5\phi$ is the increase in viscosity and ϕ is the fraction of the total volume taken up by the solid spheres.

$$\phi = n\tfrac{4}{3}\pi D^3$$

where n = number of spheres per unit volume and D is the particle diameter.

Another term that is used is the specific viscosity: $\eta_{sp} = \eta_{rel} - 1$. Since both relative and specific viscosities are dependent upon the concentration of the macromolecules in solution, an additional viscosity definition is required, the intrinsic viscosity $[\eta]$, which is the viscosity of a solution at infinite dilution.

$$(5) \qquad [\eta] = \lim_{C \to 0} \frac{\eta_{sp}}{C}$$

where C = concentration of the macromolecule.

Therefore:

$$[\eta] = \lim_{C \to 0} \frac{\eta - \eta_0}{C\eta_0}$$

where η_0 is the viscosity of the solvent and η is the viscosity of C grams of solute per ml, or the fractional increase in viscosity/unit concentration. Thus, extrapolation of a plot of $\eta_{sp/C}$ versus C to zero concentration will yield intrinsic viscosity.

The intrinsic viscosities of spherically shaped polymers (globulins) are relatively low, but those like myosin, which are very elongated, have high intrinsic viscosities. Elongated molecules also appear to effectively occupy a larger volume of solvent as a result of Brownian movement and solvent bombardment, both of which contribute to high $[\eta]$.

Another class of polymers have no definite shape. The backbones of these molecules are in constant thermal agitation and assume an infinite number of configurations. The shape of these molecules can only be quantitatively deduced from statistical treatment. This has been found to be a function of molecular weight. The statistical size or space occupied by a random polymer increases as the square root of molecular weight.

From the Einstein equation (4):

$$\frac{\eta - \eta_0}{\eta_0} = 2.5\phi$$

where $\phi = \dfrac{4n\pi D^3}{3}$. From consideration above,

$$D = K\sqrt{MW},$$

where K is a constant, and

$$\frac{\eta - \eta_0}{\eta_0} = 2.5 \, n \, \tfrac{4}{3} \pi D^3$$

$$= k_1 n D^3$$

$$= k_2 n (\sqrt{\mathrm{MW}})^3$$

where k_2 is a constant combining all previous constants.

(6) $$\frac{\eta - \eta_0}{\eta_0} = k_2 n [\mathrm{MW}]^{3/2}.$$

Since the intrinsic viscosity $= \lim\limits_{C \to 0} \dfrac{\eta - \eta_0}{C \eta_0}$, by dividing Equation (6) by C we obtain

(7) $$[\eta] = \frac{\eta - \eta_0}{C \eta_0} = \frac{k_2 n (\mathrm{MW})^{3/2}}{C}.$$

Since

(8) $$C = (\mathrm{MW}),$$

we have

(9) $$[\eta] = \frac{k_2 n (\mathrm{MW})^{3/2}}{n (\mathrm{MW})}$$

$$= k_2 (\mathrm{MW})^{1/2}$$

Therefore the intrinsic viscosity for a random polymer is equal to a constant *times* the molecular weight to the 0.5 power.

The viscosity of compact molecules with definite structure is independent of molecular weight. However, in reality, the intrinsic viscosity of random polymers increases even more rapidly than the 0.5 power of molecular weight. This could be due to particle charge, solvent layering around the molecules, and special restrictions, all of which effectively increase the diameter of a random polymer; a correct statement would be

$$[\eta] = k_2 (\mathrm{MW})^{0.5 \text{ to } 1}$$

Conventional viscosity measurements cannot be used to determine the molecular weight and shape of carefully prepared DNA samples containing long polynucleotide chains. However, routine viscometry can give a qualitative index of some of the changes caused by temperature, ionic strength, and enzymic attack. Viscosity measurement is nonetheless one of the simplest and cheapest to carry out, and for this reason alone it is most intensively studied as an index of conformational changes of nucleic acids and proteins.

Determination of the Intrinsic Viscosity of DNA. Pipet 4.00 ml of a 0.15 M NaCl solution into a dry, clean Ostwald viscometer and measure the outflow time and the temperature of the bath. Remove 1 ml of solution from the viscometer and replace with 1 ml of a DNA (1 mg/ml in 0.15 M NaCl) solution. Mix the solution gently by blowing air through the rubber tubing attached to the small arm of the viscometer, and then measure the outflow time repeatedly until uniform results are obtained. Remove 2 ml of solution from the viscometer and replace it with 2 ml of the 0.15 M NaCl. Mix thoroughly and measure the outflow time. Remove a further 2 ml of the solution, replace it with the salt solution, and after mixing determine the outflow time.

Calculate the DNA concentration and specific viscosity for each dilution. Determine the intrinsic viscosity by plotting η/C_{sp} versus C and extrapolating to zero concentration.

Dependence of Viscosity of Native DNA Upon Ionic Strength. Pipet 4.00 ml of 0.001 M NaCl solution into the viscometer and determine the outflow time. Remove 0.40 ml of solution from the viscometer and replace it with 0.40 ml of DNA solution (1 mg/ml in 0.15 M NaCl). Mix the solution and take readings of the outflow time. Record the temperature at each reading. Now pipet exactly 0.01 ml of 4.00 M NaCl into the viscometer and repeat measurements of the outflow time. Next pipet exactly 0.10 ml of 4.0 M NaCl solution into the viscometer and take a final set of readings.

Dependence of Viscosity of Single-Stranded DNA Upon Ionic Strength. Prepare some single-stranded DNA by heating the DNA solution used above in a 100° water bath for 15 minutes and then cooling quickly in an ice bath. Wash the viscometer with copious amounts of distilled water and shake dry. Measure the viscosity of 4.0 ml of distilled water in the viscometer. Then remove 0.40 ml of water from the viscometer and add 0.40 ml of the heat-denatured DNA solution. After determining the viscosity in the absence of added salt, proceed with viscosity measurements with the following consecutive additions of NaCl:

0.04 ml of 0.1 M NaCl

0.01 ml of 4 M NaCl

0.10 ml of 4 M NaCl

Plot relative viscosity versus ionic strength for both native and denatured DNA.

Part C.   The Kinetics of Degradation of Deoxyribonucleic Acid by Deoxyribonuclease

The purpose of this part of the experiment is to determine the number of strands in the DNA solution as a function of time during enzymatic degradation.

DNA, when carefully isolated, exists as a highly polymerized, highly ordered, double-stranded helical structure. Each strand is held together along its length by diester phosphate bonds linking together the sugar moieties of the deoxyribonucleosides. The enzyme DNAase I, isolated from pancreas, will cleave phosphodiester bonds, creating breaks located at random along the strands. As long as the enzyme does not create breaks in both strands at the same location in the double-stranded molecule, the DNA will not break down into smaller molecular weight fragments. When a break does occur in both strands at the same location, the molecule scissions, and a decrease in viscosity occurs, as a result of the decrease in molecular weight.

A plot of viscosity change of DNA, as a function of time of enzyme degradation, will exhibit a lag period before the viscosity begins to drop rapidly. This is due to the fact that a number of breaks must accumulate before two breaks will occur across from each other in opposite strands. The appearance of this "lag-phase" is indicative of a multistranded structure. If the molecule were single-stranded, each attack by the enzyme would cause a scission, and instead of showing a "lag-phase" the viscosity curve would drop immediately. On the other hand, if the DNA molecule were three- or four-stranded, then a very long "lag-phase" would be observed.

It is possible to plot the experimental data in such a way that the number of strands in a DNA molecule may be determined from the slope of the resulting curve. As indicated in the discussion, it may be shown for DNA that the intrinsic viscosity of the DNA solution is directly proportional to the molecular weight of the DNA. Also, the average molecular weight of the DNA at any time after the start of the experiment will be equal to the initial molecular weight divided by $(1 + S)$, where S is the number of scissions. If the probability of breaking a single bond is proportional to t, where t is time of enzyme action, then the probability that both strands will be broken, creating a scission, is proportional to t^2 for a double-stranded molecule, t^3 for a triple-stranded molecule, or t^m for a molecule containing m strands. Therefore, the following four equations may be written (where A and K are constants, MW is the molecular weight at any time, MW_0 is the initial value of the molecular weight, and $[\eta]$ is the intrinsic viscosity).

(1) $[\eta] = A[MW]$

(2) $[\eta_0] = A[MW_0]$

(3) $MW = MW_0/(1 + S)$

(4) $S = Kt^m$ (experimental form of the general rate equation).

Combining these four equations to eliminate A, MW, MW_0, and S, and taking the logarithm, yields:

(5) $\log\left(1 - \dfrac{[\eta]}{[\eta]_0}\right) = m \log t - \log(1 + Kt^m) + \text{constant.}$

In this experiment we may assume that K is so small that Kt^m is negligible compared to 1, and that $[\eta] = \dfrac{\eta_{sp}}{C}$, so that Equation (5) becomes

$$\log\left(1 - \frac{\eta_{sp}}{\eta_{sp,0}}\right) = m \log t + \text{constant.}$$

Therefore, a plot of the experimentally measurable quantity on the left hand side as a function of the logarithm of time should yield a straight line with a slope equal to m, the number of strands in the DNA molecule. The correspondence between theory and experiment is not exact, however, because the efficiency of the enzyme in causing a decrease in the viscosity contribution of the macromolecules is greater than the theory predicts.

Fill a large beaker with water and leave standing until it has reached room temperature equilibrium. Put about 4 ml of the buffer (0.03 M Tris, 0.15 M NaCl, 0.03 M $MgCl_2$) solution into the viscometer and immerse in a water bath at 23° C. Mount the viscometer firmly and do not disturb its position during the remainder of the experiment. Wait ten minutes so that it is completely equilibrated. Record the temperature. At this point, using a piece of rubber tubing, suck the fluid into the upper bulb of the viscometer so that the meniscus of the fluid is above the upper mark. Then allow the fluid to fall and start timing when the meniscus reaches the upper mark. Record the time taken for the meniscus to reach the lowest mark on the viscometer (just below the upper bulb). Repeat this operation several times, recording the temperature of the water bath during each reading. It should be possible to measure the outflow times with a reproducibility of several tenths of a second. Now remove as precisely as possible 0.4 ml of solution from your viscometer and replace it with 0.4 ml of the DNA solution, 0.5 mg/ml in 0.10 M NaCl. Measure the outflow time repeatedly until uniform results are obtained. You are now ready to begin the enzyme kinetics experiment. Add 0.03 ml of the 40 mg/ml in 0.01 M $MgCl_2$ DNAase solution. Record the time at which this addition occurs and then begin to measure viscosity repeatedly as a function of time. Record the temperature with each measurement.

Even small variations in temperature can cause relatively large errors in the viscosity measurement. The variation is approximately 2 per cent per degree, with the viscosity increasing as temperature decreases. All data should be corrected to the same temperature value using the 2 per cent correction factor. For example, an outflow time of 110.7 sec at 25.2° C is corrected to 25.0° C by the formula $110.7 + [0.02 \times (25.2 - 25.0) \times 110.7] = 110.70 + 0.44 = 111.1$ sec.

Present your results as follows:

1. Plot the relative viscosity on the ordinate of a graph against time on the abscissa.

2. Prepare a logarithmic plot according to Equation (6) and determine the number of strands in the DNA helix from the initial slope of the line.

REFERENCES

1. Courtesy of Profs. R. A. Smith and M. Konrad, Dept. of Chemistry, U.C.L.A., 1969.
2. *Experimental Biochemistry Laboratory Manual.* Department of Biochemistry, University of Wisconsin, 1967.

CHAPTER II

CHEMICAL PROPERTIES OF CARBOHYDRATES

BASIC CARBOHYDRATE CHEMISTRY

Sugars behave as very weak acids and have the chemical characteristics of aldehydes or potential aldehydes. For this reason the effects of alkali, which enhances enolization in sugars, are similar to those in aldehydes. However, in the absence of alkali an equilibrium between the enol and keto forms probably exists, but usually the equilibrium favors the keto form. When alkali is added, the acidic enols of aldehydes form a salt which directs the reaction to the right by mass action effects. Addition of acid to an alkali equilibrium mixture will cause the restoration of the tautomeric equilibrium toward the keto form:

$$
\underset{\text{keto acetaldehyde}}{
\begin{array}{c} H \quad O \\ | \quad \| \\ H-C-C-H \\ | \\ H \end{array}}
\rightleftharpoons
\underset{\text{enol acetaldehyde}}{
\begin{array}{c} OH \\ | \\ HC=C-H \\ | \\ H \end{array}}
\underset{H^+}{\overset{NaOH}{\rightleftharpoons}}
\underset{\text{enol salt}}{
\begin{array}{c} ONa \\ | \\ HC=C-H \\ | \\ H \end{array}}
$$

Tautomerization is a typical reaction of sugars containing a free aldehyde or keto group. Another important reaction of carbonyl compounds is with alcohols:

$$
2R-OH + \underset{O}{H-C-R'} \rightleftharpoons R-OH + \underset{\text{hemiacetal}}{R \quad O-\overset{H}{\underset{OH}{C}}-R'}
$$

$$
\rightleftharpoons \underset{\text{acetal}}{R-O-\overset{H}{\underset{RO}{C}}-R'} + H_2O
$$

133

Monosaccharides have both functional groups in the molecule and may react with another sugar or with itself in a similar way to produce cyclic hemiacetals.

There are many properties of simple sugars which can only be explained by assuming ring formation. This is the more stable form of the molecules. The effect of alkali, however, is to depress ring formation via formation of enol salt.

Conventional linear representation:

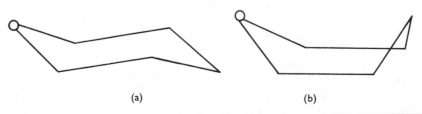

β-D-glucose *(hemiacetal)*   D-glucose *(open chain form)*   α-D-glucose *(hemiacetal)*

Haworth formulas:

While the Haworth structures of sugars such α-D-glucopyranose and β-D-fructopyranose will illustrate the fine points of structure, it must be kept in mind that the actual sugar models have a more complex conformation. Haworth structures represent the pyramose ring simply as being flat, whereas in reality the conformation is best represented by the "boat" or "chair" forms.

(a)   (b)

Figure 9. "Chair" (a) and "boat" (b) configurations.

EXPERIMENT 11

In this experiment each student will be given two unknowns which he is to identify unequivocally at the end of the designated laboratory period. Known carbohydrate solutions will be made available, which he can use to perform the various tests along with his unknowns. By entering the results he has obtained in Table 22, he will be able to identify his unknowns. At this point he will have to obtain confirming evidence by means of thin-layer

Table 22

Carbohydrate	Molish Test	Benedict's Test	Seliwanoff's Test	Bial's Test	Mucic Acid Test	Fermentation Test	Osazone Test	Starch–Iodine Test
glucose								
fructose								
galactose								
mannose								
arabinose								
ribose								
xylose								
maltose								
lactose								
sucrose								
starch								
dextrin								
inulin								
gum arabic								
glycogen								
unknown 1								
unknown 2								

chromatography, the last experiment described in this series. He should then present the record of all of his results and indicate how he arrived at his conclusions. Finally, he should be able to submit a general scheme for the identification of any unknown believed to be a common carbohydrate.

GENERAL TESTS FOR CARBOHYDRATES

Molish Test. This is a general test for carbohydrates. Carbohydrates are dehydrated to a furfural derivative in the presence of concentrated acid. The furfural then condenses with α-naphthol to give a red to violet color. A negative reaction of a given material is good evidence for the absence of carbohydrate.

Place 5 ml of a 1:10 dilution of the available carbohydrate solution in a test tube; add 2 drops of Molish reagent (5 per cent α-naphthol in alcohol) and mix. Carefully pour 3 ml of concentrated H_2SO_4 along the side of the tube so as to form a layer under the water. A reddish-violet color at the junction of the two liquids denotes a positive test. Solutions and materials available for testing are: monosaccharides, disaccharides, polysaccharides, and such material as cotton, paper, nail clippings, ovalbumin, saliva, or any other material you wish to choose.

Benedict's Test for Reducing Sugars. The reducing properties of sugars are dependent upon the presence of actual or potential aldehyde or ketone groups. When some sugar solutions are heated in the presence of certain metallic ions, the carbonyl group is oxidized and the metal ion is reduced.

The enolization of sugars under alkaline conditions is an important consideration in reduction tests. The ability of a sugar to reduce alkaline test reagents depends on the availability of an aldehyde or keto group for reduction reactions. A number of sugars, especially disaccharides or polysaccharides, have glycosidic linkages which involve bonding between each group, and hence there is no reducing group on the sugar. Such is the case for sucrose, trehalose, inulin, glycogen, starch, and dextrin. In the case of reducing sugars, the presence of alkali causes extensive enolization, especially at high pH and temperature. This leads to susceptibility to oxidation reactions, much more so than at neutral pH or acid pH. These sugars, therefore, become potent agents capable of reducing Cu^{+2} to Cu^+, Ag^+ to Ag, and so forth. Reducing sugars can react with many different oxidizing agents. Reactions such as Fehling's test, Benedict's test, and Barfoed's test have been used to distinguish between monosaccharides and disaccharides on the basis of differential reaction times.

Of the various reagents, Benedict's is the most reliable and most frequently used. Benedict's solution consists essentially of a mixture of $CuSO_4$, Na_2CO_3 and sodium citrate. In this alkaline mixture, the Cu^{+2} is prevented from precipitating as the insoluble $Cu(OH)_2$ by the citrate ion, which forms a soluble complex with it. During the reaction the enolized sugar is oxidized, but this reaction is not simple. A variety of oxidation products are formed from the sugar. However, under controlled conditions the reduction of Cu^{+2} can serve as a quantitative measure of reducing sugar.

To a series of test tubes add 5 ml of Benedict's reagent and 1 ml portions of 0.1 M solutions of the following: glucose, maltose, lactose, arabinose, sucrose, phenol, hydroquinone, acetaldehyde, starch, mannose, xylose, inulin, and dextrin. Place them in a boiling water bath for 5 to 50 minutes. A positive test is denoted by the formation of a yellow, green or red precipitate. In general, large particles, formed by slow reduction, are brick red, whereas very small particles formed by fast reactions may be greenish. The amount of precipitate, and not the color, is the primary consideration.

THE ACTION OF ACIDS ON SUGARS

The Inversion of Sucrose. Action of Dilute Acids. Add 5 ml of 0.1 M sucrose solution to each of two tubes. Add five drops of concentrated HCl to one tube and heat both tubes in a boiling water bath for 10 minutes. Test both solutions for the presence of reducing sugar with 5 ml of Benedict's reagent. Explain the result. What is invert sugar? Explain the term inversion. What can you expect from a similar experiment with starch or gum arabic? How would you identify the products of hydrolysis? What enzyme have you recently tested that has a similar action?

Condensation Reactions of Carbohydrates Dehydrated with Concentrated Acids. As most of you know, carbohydrates can be completely dehydrated to carbon by such agents as concentrated H_2SO_4. Simple sugars are dehydrated by hot 10 to 20 per cent HCl. Pentoses yield furfural; ketohexoses yield levulinic acid, hydroxymethyl furfural, and other decomposition products. Aldohexoses yield the same products as ketohexoses, except that only small amounts of hydroxymethyl furfural are formed from the former. These differences make it possible to distinguish between the three types of monosaccharides. Furfural and hydroxymethyl furfural are colorless compounds which are freely soluble in water. They can be readily detected because they react with certain aromatic amines and phenols to produce colored condensation products.

(1) The Resorcinol Test (Seliwanoff's Test) for Ketoses

fructose

hydroxymethyl
furfural

$+ 3H_2O \xrightarrow[\text{(resorcinol)}]{}$ *red condensation products*

Add 5.0 ml of Seliwanoff's reagent to 1.0 ml aliquots of 0.01 M solutions of glucose, fructose, sucrose, galactose, and arabinose. Put the tubes in a boiling water bath and observe every five minutes for 15 minutes. Record the time of appearance of color in each tube. Note the effect of continued heating. The aldoglucose gives some color especially on longer heating but in the same time interval fructose forms a deep red color and red precipitate. Why does sucrose give a positive test?

Other substances which give a positive test are sorbose, ketotriose, and oxygluconic acid.

Pentoses give a green color with this reagent, and it can be used to detect them. There are several additional tests for detecting pentoses: the aniline-acetate test, the phyloroglucin test and the orcinol test (Bial's test). The most useful one is the Bial's test, which is frequently employed to determine nucleic acids by reaction with the ribose moiety of the nucleotides.

(2) Bial's Test for Pentoses

Add 5 ml of Bial's reagent to 1 ml aliquots of 0.01 M glucose, fructose, arabinose, ribose, and sucrose. Follow the directions above for heating. Pentoses give a bluish-green or olive-green color. The condensation product with hydroxymethyl furfural from hexoses is yellow to brown in color.

Hydrolysis of Polysaccharides by Acids. Place 10 ml of starch, glycogen, inulin, gum arabic, dextrin, and a piece of cotton into three separate large test tubes. Add 2 ml of 20 per cent H_2SO_4 and place all tubes in a boiling water bath for 15 minutes while maintaining the same volume of liquid with water. Cool the tubes and neutralize with 10 per cent NaOH. Perform the Benedict's test on each hydrolysate. Compare to the original solutions previously tested for Benedict's reaction. Decolorize the gum arabic solution with charcoal if necessary and perform a Bial's test on both the hydrolyzed solution and the unhydrolyzed solution.

What are the products of acid hydrolysis of glycogen, starch and inulin?

Is the result of the Benedict's test proof of the production of a monosaccharide end product?

How would you prove this?

Why must you neutralize the solution before testing with Benedict's reagent?

THE REACTION OF I₂ AND THE NONDIFFUSIBILITY OF POLYSACCHARIDES

A. Dissolve a few grams of glucose in 20 ml of 1 per cent starch, glycogen, inulin, or gum arabic. Place each solution in dialysis tubing, and suspend it in no more than 80 ml of water. Agitate from time to time, and after about $1\frac{1}{2}$ hours, test aliquots of the solution inside and outside the bag for both glucose and polysaccharide by means of reaction with I_2. (See part B).

How can you test for glucose and starch when both may be present?
What difference between glucose and starch is revealed by this test?

B. The iodine test is useful in distinguishing polysaccharides. To 5 ml of starch, glycogen, dextrin, and inulin, add a few drops of 0.1 N iodine solution; allow to stand for a few minutes, and observe the colors formed.

These complex compounds of iodine with starch, dextrin, and glycogen are very unstable. The iodine is readily removed from these compounds by alcohol, NaOH, and $Na_2S_2O_3$. Heating also decolorizes them. This behavior shows that I_2 is very loosely bound and that it is still in the unreduced form when combined with polysaccharide. The color of iodo-glycogen can be appreciably intensified by the addition of NaCl.

The Action of an Oxidizing Acid such as HNO_3 on Sugars (Mucic Acid Test). Nitric acid oxidizes both the aldehyde group and the terminal primary alcohol group of monosaccharides to yield mono- and dicarboxylic acids which have been given the general name saccharic acids.

D-glucose   *D-gluconic acid*   *D-glucosaccharic acid*

D-galactose   *D-galactosaccharic acid*
(mucic acid)

Mucic acid is the least soluble of the saccharic acids in aqueous solution. This characteristic facilitates the identification of galactose or galactose-containing sugars. Although mucic acid contains four asymmetric carbon atoms it is optically inactive because one-half of the molecule is the mirror image of the other half.

The importance of the mucic acid test lies in the identification of galactose or lactose. Both galactose and lactose occur in milk. Its appearance in urine can be attributed to metabolic abnormalities as in lactosuria, which is often found in lactating women, and in galactosemia, a severe genetic disorder leading to mental retardation.

To 10 ml of 0.1 M galactose (or lactose) solution, add 1 ml of H_2O and 3 ml of concentrated HNO_3. Heat in a boiling water bath for 1 to $1\frac{1}{2}$ hours. Let the tube stand overnight or put it in an ice bath. Examine the crystals under the microscope.

THE ACTION OF ALKALI ON CARBOHYDRATES

As a result of tautomerization in alkali there is a rearrangement of aldoses and ketoses which can be expressed as a series of equilibrium reactions (the Lobry de Bruyn reaction).

$$
\begin{array}{c}
\text{H—C=O} \\
|\\
\text{H—C—OH} \\
|\\
\text{HO—C—H} \\
|\\
\text{H—C—OH} \\
|\\
\text{H—C—OH} \\
|\\
\text{H—C—OH} \\
|\\
\text{H}
\end{array}
$$

D-*glucose* (63%)

$^-$OH ⇅

$$
\begin{array}{ccc}
\text{H—C=O} & \text{H—C—OH} & \text{H—C—OH} \\
| & \| & | \\
\text{HO—C—H} & \text{C—OH} & \text{C=O} \\
| & | & | \\
\text{HO—C—H} & \text{HO—C—H} & \text{HO—C—H} \\
| & | & | \\
\text{H—C—OH} \rightleftharpoons & \text{H—C—OH} \rightleftharpoons & \text{H—C—OH} \\
| & | & | \\
\text{H—C—OH} & \text{H—C—OH} & \text{H—C—OH} \\
| & | & | \\
\text{H—C—OH} & \text{H—C—OH} & \text{H—C—OH} \\
| & | & | \\
\text{H} & \text{H} & \text{H}
\end{array}
$$

D-*mannose* (2 to 4%)   *enediol of glucose*   D-*fructose* (31%)

To three test tubes add 1.0 ml of 0.1 M fructose, 0.1 M glucose and 0.1 M lactose respectively. Add 0.5 ml of saturated $Ba(OH)_2$ to each tube.

Prepare a similar series of tubes substituting 1.0 ml of water for the $Ba(OH)_2$ solution.

Add a layer of toluene, 1 to 2 mm deep, to each of the six tubes and allow them to stand overnight. Add 5 ml of Seliwanoff's reagent to each tube; mix, then immerse in a boiling water bath. Record the development of color. What is your interpretation?

FERMENTATION OF SUGARS

Test *unpreserved* solutions of 0.1 M glucose, fructose, sucrose, inulin, lactose, glycogen, xylose, and ribose in a yeast suspension. Add 0.5 gm of Baker's yeast and enough of each sugar solution to fill the closed arm of a fermentation tube. Shake the tube long enough for most of the air bubbles to escape and *to uniformly suspend the yeast in fluid*. Place the tubes upright in a water bath maintained at 37°. Which tubes form gas and approximately how much? Is the gas CO_2? Test this possibility by adding some dilute NaOH.

THE FORMATION OF OSAZONES (PHENYLHYDRAZINE REACTION)

The sugars with a free carbonyl group will react with phenylhydrazine to form phenylhydrazones. With the exception of mannose phenylhydrazones, this derivative of most sugars is soluble, and this reaction alone cannot be used to identify sugars. In an excess of phenylhydrazine, however, the hydroxyl carbon adjacent to the phenylhydrazone adduct is oxidized to a carbonyl group and a second condensation takes place to form yellow insoluble compounds known as osazones. Why?

Osazones are important in the identification of sugars because they possess characteristic crystalline forms and specific optical rotations. The proximity of their melting points, however, serves as one limitation. Only *reducing* sugars react with phenylhydrazine in slightly acid medium to give osazones.

D-*Glucose*   *Glucose phenyl hydrozone*

$$
\begin{array}{c}
H \\
| \\
H-C=N-N-\phi \\
| \\
C=O \\
| \\
HO-C-H \\
| \\
H-C-OH \\
| \\
H-C-OH \\
| \\
H-C-OH \\
| \\
H
\end{array}
\;+\;\phi NH_2 \;+\; CH_3COONH_4
\;\xrightarrow{\;\overset{H}{\underset{}{\phi N-NH_2}}\;}\;
\begin{array}{c}
H \\
| \\
H-C=N-N-\phi \\
| \\
H \\
| \\
C=N-N-\phi \\
| \\
HO-C-H \\
| \\
H-C-OH \\
| \\
H-C-OH \\
| \\
H-C-OH \\
| \\
H
\end{array}
$$

Keto phenyl hydrazone   *Osazone*

Since the asymmetry of C2 in the aldoses is destroyed in osazone formation, D-glucose, D-mannose, and D-fructose all yield the same osazone.

Introduce 1 ml of 0.01 M solutions of glucose, mannose, fructose, galactose, lactose, sucrose, and xylose into labeled test tubes. To each tube add 2 ml distilled water and 0.7 gram of freshly prepared phenylhydrazine hydrochloride-sodium acetate reagent (ratio of 2:3 grams). Mix well, stopper each tube with cotton, and place all in a beaker with boiling water. Observe for 15 minutes. Continue heating for another 10 minutes; remove the tubes; allow to cool. Examine the crystals with a microscope.

THIN-LAYER OR PAPER CHROMATOGRAPHY OF CARBOHYDRATES

Principles. The same principles involved in paper chromatography apply to thin-layer chromatography. The substance to be separated is extracted from a flowing liquid phase and retained in a solid phase. The process is concerned with absorption by van der Waals forces or hydrogen bonds or with ion exchange. The experiment itself can be carried out either with paper or with prefabricated thin layer sheets.

In thin-layer chromatography, absorption plays a significant role, which is sometimes of equal importance to partition. All chromatographic processes can be classified as column chromatography or thin-layer chromatography, of which paper chromatography is a special case of thin-layer chromatography because the paper may be regarded as a thin film of long cellulose fibers. The principle of thin-layer or plate chromatography is that a suitable absorbant is spread in a thin layer on a glass plate, or other suitable support plate, and in some cases is dried and activated. The drop of solution to be analyzed is applied as in paper chromatography at a known starting point. The plate is placed in a sealed chromatographic chamber with a suitable solvent system.

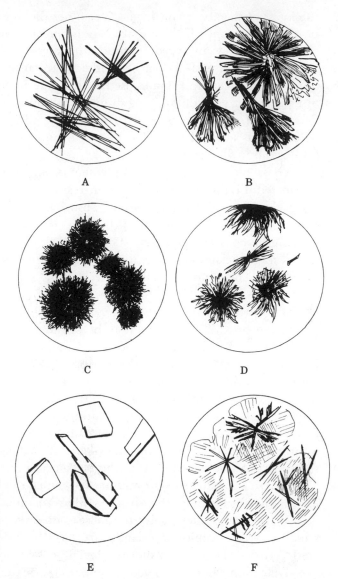

Figure 10. Appearance of phenylhydrazine derivatives under low power of the microscope. A, glucosazone; B, galactosazone; C, lactosazone; D, maltosazone; E. mannose phenylhydrazone; F, xylosazone.

By capillary action the solvent creeps up through the stationary layer and separates the components of the sample into a number of spots. After separation, the plates are dried and spots can be individually identified either by characteristic color in daylight or by being made visible by addition of reagents or by other means.

TLC has several advantages over paper chromatography: 1) it is faster, taking one to two hours versus 12 to 16 for paper; 2) it is 10 to 100 times more sensitive; 3) it can utilize a greater variety of adsorbent materials and is particularly superior for lipophilic substances; and 4) it can use more drastic chemical detection methods.

Thin-Layer Chromatography of Carbohydrates. Square glass plates 20 × 20 cm have been used in the past as the support for absorbant. The thickness of the absorbant layer determines the amount of material which can be chromatographed. There are a number of commercial devices which can be used to apply any predetermined thickness of absorbant. The glass plates are first cleaned with 95 per cent ethanol to ensure adhesion of absorbant on the glass. For example, plates used in this experiment are made from a mixture of 1 part silica gel, containing gypsum as a binding agent, and 2 parts of water. The special apparatus is used to spread the silica uniformly over the plates. The silica is allowed to set for 10 to 15 minutes and put in an oven at 110° for 45 minutes for activation.

Glass plates have been supplemented and successfully replaced by pre-fabricated TL sheets having a solvent-resistant plastic backing or impregnated fiber glass backing for support of absorbant. Prefabricated sheets come in several basic sizes, 5 × 20 cm or 20 × 20 cm for ordinary use, and 20 × 40 cm for thin-layer electrophoresis. The prefabricated sheets can also be obtained in several thicknesses as required. For better carbohydrate separation the sheets can be dipped in 0.1 M KH_2PO_4 and activated for one hour at 110° C.

The prefabricated sheets offer a more convenient method, particularly for student use. These sheets may be cut to 5 × 10 cm pieces. These can be spotted with 0.1 M sugar solutions and unknowns. At least four spots 0.5 cm in diameter and 2 cm from the bottom can be placed on these sheets with capillary tubing. After drying (use a heat gun if necessary), two sheets can be placed in a 400 ml beaker filled to the 0.5 cm level with solvent. The sheets are put in a ∨ or inverted ∧ configuration so as to support each other. The beaker is covered with a watch glass or a piece of aluminum foil. In about 20 minutes the developer solvent will rise to a height of 7 cm, at which time the sheets can be removed from the beaker and analyzed.

There are many solvent mixtures which can be employed. The references cover this topic comprehensively. Ethyl acetate, pyridine, and water (v/v 8:2:1) is a convenient solvent to use with sugars. Other solvent mixtures which may be employed usefully are: water, n-butanol, and glacial acetic acid (v/v 5:4:1) and chloroform, glacial acetic acid and water (30:35:5).

Detection can be involved. Some prefabricated sheets include fluorescent dyes in the absorbant which are quenched by absorbents. These substances can be detected by examination under a U.V. lamp, and will appear as dark spots against a green background. There are also many sprays in use for detection. For instance, a TL sheet can be sprayed with concentrated sulfuric acid, followed by heating at 110° C in an oven for 15 minutes. Charred spots will appear at any locus of organic molecules. This technique is best suited for use with glass support plates or with prefabricated sheets having inert binder.

There are a variety of spray reagents employed for the selective detection of carbohydrates.

(a) Freshly prepared 2 per cent $NaIO_4$; 1 per cent $KMnO_4$ in 2 per cent Na_2CO_3 4:1. The yellowish spots will appear immediately against a pink background.

(b) 1.23 g of p-anisidine, 1.66 g of phthalic acid in 100 ml of methanol.

(c) 0.93 g of aniline and 1.66 g of phthalic acid in 100 ml of H_2O-saturated n-butanol.

(d) H_2SO_4 spray: spray concentrated or dilute H_2SO_4 and heat to 110°.

Paper Chromatography of Carbohydrates. For paper chromatography, use 36 × 36 cm sheets of Whatman #1 paper. After spotting, roll the filter papers into cylinders and develop them in a cylindrical chromatography jar, using the same solvent system. As previously, pin the $\frac{1}{2}$ inch overlap with capillary tubing, and cut a 1 × 1 inch rectangle of paper from one of the lower corners to eliminate deformation of the solvent front.

REFERENCES

1. Daniel, L. J., and Leslie, N. A. *Laboratory Experiments in Biochemistry.* Academic Press Inc., New York, 1966.
2. Bobbitt, J. M., et al. *Introduction to Chromatography.* Reinhold, New York, 1968.
3. Kirchner, J. G. *Thin-Layer Chromatography.* Interscience, New York, 1967.
4. Bobbitt, J. M. *Thin-Layer Chromatography.* Reinhold, New York, 1963.
5. Stahl, E. (ed.) *Thin-Layer Chromatography: A Laboratory Handbook.* Academic Press, New York, 1965.
6. Heftman, E. *Chromatography.* Reinhold, New York, 1961.
7. Whistler, R. L., and Wolfrom, M. L. *Methods in Carbohydrate Chemistry*, Vol. I. Academic Press, New York, 1962.

CHAPTER 12

THE USE OF POLARIMETRY FOR CONFIGURATIONAL AND QUANTITATIVE ANALYSIS OF CARBOHYDRATES

THEORY

1. Stereoisomers. Stereoisomers are substances having the same elements, the same empirical structural formulae, and the same substituent groups about each carbon atom. However, these substituent groups are oriented differently in space. Optical isomerism is a specific type of stereoisomerism involving the general formula Y about a carbon atom where X, Y, and Z are different

$$X-\overset{\displaystyle Y}{\underset{\displaystyle |}{C}}-Z$$

substituent groups. These molecules all have identical chemical and physical properties except one. Pairs of optical isomers will rotate the plane of polarized light in opposite directions. This is a problem of symmetry or "handedness." Such compounds can have two asymmetrical conformations in space. They are mirror images of one another and are non-superimposable. Optically isomeric pairs are enantiomorphs of one another; one rotates the plane of polarized light to the right and the other to the left.

2. The Polarimeter. The polarimeter is an instrument which has an optical system capable of producing plane polarized light. Its construction and design also allow an accurate measurement of the optical rotation of enantiomorphic pairs.

Ordinary light is a wave motion in which the vibrations take place in all planes perpendicular to the direction of propagation. When light passes

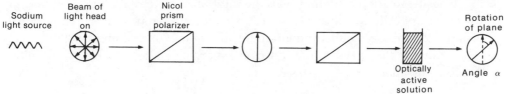

FIGURE 11.   Polarization of light.

through certain substances it emerges in a single plane rather than a multiplicity of possible planes. In this circumstance it is regarded to be plane polarized. The figures indicate what happens in a polarimeter.

Substances which rotate the beam of polarized light to the right or clockwise are called dextrorotating, and those rotating it to the left or counter clockwise, levorotating. These terms are not to be confused with the D and L forms of compounds, which refer to a specific structural orientation in space of the residual groups about the carbon atom, and not particularly to the observed rotations obtained by polarimetric analysis.

Sugars such as glucose have five asymmetric carbon atoms; therefore, there are 32 stereoisomers. However, most naturally occurring sugars are D with respect to the configuration about carbon atom 4 or 5, depending upon whether the sugar is a pentose or hexose. For hexoses, consequently, there can only be 16 naturally occurring optical isomers. The three most important and abundant are glucose, mannose, and galactose. However, they are not mirror images of one another since the differences in geometry involve carbon atoms other than number 5. The D series of hexoses is determined by the configuration about carbon 5, and in the case of these three sugars the orientation is the same.

The geometry about carbon 1 has the designation alpha or beta. For instance, crystals containing only one of these optical forms can be obtained with most reducing sugars. The commercial anhydrous reagent grade D-glucose is usually the pure α-pyranose ring form. When it is dissolved in water it undergoes a molecular rearrangement, called mutarotation, involving the hemiacetal linkage between carbons 1 and 5. This results in an equilibrium mixture consisting of α- and β-D-glucopyranose.

THE DETERMINATION OF THE SPECIFIC ROTATION, INVERSION, AND KINETICS OF MUTAROTATION OF SUGARS

Principles.   If the rotation of an optically active substance is measured under standard conditions, the specific rotation is given by:

$$[\alpha]_D^T = \frac{[\alpha]_{obs}}{1 \times c} \text{ or } = \frac{[\alpha]_{obs}(100)}{1 \times c}, \text{ depending on the units of } c.$$

α = observed rotation in circular degrees

T = temperature

D = the line for sodium vapor (5890 Å)

l = length of the tube in decimeters

c = concentration in g/ml or g/100 ml

Specific rotation by definition is the rotation caused by a solution containing 1 g/ml of optically active solute in a 1 decimeter tube at a given temperature (usually 20°) using monochromatic light, the D line of sodium at 5890 Å.

Knowing the specific rotation of a sugar, it is also possible to calculate the concentration.

$$c = \frac{[\alpha]_{\text{obs}}}{1 \times [\alpha]_D^T}$$

Every optically active compound has a characteristic specific rotation which may be used as a physical constant to distinguish it from other compounds. Identification is facilitated by obtaining a characteristic initial specific rotation when a sugar is first dissolved as well as a characteristic equilibrium specific rotation after the sugar has mutarotated. These two values provide information on the original anomeric form of the sugar as well as help to identify it.

EXPERIMENT 12

Work in pairs.

THE SPECIFIC ROTATION AND MUTAROTATION OF GLUCOSE: CALCULATION OF THE MUTAROTATION CONSTANT

A number of different polarimeters are available. These are equipped with a circular scale with a vernier for fine readings. Before these experiments are performed each student should become familiar with the operation of the polarimeter. An instructor will assist you. The polarimeter tubes themselves are expensive precision items which must be handled with care. Chipping of the end plates or tube ends will produce leaks. The end plates must be handled with extreme care. They are polished optically flat and must not be scratched. Use only lens paper to wipe them.

Before any measurements are taken, a zero reading of the polarimeter must be obtained. Rinse the polarimeter tube twice with 5 to 10 ml portions of distilled water, and then fill to the top with water and stopper. Use a medicine dropper to complete the filling. Place in the polarimeter and rotate the analyzer of the polarimeter until both halves of the circular field, as seen through the eye piece, are equally illuminated. Change the analyzer setting and bring it once again to the point where the field is uniform, and read again. Take three readings; calculate the average and use this as the zero point.

After obtaining a zero reading, thoroughly dry the polarimeter tube. Work rapidly and make up a fresh solution of 8 per cent D-glucose. Add 8 g of solid D-glucose to a dry 100 ml volumetric flask. Slowly add, with swirling, about 90 ml of distilled water. Add a drop of concentrated NH_4OH; bring to volume with water and mix thoroughly. Quickly add the solution to the dry polarimeter tube. Heap the solution above the end of the tube so that when the tube is closed off no bubbles are trapped inside. Obtain a reading as soon as possible. Designate this first reading as the zero time reading, and thereafter, take readings every 2 minutes for 10 to 20 minutes.

Repeat this experiment but omit the NH_4OH. In this case, however, take readings every 5 minutes after the initial reading for 30 minutes, then every 20 minutes for the next 60 minutes.

Calculation of the Specific Rotation. Plot in each case time versus observed rotation. Calculate the specific rotation of α-D-glucose by extrapolating the curve of the untreated glucose to zero time. How does this compare to the data in Table 23?

Determine the specific rotation of the equilibrium mixture of α- and β-D-glucose; compare to the data of Table 23.

Why is it necessary to add NH_4OH to the reducing sugar before reading the polarimeter?

FIGURE 12. (Top) Polarimeter, half-shade field, Rudolph. (Courtesy Arthur H. Thomas Co., Philadelphia, Pa.) (Bottom) Diagram of a polarimeter. **S,** Source of light; **D,** bichromate light filter, giving the effect of sodium light; **L$_1$,** collimator lens, from which the light emerges as a parallel beam, **R$_1$, R$_2$; N$_1$,** Nicol prism, polarizer; **N$_2$,** Nicol prism, analyzer—here it is assumed that their principal planes are parallel; **E,** eyepiece with lens, **L$_2$.** The courses of two rays are shown. **R$_1$** is split into two rays, vibrating in planes at right angles to each other. **O$_1$** is the "ordinary" ray, vibrating in a plane at right angles to the page and is reflected (by the surface **AB**) and eliminated. The "extraordinary" ray, **E$_1$,** passes out of the polarizer, vibrating in a plane parallel to the page. Similarly, **O$_2$** is eliminated and **E$_2$** passes out of the polarizer, vibrating in a plane parallel to that of **E$_1$.** A quartz plane, **Q,** covering half the field, is here interposed. Actually this covers the left or right half, but in the figure it is shown as covering the lower half. **Q** turns the plane of polarized light of **E$_1$** at a small angle to **E$_2$.** Both now enter the tube, **T,** which contains the sugar solution. Here the planes of both **E$_1$** and **E$_2$** are again twisted. Now **E$_1$** is vibrating in a plane at right angles to the plane of the page, and hence, when it enters **N$_2$,** it becomes eliminated because the principal planes of **N$_1$** and **N$_2$** are parallel. Therefore all the light passing through **Q** will be lost and half the field will appear dark as shown at **F$_2$** or **F$_3$.** **E$_2$,** although turned just as much as **E$_1$,** is vibrating in a different plane from **E$_1$** and therefore can pass through **N$_2$** to a slight extent. The analyzer may be rotated and thus permit **E$_1$** and **E$_2$** to pass through to an equal extent. The field, **F$_1$,** would then appear, and the amount of rotation necessary to produce this field, i.e., the compensation for the twisting due to the sugar solution in **T,** may be read off on a scale. (From Kleiner, I. S., and Dotti, L. B.: Laboratory Instructions in Biochemistry, ed. 4, St. Louis, 1954, The C. V. Mosby Co.; slightly modified with the aid of the staff of the Bausch & Lomb Optical Co.)

Calculation of the Mutarotation Constant. A first order equilibrium reaction such as the mutarotation of α-D-glucose to β-D-glucose is given as:

$$\alpha \underset{k_2}{\overset{k_1}{\rightleftharpoons}} \beta$$

The rate equation for this reaction may be written:

(1)
$$-\frac{d\alpha}{dt} = k_1\alpha - k_2\beta.$$

Table 23

Sugar	α Form	Equilibrium Mixture	β Form
D-ribose	−23.1	−23.7	—
L-arabinose	+54.0	+104.5	+175.0
D-xylose	+92.0	+19.0	−20.0*
D-glucose	+113.4	+52.2	+19.0
D-galactose	+144.0	+80.5	+52.0
D-fructose	−21.0*	−92.0	−133.5
D-mannose	+34.0	+14.6	−17.0
L-rhamnose	−7.7	+8.9	+54.0*
L-sorbose	—	−43.4	—
lactose	+90.0	+55.3	+35.0
maltose	+168.0*	+136.0	+118.0
sucrose	—	+66.5	—
raffinose	—	+105.2	—
trehalose	—	+178.3	—

* Calculated value

This differential equation states that the rate of disappearance of the α form is a function of the rate of the forward reaction less the rate of the reverse reaction.

Let α_0 be the original concentration of D-glucose. Then $\alpha + \beta = \alpha_0$; that is, at any time t, α_0, the original amount of D-glucose, is equal to the sum of α-D-glucose and β-D-glucose in solution. Consequently,

$$(2) \qquad \beta = \alpha_0 - \alpha.$$

Substituting in (1),

$$(3) \qquad \frac{-d\alpha}{dt} = k_1\alpha - k_2(\alpha_0 - \alpha)$$

$$= \alpha(k_1 + k_2) - k_2\alpha_0.$$

Rearranging terms,

$$(4) \qquad \frac{-d\alpha}{\alpha(k_1 + k_2) - k_2\alpha_0} = dt,$$

and integrating,

$$(5) \qquad \frac{-1}{k_1 + k_2} \ln\left[(k_1 + k_2)\alpha - k_2\alpha_0\right] = t + C.$$

Evaluating for the integration constant at a specific value (when $t = 0$ and $\alpha = \alpha_0$),

$$(6) \qquad C = -\frac{1}{k_1 + k_2} \ln\left[(k_1 + k_2)\alpha_0 - k_2\alpha_0\right]$$

$$(7) \qquad C = -\frac{1}{k_1 + k_2} \ln\left[k_1\alpha_0\right]$$

Substituting for C in Equation (5),

(8) $\quad -\dfrac{1}{k_1 + k_2} \ln \left[(k_1 + k_2)\alpha - k_2\alpha_0\right] = t - \dfrac{1}{k_1 + k_2} \ln \left[k_1\alpha_0\right]$

(9) $\quad -\ln \left[(k_1 + k_2)\alpha - k_2\alpha_0\right] = (k_1 + k_2)t - \ln \left[k_1\alpha_0\right]$

(10) $\quad \ln \left[k_1\alpha_0\right] - \ln \left[(k_1 + k_2)\alpha - k_2\alpha_0\right] = (k_1 + k_2)t$

The integrated rate equation is thus

(11) $\qquad \ln \dfrac{k_1\alpha_0}{(k_1 + k_2)\alpha - k_2\alpha_0} = (k_1 + k_2)t$

At equilibrium, however,

(12) $\qquad \dfrac{d\alpha_e}{dt} = \dfrac{d\beta_e}{dt} = 0$

where α_e and β_e are the concentrations of the alpha and beta forms at equilibrium.

(13) $\qquad K_{\text{equil}} = \dfrac{\beta_e}{\alpha_e} = \dfrac{k_1}{k_2}$

(14) $\qquad \beta_e k_2 = \alpha_e k_1$

But since β_e is $\alpha_0 - \alpha_e$, then

(15) $\qquad (\alpha_0 - \alpha_e)k_2 = \alpha_e k_1$

Solving for k_1 in Equation (15), and substituting into the logarithmic expression of Equation (11),

(16) $\qquad \ln \dfrac{\alpha_0 - \alpha_e}{\alpha - \alpha_e} = (k_1 + k_2)t$

where $t =$ time

$\qquad \alpha =$ concentration of α-D-glucose at time t

$\qquad \alpha_0 =$ concentration of α-D-glucose at zero time

$\qquad \alpha_e =$ concentration of α-D-glucose at equilibrium.

Since concentration is directly proportional to specific rotation, Equation (16) may be written as follows:

(17) $\quad \log \left([\alpha]_D^T - [\alpha_e]_D^T\right) = \dfrac{-k_1 + k_2}{2.303} \, [t] + \log \left([\alpha_0]_D^T - [\alpha_e]_D^T\right)$

This is a straight line equation where $y = ax + b$; $y =$ the log of the specific rotation of D-glucose at time t minus the specific rotation of D-glucose at equilibrium; $x =$ time;

$$a = \text{slope} = -\frac{k_1 + k_2}{2.303}\,,$$ which is now called the mutarotation constant k_m; and b is the y intercept $=$ log of the specific rotation at $t = 0$ minus the specific rotation at equilibrium.

Plot graphs of log $[\alpha]_D^T - [\alpha_e]_D^T$ versus time. From the slope determine k_m. Use only the data obtained with NH_4OH.

THE SPECIFIC ROTATION AND INVERSION OF SUCROSE

The Specific Rotation of Sucrose.   Since sucrose is not a reducing sugar it does not exhibit mutarotation. Carefully and accurately weigh about 10 g of sucrose and transfer it to a 100 ml volumetric flask. The weight does not have to be exactly 10 g, but the weight should be accurate. Dissolve the sugar in distilled water, make it up to volume, and mix thoroughly.

Obtain a zero reading on the polarimeter with water. Then, using either a dry polarimeter tube or one rinsed out with sucrose solution, obtain a reading for the sucrose solution.

Calculate the specific rotation of sucrose.

The Inversion of Sucrose.   Because fructose is strongly levorotating as compared to sucrose, which is dextrorotating, it is possible to demonstrate inversion of rotation when sucrose is hydrolyzed. Transfer some of the sucrose solution to a second 100 ml volumetric flask; add 25 ml of water and 5 ml of concentrated HCl. Place the flask in a beaker of water heated to 90°. When the sucrose solution has reached 69°, lower the heat of the bath to 68–70° and maintain the sucrose solution at 69° for 10 minutes. Cool the contents to 20°, rinse the thermometer with water, making sure the rinse goes back into the flask, and bring up to volume. Mix thoroughly and read in the polarimeter.

Since the sucrose concentration has been halved, the observed optical rotation has to be multiplied by two in order to calculate specific rotation. How does this compare now to the specific rotation of the untreated sample?

From the readings before and after inversion, calculate the percentage of sucrose in the inverted sample, which by subtraction from the original concentration will give the per cent inversion. Since the degree of rotation is directly proportional to the concentration of the solution, the following proportionality can be used to calculate the amount of sucrose remaining.

$$\frac{45.98}{P - I} = \frac{26}{x}$$

where $x =$ concentration of sucrose in grams per cent. The proportion is based on the fact that 26 grams of sucrose in 100 ml gives a positive reading, P, of $+34.06°$ before inversion, and an invert reading of $-11.92°$ after inversion. The algebraic difference is $45.98°$. I is the experimental specific rotation.

ISOLATION AND CHARACTERIZATION OF TREHALOSE BY POLARIMETRY

Work in pairs.

Trehalose, α-D-glucopyranosyl α-D-glycopyranoside, was discovered in rye ergot over 100 years ago. It has been isolated from many sources. It was isolated from yeast in 1925 by Koch and Koch while they were attempting to isolate "bios," a yeast growth factor. They allowed alcoholic yeast extract to stand for several months and obtained crystals which eventually were identified as trehalose. It had no bios activity, however.

Trehalose

Make a paste of 16 g of dried baker's yeast with 34 ml of water. Add 125 ml of 95 per cent ethanol and allow the mixture to stand for 30 minutes with occasional stirring. Filter the mixture in a Büchner funnel and wash the precipitate with three 15 ml portions of 70 per cent ethanol. Combine the washes and the original extract. Add 10 ml of 20 per cent $ZnSO_4$, 0.5 ml of 1 per cent phenolphthalein, and sufficient saturated $Ba(OH)_2$ to make the mixture pink to the dye. Add 1 g of activated charcoal. Heat the entire mixture to 70° C in a steam bath, and filter while hot through a Büchner funnel precoated with Filter Cel. Adjust the deproteinized solution to about pH 7 with 0.1 M HCl and concentrate it by gentle heating under vacuum in a round bottom flask until 5 ml of syrup remain.

Slowly mix 40 ml of 95 per cent ethanol into the syrup; stopper the flask and store in the desk from 24 hours to as long as one week in order to assure crystallization. Slow crystallization often produces crystals over $\frac{1}{2}$ inch long. Crystallization can be hastened by cooling the sample to 0° C or by adding more 95 per cent ethanol. Filter the crystals and allow them to dry.

(1) Chromatograph the sample along with an authentic sample and other sugars by thin-layer chromatography and obtain a distinctive R_f for trehalose. Some judgement of purity can be made from chromatography (cf. page 143).

(2) Check the specific rotation of a 1% concentration of authentic trehalose. It has an $[\alpha D]_D^{20} = +178.3°$. What is the purity of your product according to this criterion?

(3) Is this a reducing sugar? Explain.

REFERENCES

1. Clark, J. M. *Experimental Biochemistry*. W. H. Freeman and Co., San Francisco, 1964.
2. Pritham, G. H. *Anderson's Laboratory Experiments in Biochemistry*. The C. V. Mosby Co., St. Louis, 1968.
3. Harrow, B., Borek, E., Mazur, A., Stone, G. C. H., and Wagreich, H. *Laboratory Manual of Biochemistry*. W. B. Saunders Co., Philadelphia, 1960.
4. Daniel, L. J., and Leslie, N. A. *Laboratory Experiments in Biochemistry*. Academic Press Inc., New York, 1966.
5. *Experimental Biochemistry Laboratory Manual*, Department of Biochemistry, University of Wisconsin, 1967.

CHAPTER 13

THE ISOLATION OF GLYCOGEN AND THE DETERMINATION OF THE DEGREE OF BRANCHING BY PERIODIC ACID OXIDATION

CHEMISTRY

Periodic acid oxidation is limited to compounds having the following groups on *adjacent* (vicinal) carbon atoms: OH and OH; OH and NH_2; C=O and OH; CO and CO; CO and CHO. There are other examples.

$$\begin{array}{c} R \\ | \\ H-C-OH \\ | \\ H-C-OH \\ | \\ R' \\ \textit{glycol} \end{array} + HIO_4 \rightarrow \begin{array}{c} R \\ | \\ H-C=O \end{array} + \begin{array}{c} H-C=O \\ | \\ R' \end{array} + H_2O + HIO_3$$

mixed aldehydes

$$\begin{array}{c} R \\ | \\ H-C-OH \\ | \\ H-C-NH_2 \\ | \\ R' \\ \textit{Hydroxyamino} \\ \textit{compounds} \end{array} + HIO_4 \rightarrow \begin{array}{c} R \\ | \\ H-C=O \end{array} + \begin{array}{c} H-C=O \\ | \\ R' \end{array} + NH_3 + HIO_3$$

mixed aldehydes

156

$$
\begin{array}{l}
\text{H} \\
| \\
\text{H}-\text{C}-\text{OH} \\
| \\
\text{C}=\text{O} \\
| \\
\text{H}-\text{C}-\text{OH} \\
| \\
\text{H}
\end{array}
+ \text{HIO}_4 \rightarrow
\begin{array}{l}
\text{H} \\
| \\
\text{H}-\text{C}=\text{O} \\
\textit{formaldehyde} \\
\textit{from primary} \\
\textit{alcohol}
\end{array}
+
\begin{array}{l}
\text{O} \\
\| \\
\text{C}-\text{OH} \\
| \\
\text{H}-\text{C}-\text{OH} \\
| \\
\text{H}
\end{array}
+ \text{HIO}_3
$$

Hydroxyl carbonyl

$$
\begin{array}{l}
\text{R} \\
| \\
\text{C}=\text{O} \\
| \\
\text{H}-\text{C}-\text{OH} \\
| \\
\text{R}'
\end{array}
+ \text{HIO}_4 \rightarrow
\begin{array}{l}
\text{R} \\
| \\
\text{C}=\text{O} \\
| \\
\text{OH} \\
\textit{acid}
\end{array}
+
\begin{array}{l}
\text{H}-\text{C}=\text{O} \\
| \\
\text{R}' \\
\textit{aldehyde}
\end{array}
+ \text{HIO}_3
$$

carbonyl-hydroxyl

$$
\begin{array}{l}
\text{R} \\
| \\
\text{C}=\text{O} \\
| \\
\text{C}=\text{O} \\
| \\
\text{R}'
\end{array}
+ \text{HIO}_4 + \text{H}_2\text{O} \rightarrow
\begin{array}{l}
\text{R} \\
| \\
\text{C}=\text{O} \\
| \\
\text{OH}
\end{array}
+
\begin{array}{l}
\text{OH} \\
| \\
\text{C}=\text{O} \\
| \\
\text{R}'
\end{array}
+ \text{HIO}_3
$$

acids

α-diketone

$$
\begin{array}{l}
\text{CH}_3 \\
| \\
\text{C}=\text{O} \\
| \\
\text{H}-\text{C}=\text{O}
\end{array}
+ \text{HIO}_4 + \text{H}_2\text{O} \rightarrow
\begin{array}{l}
\text{CH}_3 \\
| \\
\text{C}=\text{O} \\
| \\
\text{OH}
\end{array}
+
\begin{array}{l}
\text{HO}-\text{C}=\text{O} \\
| \\
\text{H}
\end{array}
+ \text{HIO}_3
$$

α-ketonic aldehyde

The end product is always HIO_3 and mixtures of aldehydes or acids. The aldehydes, particularly formaldehyde, can be determined gravimetrically by precipitation with dimedon:

formaldehyde   *dimedon*

HIO_3 can be determined titrimetrically by several methods. The one most suitable is to determine the difference in amount of I_2 released. Iodine is released from IO_4^- and IO_3^- according to the following equations and it is determined by titration with standard $Na_2S_2O_3$. A difference of one mole of I_2 is obtained for each mole of IO_4^- reduced, and two equivalents of $Na_2S_2O_3$ are needed to titrate the mole of I_2.

$$IO_4^- + 8H^+ + 7I \rightarrow 4I_2 + 4H_2O$$

$$IO_3^- + 6H^+ + 5I \rightarrow 3I_2 + 3H_2O$$

$$IO_4^- - IO_3^- = \Delta I_2, \text{ 1 mole } I_2 \text{ released/mole } IO_4^- \text{ reduced}$$

and $I_2 + 2S_2O_3^= \rightarrow 2I^- + S_4O_6^=$

This means that 1 mole of IO_4^- reduced $=$ two equivalents of $Na_2S_2O_3$.

When formic acid is the end product, it may be determined by titrating an aliquot of the reaction mixture after decomposing the excess IO_4^-, provided buffers have not been used. Decomposition of the IO_4^- is usually accomplished by adding an excess ethylene glycol.

The simple sugars, such as glucose, galactose, and mannose, react with periodic acid to yield 1 mole of formaldehyde and 5 moles of formic acid.

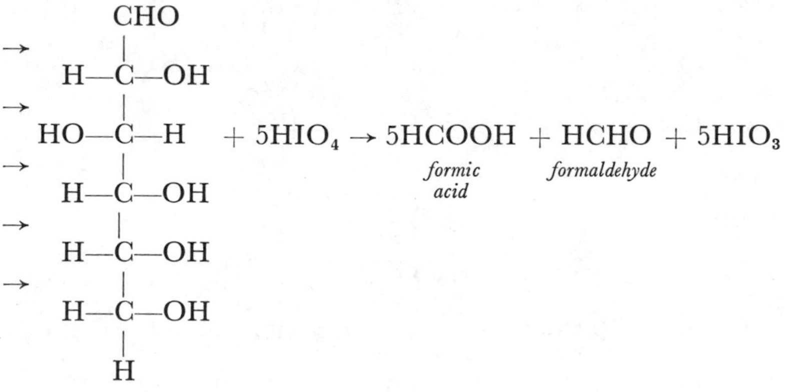

glucose

Application of this method to substituted sugars, such as glycosides, is very valuable in determining chemical structure.

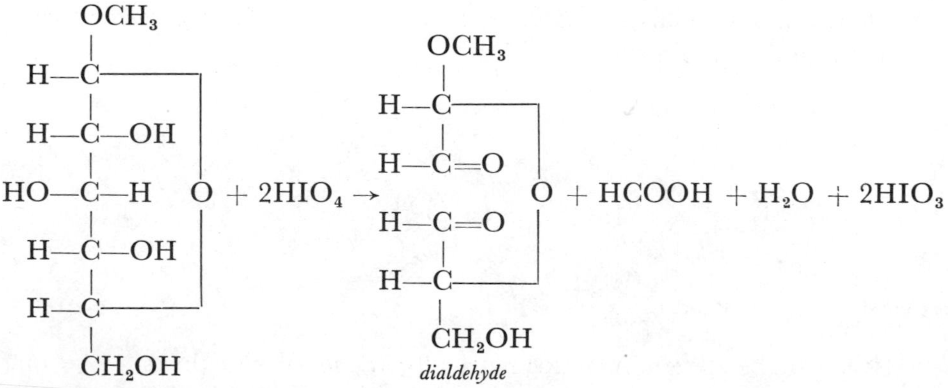

D-*Methylglucopyranoside*

PRINCIPLE

The structure of glycogen and similar polysaccharides can be described as a branched, fanlike arrangement of glucopyranose residues with 1–4 linkages along the main chains and 1–6 linkages at the branch points. As a result of this structure, only the terminal reducing end and all of the non-reducing ends yield formic acid on periodate oxidation. Because of this unique structure, the amount of formic acid formed from the *single* reducing end is insignificant as compared to that formed by the many non-reducing ends. In this procedure, then, it is presumed that the number of non-reducing ends of glycogen is approximately equal to the number of molecules of formic acid formed. The classical methylation procedure for end-group analysis is extremely tedious, but the periodate method makes routine determinations possible, although subject to errors.

The four types of linkages of glucose residues found in a molecule of glycogen are illustrated:

At the non-reducing end groups there are two pairs of vicinal hydroxyls. Two moles of periodate will be consumed with the liberation of C-3 as formic acid. At the reducing end, cleavage takes place between C-3 and C-2, and also between C-2 and C-1; C-2 is liberated as formic acid; C-1, which is attached to the ring oxygen as the formic ester, eventually is hydrolyzed to yield another molecule of formic acid and an OH group on C-5. This is followed by a C-5 and C-6 split freeing C-6 as formaldehyde. The branch point glucose residues and the 1,4-linked residues are split only at the C-2 and C-3 carbons to yield dialdehydes, and no one-carbon fragments. The total periodate

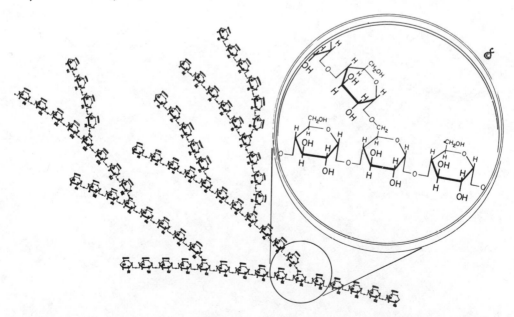

FIGURE 13. Glycogen structure. (From *Experimental Biochemistry*, John M. Clark, Jr., ed. W. H. Freeman and Company. Copyright © 1964.)

consumption for the reducing end group is 3 moles, along with the formation of 2 moles of formic acid. This makes a very small contribution and can be ignored in a molecule having a molecular weight of about 1×10^6 with at most one or two reducing end groups in the entire molecule. The residues in the inner part of the chain consume one mole of periodate. The non-reducing end group consumes 2 moles of periodate, with the liberation of 1 mole of formic acid. Therefore, for all practical purposes *the number of non-reducing ends equals the number of moles of formic acid released* and this fact is used to determine the degree of branching.

EXPERIMENT 13

ISOLATION OF RAT LIVER GLYCOGEN

Each student will isolate glycogen from the liver of a single rat. There is just enough glycogen in one rat to perform the experiment properly.

(a) Decapitate a well fed 250 g rat and immediately excise the liver; weigh it quickly, slice it, and transfer the slices without delay to a 100 ml flask containing 12.0 ml of hot 50 per cent aqueous KOH.

(b) Heat the liver mixture in a boiling water bath for one hour or until digested. Saponification will be hastened by swirling the flask. *Wear safety goggles during this procedure.*

(c) Dilute this mixture to 30 ml with water and transfer it to a 500 ml flask; add 135 ml of 95 per cent ethanol and mix by swirling. The flocculent precipitate formed is principally glycogen, contaminated with some protein. Stopper the flask and allow it to stand for 30 minutes.

(d) Collect the precipitate by centrifugation for 5 minutes at 3000 rpm using 50 ml polyethylene centrifuge tubes. Collect the glycogen precipitate quantitatively with the aid of a rubber policeman.

(e) Dissolve the glycogen in 15.0 ml of cold 10 per cent trichloroacetic acid, and remove the brown protein contaminating material by centrifugation.

(f) To the glycogen supernatant add approximately 60 ml of 95 per cent EtOH. Wait until the glycogen flocculates, and then collect it quantitatively by centrifugation as previously.

(g) Suspend the glycogen in 15 ml of cold 70 per cent EtOH; centrifuge again and discard the supernatant.

(h) Repeat step (g) but use 10 ml of absolute ethanol.

(i) Recover the precipitate and evaporate it to dryness on a tared watch glass. Place it about 9 inches from an infra-red lamp in the hood.

(j) Weigh the glycogen, and determine the yield on a wet weight basis.

(k) Weigh very accurately a 100 to 200 mg sample of this glycogen. (Be sure to save about 20 mg of glycogen for purity determination.) Transfer the glycogen to a 50 ml volumetric flask. Add 20 ml of 3 per cent NaCl, and heat gently on a steam bath to facilitate solution. Add 10 ml of 0.350 M $NaIO_4$, and adjust the volume to 50.0 ml with distilled water.

(l) Make up a blank in a similar fashion but omit the glycogen. Place all solutions in the dark in the cold till the next laboratory period.

ANALYSIS OF GLYCOGEN

Work as individuals.

Determination of Glycogen Purity by the Anthrone Reaction. For a reliable determination, the reaction must be carefully controlled; i.e., the concentration of anthrone and acid in the reagent, time and temperature of the reaction, and the presence of interfering substances, such as lint, skin, and paper fibers, which may contaminate the tubes must be taken into account.

Quantitatively weigh out two 9.00 mg samples of glycogen; transfer them to 100 ml volumetric flasks and dissolve in 100 ml of distilled water.

Add 4.0 ml of anthrone reagent, slowly and using a propipette, to each of the following: duplicate 1.0 ml aliquots of the glycogen solution; duplicate 1.0 ml aliquots of a standard 9.0 mg per cent (0.5 μmoles/ml) glucose solution; and 1.0 ml of distilled water. Cool all tubes to 0° in an ice bath, place a marble in each tube, and then heat in a boiling water bath for 10 minutes. After boiling is completed, place the tubes in cold water to stop the reaction and bring the tubes to room temperature.

Remove the tubes from the water bath and wipe the outsides dry.

Carefully transfer a 3 ml portion from each tube to individual cuvettes and read the absorbances at 620 nm.

The micromoles of glucose in 9.0 mg per cent glycogen will be equal to the absorbance equivalent for a 9.0 mg per cent glucose solution corrected for the loss of water in glycogen, which is a polymer of glucose. The molecular weight of a glucose residue in glycogen is $180 - 18$ or 162, and $\frac{180}{162} = 1.11$. Therefore:

$$\mu\text{moles of glucose per 1 ml glycogen sample} = \frac{A_{620} \text{ of glycogen } (A_u)}{A_{620} \text{ of glucose } (A_s)} \times \text{glucose}$$

$$\text{standard concentration } (C_s)$$

In reality, both solutions should exhibit the same absorbance, since they are both glucose; glycogen has an effective molecular weight of 162. Therefore:

$$\frac{A_{620} \text{ of glycogen}}{A_{620} \text{ of glucose}} \times 1.11 \times 100 = \text{purity}$$

or if you prefer:

$$\frac{\text{Total } \mu\text{moles of glucose in glycogen}}{\mu\text{moles of glucose in the standard}} \times 1.11 \times 100 = \text{purity.}$$

Determination of Per Cent Branching and the Number of Glucose Units per Segment in the Glycogen Sample. After completing the absorbance readings, transfer 5 ml aliquots of the glycogen–periodate mixture to 50 ml flasks; add two or three drops of ethylene glycol to each and place them in the dark for 15 minutes. Titrate each to a phenolphthalein end point with 0.010 N NaOH, using an ascarite tube over the top of the burette to exclude CO_2. Do three titrations and determine the average value.

CALCULATIONS

If we were to look at the structure of glycogen we would see that it is fan-like and begins at a single glucose residue linked by α-1,4 bonds to a second

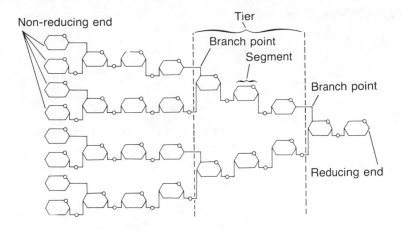

FIGURE 14. Schematic glycogen structure.

glucose residue and so on. At each branch point there is an α-1,6 linkage. The first glucose residue is a reducing sugar because carbon 1 is still free. For every branch point, however, there are two non-reducing glucose residues. Why? The total number of branching α-1,6 bonds in the molecule of Figure 14 is one less than the number of non-reducing ends (N) of glycogen. According to the diagram there are seven points in this representation of a glycogen molecule, and consequently there are eight non-reducing end groups and the number of chains is eight. However, in an authentic sample of glycogen with a molecular weight of the order $> 1 \times 10^6$, N would be a large number, and for all practical purposes the number of α-1,6 bonds is equal to the number of non-reducing ends. Therefore the per cent branching or per cent α-1,6 glucose residues in a glycogen sample is determined by:

$$\text{per cent branching} = \frac{\alpha\text{-1,6 bonds in glycogen sample}}{\text{total glucose residues in glycogen sample}} \times 100.$$

Since we have already established that the number of non-reducing glucose residues is equal to the number of molecules of formic acid released, then:

$$\text{per cent branching} = \frac{\text{total moles of HCOOH formed from glycogen sample}}{\text{total moles of glucose residues in glycogen sample}} \times 100.$$

Since there are N chains present,

$$\text{average chain length in glycogen} = \frac{\text{moles of glucose in sample}}{\text{number of chains}}$$

$$= \frac{\text{moles of glucose in tared glycogen sample}}{\text{moles of formic acid formed from tared glycogen}}$$

For these calculations use data obtained under "Analysis of Glycogen" and correct for the state of purity.

$$\frac{\text{purity per cent} \times \text{g of glycogen dissolved in 50 ml}}{}= \text{total moles of glucose in tared glycogen sample}$$

and:

$$\frac{\text{Vol} \times N_{\text{NaOH}} \times 3.33}{1000} = \text{total moles of HCOOH formed from the tared glycogen sample (The factor of 3.33 is used to correct for the fact that 15 ml out of 50 ml was titrated.)}$$

N = the number of non-reducing residues = the number of chains present. If we look at the schematic representation of glycogen we find that N, the number of non-reducing residues, is equal to 8 and that the number of segments (portion between branch points) is equal to 15 or $2N - 1$. Since $2N$ also is a relatively large number it is possible to calculate the average number of glucose residues in a segment of glycogen from the relationship:

$$\text{average glucose units in a segment} = \frac{\text{total moles of glucose in glycogen sample}}{\text{number of segments in glycogen sample}}$$

Now, it is also possible to conclude that if the number of segments in a glycogen molecule is equal to $2N - 1$ or $\cong 2N$, and N = the total number of branching α-1,6 linkages, which in turn is equal to the total number of non-reducing groups, and finally equivalent to the total moles of HCOOH formed, then it follows that:

average number of glucose units per segment

$$= \frac{\text{total moles of glucose in glycogen sample}}{2 \times \text{moles of HCOOH formed}}$$

Assuming a molecular weight of 1.62×10^6 for glycogen assayed, use your data to calculate, in one gram molecular weight of glycogen, the total number of glucose residues, the total number of non-reducing end groups (N), the total number of segments ($2N - 1$), and the number of tiers in your glycogen, where $2N - 1 = 2^T - 1$, and T is the number of tiers.

In order, however, to obtain a real number for N and T, something has to be known about the molecular weight of the sample.

You have obtained the number of moles of glucose in your glycogen sample, but you cannot ascertain from this how many units of glucose are in the glycogen. By assigning a molecular weight of 1.62×10^6, you can easily calculate that there are 10,000 glucose units in a mole of glycogen, since:

$$\frac{1.62 \times 10^6}{162} = 1 \times 10^4$$

But because you have determined the per cent branching, let us say, to be 8.35 per cent, which means that 8.35 per cent of the glucose residues are branch points, then $10,000 \times 0.0835$, or 835 out of 10,000, are branching glucose residues.

This number is also approximately equal to the number of non-reducing residues (N); hence, since we have previously stated that N is 1 more than the number of branch points, N now also represents the number of chains in the sample as well.

N can be derived from the titration data directly. Suppose in this example you found 1.17×10^{-4} moles of end groups from the formic acid titration; your sample of glycogen weighed 0.250 gm and was 0.91 per cent pure.

The moles of glycogen in the sample would be:

$$\frac{0.91 \times 0.250}{1.62 \times 10^6} = 1.4 \times 10^{-7} \text{ moles in the sample.}$$

Since

$$N = \frac{\text{moles of end groups in sample}}{\text{moles of glycogen in the sample}} = \frac{1.17 \times 10^{-4}}{1.4 \times 10^{-7}} = 835,$$

the total number of segments equals

$$(2 \times 835) - 1 = 1669$$

We can now calculate the average number of glucose residues in a segment because we know there are a total of 10,000 glucose residues in the molecule of glycogen. Therefore,

$$\frac{10,000}{1669} = 6.$$

PROBLEMS

1. How would you go about correcting for the contribution of the reducing end groups to formic acid formation?

2. What additional information, if any, would be provided by performing a periodate titration in order to obtain the net amount of IO_4^--reduced to IO_3^-?

3. Assuming a fixed length of 5 residues in a glycogen sample, show that the amount of HCOOH produced from the reducing end of a symmetrical 10 tier molecule of glycogen by IO_4^- oxidation is very small compared with that produced from the non-reducing end.

REFERENCES

1. Clark, J. M. *Experimental Biochemistry.* W. H. Freeman and Co., San Francisco, 1964.
2. *Experimental Biochemistry Laboratory Manual.* Dept. of Biochemistry, University of Wisconsin, 1967.
3. Montgomery, R., and Smith, F. "End Group Analysis of Polysaccharides. Part IV. End Group Determination by Periodate Oxidation." *Methods of Biochemical Analysis,* D. Glick, ed. Vol. 3 (1956).
4. Dyer, J. R. "Use of Periodate Oxidation in Biochemical Analysis." *Methods of Biochemical Analysis,* D. Glick, ed. Vol. 3 (1956).
5. Boureng, H. G., and Lindberg, B. "Methods in Structural Polysaccharide Chemistry." In *Advances in Carbohydrate Chemistry,* M. L. Wolfrom, ed. (1960).
6. Whistler, R. L., and Wolfrom, M. L. *Methods in Carbohydrate Chemistry.* Academic Press, New York, 1962.
7. Carroll, N. V., Longley, R. W., and Roe, J. H. *The Determination of Glycogen in Liver and Muscle by Use of Anthrone Reagent.* J. Biol. Chem. **220** 583 (1956).

CHAPTER 14

THE SEPARATION AND PHYSICAL AND CHEMICAL PROPERTIES OF LIPIDS

INTRODUCTION

Lipids are classified either as simple or as complex. The simple lipids comprise fats and oils, waxes, sterolesters, and certain vitamins. Upon hydrolysis they yield fatty acids and alcohols. The complex lipids contain not only fatty acids and alcohols but also phosphate, carbohydrate, and nitrogenous components. They are divided into four major classes: the phospholipids, the sphingolipids, the glycolipids, and the sulpholipids. Each of these are divided into several subgroups.

Plant pigments have lipoidal properties and are principally green or yellow colored unsaturated polyalcohols and hydrocarbons. The yellow carotenoid pigments comprise two groups, the carotenes and the xanthophylls. There are three main subgroups of carotenes, all of which are non-polar: α, β, and γ. The xanthophylls are polar oxygen derivatives of the carotenes. Both groups are polyene isometric hydrocarbon compounds. The compounds like β-carotene are converted in mammalian tissue to Vitamin A, and are termed provitamins.

The green plant pigments are found in two chief forms, as chlorophylls a and b. They are tetracyclic pyrollic compounds, a structural feature which is also found in hemoglobins and cytochromes. Although classifying chlorophylls as lipoidal may be ambiguous, the main reasons for placing them here is that they are extractable by lipid solvents, and they are substituted esters which yield phytylalcohol and a porphyrin pyrole ring system on hydrolysis.

Look up the structure of cholesterol, ergosterol, deoxycholic acid, estradiol, progesterone, Vitamin A, D_3, and tocopherol. What do all of these important biochemicals have in common structurally?

EXPERIMENT 14

The tasks in this experiment are divided into two major parts. Student pairs are to begin the procedure described for separating plant pigments in Part 1. In their spare time, as the extractions proceed, they are to do as many experiments as are possible in Part B.

Part A. Extraction and Adsorption Column Chromatography of Plant Pigments

In 1906 Michael Tswelt, a Russian botanist, percolated a solution of chlorophyll through a column packed with pulverized $CaCO_3$ and fractionated the chlorophyll. He called the process chromatography. Solvent extracts of plant material can still serve as useful illustrations of techniques of column chromatography primarily based in selective adsorptive mechanisms. The separation of plant pigments by chromatography often reflects small differences in the chemical composition and molecular structure of these compounds. Many kinds of fresh plant material can be used for this purpose, the most common being fresh spinach leaves and carrots obtained from a market. In this experiment, pairs of students in half of the class will extract pigments from spinach leaves and the other half from carrot leaves.

SPINACH LEAF EXTRACTION

Two grams of tender green spinach leaves, with large veins and petiols removed, are cut up and placed in 100 ml of boiling water in a beaker. After 2 minutes, the vessel is placed in an ice bath and the water is poured off when cool. The leaves are then extracted with 100 ml of 90 per cent methanol–10 per cent diethyl ether followed by 100 ml of 70 per cent methanol–30 per cent diethyl ether. The extracts are combined in a separatory funnel, and 100 ml of petroleum ether (20 to 40° C) is added; 100 ml of saturated aqueous NaCl is added, and the mixture is shaken. The layers are allowed to separate. The lower layer is discarded, and the upper layer is washed with two 100 ml portions of distilled water. The upper green layer is poured out through the top of the funnel into a round bottom flask and evaporated to dryness below 40° C under vacuum. Be sure that the sample is dry and remains dry after this step. Use a filter pump with a water trap. Store the evaporated extract under vacuum in the dark and cold in order to avoid decomposition. However, should time permit, the residue may be dissolved in 1 ml of petroleum ether (60 to 110° C) and chromatography can proceed. (In order, however, to examine the xanthophylls and carotenes more conveniently, the chlorophylls will have to be removed by saponification with alkali. This may be carried out by adding 10 ml of 30 per cent KOH in methanol to the methanol-diethyl ether mixture in the separatory funnel. After 30 minutes with occasional swirling, 40 ml of cold 1:1 petroleum ether (20 to 40° C)–diethyl ether plus 100 ml of 10 per cent aqueous NaCl are added. The resultant upper golden-yellow layer is washed with water and taken to dryness as above; the residue is then dissolved in 1 ml of petroleum ether in preparation for chromatography.)

100 ml of 10 per cent aqueous NaCl are added. The resultant upper golden-yellow layer is washed with water and taken to dryness as above; the residue is then dissolved in 1 ml of petroleum ether in preparation for chromatography.)

EXTRACTION OF CAROTENES FROM CARROTS

Extract 10 grams of finely lacerated carrots with 100 ml of hot 95 per cent alcohol in a 400 ml Erlenmeyer flask for 30 minutes. Decant the yellow solution and dilute it to approximately 85 per cent alcohol with water, and cool to room temperature. Shake the mixture in a separatory funnel with 50 ml of petroleum ether (20 to 40° C). Allow the layers to separate; the upper one of petroleum ether carries the carotenes, and the lower one the xanthophylls. After removing the upper layer, wash the lower alcohol layer with 2 ml portions of petroleum ether until the wash, which forms a new upper layer, remains colorless. This indicates that all of the carotenes have been extracted. Combine the petroleum ether extracts and wash with 85 per cent alcohol to remove the remaining xanthophylls. Concentrate the petroleum ether layer in vacuo below 40° C, using a rotary evaporator if available. Dissolve the oily residue in 2 ml of Skelly Solvent B in preparation for chromatography. If storage is necessary, place it in the dark, under vacuum and in the cold.

PREPARATION OF ADSORPTION COLUMN

Students extracting spinach leaves are to pack a column of commercial confectioner's sugar which contains 3 per cent starch, and which has been passed through a 60 mesh sieve. Students extracting carrots are to pack a column containing a 1:1 mixture of activated magnesia and celite filter aid.

The 1.4×20 cm chromatography tubes, fitted with fritted glass discs or plugs of cotton or glass wool, are carefully packed with adsorbant. A paper funnel with a narrow hole should be made and the adsorbant poured through uniformly and continuously while agitating the column with an electric vibrator. *Do not force pack.* The packed height of the column should be 20 cm for the sugar column and 15 cm for the MgO-celite column. After packing each column, place a 5 cm layer of anhydrous Na_2SO_4 on the top. This protects the column from traces of water and prevents mechanical erosion.

With the use of mild suction, the sample solution is introduced gently into the column with a pipette and allowed to form a narrow initial zone. The sample adhering to the inside of the tube above the absorbant bed may be washed in by adding a few milliliters of Skelly Solvent B or petroleum ether, whichever is appropriate. The development with fresh solvent may now proceed. It is important that once this point is reached *the columns should not be allowed to run dry,* nor should the vacuum be stopped or altered in intensity. The sugar column is developed with 0.5 per cent n-propanol in petroleum ether (20 to 40° C), and the MgO-celite column is developed with 5 per cent

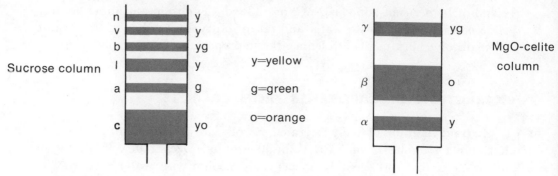

FIGURE 15.   Extraction of plant pigments by column chromatography: spinach (*left*) and carrots (*right*).

acetone in petroleum ether (20 to 40°). This development requires about 30 to 90 minutes, depending upon sorbent, solvent, and column size. It may be speeded up by applying 0.5 atmosphere of suction. A flow rate of 1 ml/min gives good separation. Favorable loading capacity, as indicated by good separation and detection of the major zones, is about 400 μl.

The development of the spinach extraction on confectioner's sugar with 0.5 per cent n-propanol in petroleum ether (20 to 40° C) yields separate zones of the leaf pigments in the sequence: Carotenes (least sorbed), chlorophyll *a*, lutein, chlorophyll *b*, violxanthin, and neoxanthin (most sorbed) (cf. Figure 15).

The carotenes are named on the basis of the rate at which they flow through such a column. Alpha carotene flows through most rapidly, followed in order by beta and gamma. The carotene from carrots is predominantly beta; hence this band is the broadest of the three. Any traces of carotenols will appear above the band of γ-carotene. The desired compound may be isolated by allowing it to flow from the bottom of the column or by extruding the column and mechanically separating the desired band, using a more polar solvent than the developing solvent to remove the compound. (cf. Figure 15 and Reference 1.)

These experiments can be performed by thin layer chromatography and paper chromatography as well as by adsorption column chromatography. All methods can also be adapted to partition chromatography, in which separation is based upon the selective partition of the pigments between two immiscible liquids, the more polar liquid being fixed in a non-sorptive porous support. (cf. Reference 1.)

Part B.   The Physical and Chemical Properties of Lipids

Historically, lipid chemistry at first was concerned with the properties of crude mixtures which were important items in commerce. Certain chemical tests were used and continue to be used to determine the suitability of lipids for human consumption and for the preparation of commercial products such as soaps, greases, cosmetics, and drying oils. Examples of the more common tests are given to provide background information.

Saponification Number: the number of milligrams of KOH required to neutralize the fatty acids in one gram of fat. Fats containing short chain acids will have more COOH's per gram. Therefore, this number becomes a criterion for the average size of the fatty acid chains in the sample.

Iodine Number: the number of grams of iodine absorbed by 100 grams of fat. The unsaturated fatty acids take up iodine and other halogens, yielding saturated halogen derivatives. This can be used as a criterion for determining adulteration in food. For example, olive oil has an I.N. = 79 to 88, and cotton seed oil has an I.N. = 175 to 202. A sample of olive oil having an I.N. greater than 88 might be presumed to be adulterated with cotton seed oil.

Reichert-Meisel Number: a measure of the amount of volatile soluble C_2 to C_8 fatty acids.

Acetyl Number: a measure of the reactivity of fats with acetic anhydride as a means of determining the presence of hydroxyl groups.

Acid Number: acidity in fats is ascribed to hydrolysis. This takes place in various ways and is a factor in ascertaining rancidity.

Peroxide Number: a measure of oxidative rancidity resulting from addition of oxygen across double bonds.

EXPERIMENTAL

Work in pairs.

1. Saponification: To 10 g of corn oil in a 250 ml flask, add 30 ml of 30 per cent alcoholic KOH and reflux gently on a water bath for one hour. Cool for a few minutes, add 50 ml of water, and cool under running water. Pour the solution into a separatory funnel and extract with 30 ml of petroleum ether. Retain the lower phase (saponifiable fraction) and treat it as follows.

(*a*) *SALTING OUT:* To about 5 ml of the soap solution obtained by saponification, add 20 ml of water. Add sodium chloride to saturate (about 5 g). Note the physical properties of the soap that separates. Test it for detergent properties

(*b*) *INSOLUBLE SOAPS:* To 5 ml portions of the soap solution add a few drops of calcium and magnesium chloride. What is the practical importance of these insoluble soaps?

(*c*) *COLORIMETRIC TEST FOR GLYCEROL:*   The bromine oxidation of glycerol to dihydroxy acetone and subsequent dehydration to methylglyoxal has been used in selective analysis for glycerol.

To an aliquot of the "saponifiable fraction" add an equal volume of fresh, saturated bromine water. (**CAUTION:** Bromine fumes are toxic. Use hood and proceed so that no fumes will be inhaled.) Heat in a boiling water or steam bath for 20 minutes and then boil to completely expel the bromine. To 0.4 ml of this solution, add 0.1 ml of 5 per cent β-naphthol in ethanol and 2 ml of concentrated H_2SO_4. Mix gently, heat for 2 minutes in boiling water, cool and note blue-green color. This assay can readily be made quantitative on as little as 0.2 mg of glycerol.

What enzymes accomplish the same end result in the digestive tract?

2. Polybromide Test: To 2 ml portions of linseed oil, cotton seed oil, olive oil, and cod liver oil in dry test tubes, add 10 ml portions of dry ether. Stopper the tubes and chill in an ice bath. When cold, add enough brominating reagent (20 per cent bromine in CCl_4) to each to produce a deep red color (about 1 ml). Allow to stand in the ice bath for one hour, then examine for a white precipitate of polybromide. Which oil yields the largest amount of polybromide and which the least? What could you infer from these data relative to the composition of the oil?

3. Isomerization (Elaidinization):   Into a test tube containing 5 ml of oleic acid and 1 ml of concentrated nitric acid, drop a small coil of copper wire. In the vigorous reaction which occurs, oxides of nitrogen rise through the oleic acid and produce cis-trans isomerization. Should the contents threaten to escape the tube, cool momentarily under the tap. Allow the reaction to proceed for 15 minutes; then add water to dilute the aqueous acid, transfer the oil phase to another test tube, and dissolve in 5 to 7 ml of alcohol. Place in an ice bath and note the appearance of crystalline elaidic acid. The elaidinization reaction is a general one for unsaturated fatty acids. Is the reaction quantitative? What other reagents will produce elaidinization? How could oleic acid be prepared from pure elaidic acid?

4. Liebermann-Burchard Test for Sterols: Acetic anhydride can condense with the C-3OH groups of cholesterol and related sterols to yield esters. If, as in the case of cholesterol, the sterol also has a C-5 unsaturation, a subsequent epimerization to the 3 form and dehydration occurs, along with the formation of a characteristic color. Hence the reaction serves as a specific test for 3-hydroxysteroid with C-5 unsaturation.

In this test, moisture must be carefully avoided. To 2 ml portions of dilute solutions of ergosterol and cholesterol (0.02 per cent) in chloroform in *dry* test tubes, add 15 drops of acetic anhydride, mix, cool, and finally add 5 drops of concentrated sulfuric acid. Mix, and note the color changes. A deep blue-green color develops in both tubes, but is preceded in the case of ergosterol by a fleeting red color. The rate of color development is a function of the degree of unsaturation of the sterol. Thus, ergosterol develops maximum color in 3 to 5 minutes, while cholesterol requires 15 to 20 minutes.

This reaction may be made quantitative by controlling temperature, time, and volume of reactants. Standards must be included; absorbance is determined at 625 mμ.

5. Digitonin Precipitation of Sterols: In nonpolar solvents, digitonin will condense with equivalent amounts of 3-β-sterols, e.g., cholesterol, to yield an insoluble digitonide. This reaction is very specific for 3-β-sterols, but can be distinguished from 5-sterols by the Liebermann-Burchard Test. Dissolve some cholesterol in acetone and absolute EtOH (1:1). Add three volumes of 0.5 per cent digitonin in 50 per cent EtOH. Warm for a few seconds on a steam bath, and allow suspension to stand 1 hour at 0°. Observe precipitate formation.

6. Salkowski Test: When chloroform solutions of 0.1 per cent cholesterol and ergosterol are layered over an equal volume of concentrated H_2SO_4 and mixed gently, a characteristic color develops. Note the yellow ring obtained with cholesterol and the dark orange-to-red ring with ergosterol. There may also be a green fluorescence in the upper layer of ergosterol. Try some other sterols. What conclusions can you draw about interpreting the results of this ring test?

7. Modified Furter-Meyer Test for Tocopherols: To 2 ml of a 0.05 per cent solution of α-tocopherol in chloroform, add 7 ml of n-butyl alcohol and 1 ml of concentrated nitric acid. Mix and place the tube in a water bath at 80° for 10 minutes. The *bronze-red* color which develops is specific for tocopherols. Carotenoids and certain sterols produce a yellow or brown color, and when present in sufficient quantity may mask the bronze-red color of small amounts of tocopherols. To satisfy yourself on this point repeat the test with carotene.

What relation exists between the keeping qualities of a fat and the tocopherol content? What other compounds act similarly to tocopherols in this respect? What factors hasten oxidative changes in fats?

8. Carr-Price Reaction for Vitamin A: Dissolve 2 drops of cod liver oil in 1 ml of chloroform. Cool in an ice bath and add 2 ml of a cold saturated chloroform solution of antimony trichloride. Note the changes in color and the subsequent fading. Repeat the test with carotene. Note that the color is much more stable. This method is commonly used for the chemical determination of Vitamin A in foodstuffs.

REFERENCES

1. Strain, H. H., and Sherman, J. *Modification of Solution Chromatography Illustrated with Chloroplast Pigments.* J. Chem. Ed. **46** 476 (1969).
2. *Experimental Biochemistry Laboratory Manual.* Department of Biochemistry, University of Wisconsin, 1967.
3. Loren, P. *The Chemistry of Lipids of Biochemical Significance.* Methuen and Co., London, 1955.
4. Deuel, H. J., Jr. *The Lipids (Vol. I to III).* John Wiley and Sons, Inc., New York, 1957.
5. Hanahan, D. J. *Lipide Chemistry.* John Wiley and Sons, Inc., New York, 1960.

CHAPTER 15

THE PURIFICATION OF EGG WHITE LYSOZYME (MURAMIDASE)

EXPERIMENTAL PRINCIPLES

During the purification of proteins the investigator must use a number of precautions not generally required for purification of simpler compounds. Most proteins are denatured by foaming, by heating above 40° C, by excess organic solvents at room temperatures, by drying at room temperature, and by concentrated acids or bases. Proteins in aqueous solution are excellent nutrient systems for microorganisms, and therefore, cleanliness of equipment and avoidance of unnecessary contamination will contribute to the success of purification attempts. When aqueous solutions or suspensions of proteins must be stored overnight or longer, toluene or other preservatives should be added to minimize the growth of microorganisms.

A single purification step is seldom adequate to purify a protein. Typically, several different principles are applied at each step, which exploit different properties of the protein. Each step often accomplishes a two- to ten-fold purification, so that the several-hundred-fold purification often needed for meaningful study can finally be accomplished. Another consideration that determines the usefulness of a given purification step is the yield of the enzyme. Some tissues are unusually rich in a particular protein, so that the selection of an appropriate tissue (or growing it in an appropriate manner) often makes a difference in the final yield of the desired enzyme.

Each enzyme undergoing purification presents an individual problem. An enzyme in any given cell is unique, having its own physiological, cytological, and chemical properties. In attempts to purify an enzyme this uniqueness has to be taken into account, and no two enzymes have been purified in exactly the same way. The overall principles and procedures are fairly uniform, however, and can serve as a guide in any new problem. The most useful techniques involve salting out, adsorption, detergent solubilization, organic solvent precipitation, and isoelectric precipitation, each step of which can be used independently or in combination with a variety of ion exchange and exclusion chromatographic procedures as well as preparative gel electrophoretic techniques, a more recent development.

Determination of the progress of a purification process is important, and whenever the protein of interest has a measurable biological activity or an unusual feature (colored prosthetic group, for instance) the "specific activity" of enzyme can be used to follow the purification process. This is one of the most useful concepts in protein chemistry, and when it is not possible to use it, purification progress can only be inferred by making indirect tests or by ultimately resorting to analysis for impurities by electrophoresis.

Purification by Ion Exchange Chromatography. The basis of ion exchange chromatography is the electrostatic attraction between oppositely charged ions, one of which is an electrolyte and the other a synthetic resin polymer. An ion exchanger is an insoluble material, usually silicates, resins, or polysaccharides, containing fixed charged groups associated with a small mobile counter-ion of opposite charge. The latter can be exchanged reversibly with other ions of the same charge without physically changing the matrix. Exchangers are termed anionic or cationic, according to whether they have an affinity for negative or positive ions. A summary of the properties of ion exchangers is given in Table 24. The exchange resins listed have been extremely useful for separating and purifying proteins, nucleotides, amino acids, peptides, and many low molecular weight polyelectrolytes and ions.

The electrostatic interactions taking place between resin and a protein are equilibrium processes involving diffusion of the charged protein to the resin surface, and then to a charged exchange site. The actual exchange takes place at this site. Finally, diffusion away from the resin takes place upon elution with an appropriate solvent. The rate of movement of a given ion down the column is a function of its ionizability, the concentration of other ions present, and their relative affinity for charged sites on the resin. By adjusting the pH and the ionic strength of the eluting solvent, the ions held by electrostatic attraction on the resin are eluted differentially to yield the desired separation.

There are other considerations which determine the suitability of an ion exchanger. These include chemical and mechanical stability, resistance to flow of liquids, and capacity. The latter is of major importance. Capacity may be expressed as total (theoretical) capacity, which is calculated from the number of functional groups per unit of resin dry weight, and as available capacity, the actual capacity under specified experimental conditions. This is usually smaller than the theoretical capacity, and depends on accessibility of the functional groups, the concentration of eluent, the nature of the counterions, and the selectivity of the resin functional groups toward them. The pH and temperature of the eluent are also important.

Table 24

Name	Type	Active Group	Exchange Capacity (me/g Dry Resin)
I—SYNTHETIC ION EXCHANGE RESINS			
Dowex 1	strongly basic anion exchanger	$-CH_2\overset{+}{N}(CH_3)_3$	3.5
Dowex 2	strongly basic anion exchanger	$-CH_2-\overset{+}{N}-(CH_3)_2$ $\quad\quad\; \vert$ $\quad\quad CH_2-CH_2-OH$	3.5
Dowex 3 Amberlite 1R-4B	weakly basic anion exchanger	$-NH_3^+$	5.5
Dowex 50 Amberlite IR-120	strongly acidic cation exchanger	$-SO_3^-$	5.0
Amberlite IRC-50 XE-64	weakly acidic cation exchanger	$-COO^-$	10.0
Biorex 63	intermediately acidic cation exchanger	$-PO_3^-$	
II—CELLULOSE AND DEXTRIN (SEPHADEX) ION EXCHANGERS			
diethylaminoethyl-cellulose (DEAE—C)	weakly basic anion exchanger	$-O-CH_2-CH_2-\overset{H}{\underset{+}{N}}-(CH_2-CH_3)_2$	0.7
DEAE-Sephadex	weakly basic anion exchanger	$-O-CH_2-CH_2-\overset{H}{\underset{+}{N}}-(CH_2-CH_3)_2$	3.5
triethylaminoethyl-cellulose (TEAE—C)	weakly basic anion exchanger	$-O-CH_2-CH_2-\overset{+}{N}-(CH_2-CH_3)_3$	0.5
ECTEOLA—C‡	weakly basic anion exchanger	$-O-CH_2-CHOH-CH_2-\overset{+}{N}-(CH_2-CH_3)_3$	0.3
carboxymethyl-cellulose (CM—C)	weakly acidic cation exchanger	$-O-CH_2-COO^-$	0.7
CM-Sephadex	weakly acidic cation exchanger	$-O-CH_2-COO^-$	4.5
phosphocellulose (P—C)	intermediately acidic cation exchanger	$-O-\overset{\textstyle O}{\underset{\textstyle O^-}{\overset{\|}{\underset{\|}{P}}}}-O^-$	0.85
sulfoethyl cellulose (SE—C)	strongly acidic cation exchange	$-O-CH_2CH_2SO_3^-$	0.2

In addition to electrostatic adsorption these resins may have specific adsorptive properties which are governed by Vander Waals forces and polar attractions, which also influence exchange properties.

Because of the limited available surface and the lability of proteins, ion exchange resins have some shortcomings. The cellulose based resins, however, have high attractive capacities, and at the same time hold proteins weakly. This means that by continuous gradient variations of pH and salt concentrations, efficient elutions of absorbed proteins can be made. CM-Sephadex-25, the cationic resin used in this experiment, is made negatively charged by addition of an appropriate buffer. A protein mixture in this same buffer takes on a positive charge and is passed onto the column. Elution of the proteins is accomplished by increasing the pH of the eluting solvent so that the number of positive charges on the protein diminish. Elution can also be accomplished by increasing salt concentration. DEAE columns are used in the opposite way. These columns are made negative and elution occurs by decreasing pH or increasing salt concentration or both.

Detection and Collection Methods.   A very useful, time saving technique employed in chromatography involves the use of automated fraction collectors in conjunction with liquid flow detection devices. As fluid is eluted from a

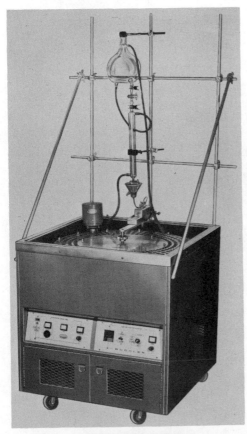

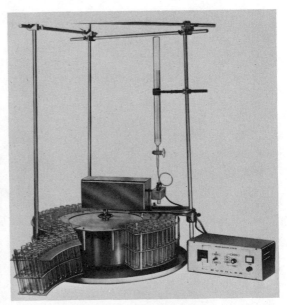

SECTIONAL ASSEMBLY

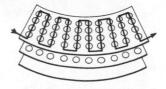

TUBE FILLING SEQUENCE
IN SECTION

REFRIGERATED ASSEMBLY

FIGURE 16. Fraction collectors. (Courtesy of Buchler Instruments Inc.)

chromatographic column, the eluate can be monitored continuously by one of several means, either fluorometrically, by ultra-violet or visible absorption spectrophotometry, or by differential refractometry if desirable and useful. After monitoring, the eluate then passes to the fraction collector, which can be operated in two ways. The eluate can be collected for a precise preset time or it can be collected on the basis of the number of drops or volume passing by. In both cases, when the eluate flow satisfies either the time or volume setting, the tray mechanism of the fraction collector is activated to place a new tube or vessel in place. The instrument is automatically operated to collect a large number of eluate fractions of precise volume. Fraction collectors are designed to contain more than 100 tubes or vessels and some have removable sections in order to facilitate replacement of filled tubes. They are also designed as single refrigerated units that help protect enzymes during a long fractionation period. By combining fractions with the eluate detection data, it is possible to identify these fractions containing the protein of interest. (See Figure 16.)

EXPERIMENT 15

Lysozyme (trivial name): IUB number 3.2.1.17. Systematic name: Mucopeptide N-acetylmuramylhydrolase. Lysozyme hydrolyzes β-1,4-linkages between N-acetylmuramic acid and 2-acetamido-2-deoxy-D-glucose residues in mucopolysaccharides or mucopeptides of a variety of microorganisms. Egg white, human milk, tears, spleen, and many other tissues, including plant sources, contain this enzyme. Its antibiotic action is attributed to the ability to destroy invading bacteria by hydrolyzing the mucopolysaccharides of the cell wall. The enzyme has been isolated in crystalline form, and the complete amino acid sequence was published in 1963. This was the first enzyme for which the complete three-dimensional structure was determined. Lysozyme has a higher isoelectric point (pH 10.5) and a lower molecular weight (14,307) than most proteins. These properties will be used to purify it by either ion exchange or size exclusion chromatography in this experiment.

Part A. Packing the Columns

The class will be divided in half. Each group, subdivided into student pairs, will purify lysozyme by either ion exchange or size exclusion chromatography. An attempt will be made to evaluate these two methods for convenience, yield, and extent of purity obtained.

Vertically mount a 0.9 × 15 cm chromatographic column. Make sure that the bottom piece is assembled with the "O" ring in place, thus holding a nylon net above the coarse nylon screen. The outlet spout is screwed in at the bottom. Remove the top end piece and prepare to pack the column. You may also want to devise a reservoir by connecting the top end piece to a small funnel by means of suitable diameter polyethylene tubing.

CM-Sephadex 25 ion exchanger has been equilibrated for 48 hours in 0.05 M TRIS-EDTA and 0.05 M NaCl buffer, pH 8.2. Place a slurry of this resin in a flask in fresh buffer; degas, and quickly pour it into the column. Allow it to settle by gravity. A bed height-to-diameter ratio of 10:1 gives good resolution. After achieving a resin height of 10 cm, assemble the top end piece and attach a reservoir to the inlet part. Put some buffer in it and allow it to gravity-feed the column so that it won't go dry. This may be accomplished by pinching off the outlet tube with a clamp.

Part B. Preparing the Enzyme

(A group of four students can collaborate in this preparation.) Place a 10 × 10 cm square of a double layer of cheese cloth over a 100 ml beaker. Gently filter the white of an egg by stroking the balled cheese cloth against the side of the beaker. Do not force the egg white through. When more than 5 ml has been collected, transfer 5 ml of the filtered egg white to a second beaker. Add 30 ml of 0.05 M TRIS buffer (a 1:8 dilution), and pass this mixture through a large glass wool plug. Centrifuge the filtered egg-white-buffer mixture for three minutes in a table model centrifuge at top speed, using polyethylene tubes; 4 ml will be used for the ion exchange purification, and 2 ml for the gel filtration purification. (Cf. page 190.)

Carefully and slowly pour 4 ml of clear supernatant along the side of the column. *Save the rest of the supernatant for assay.* Continuously wash the column with TRIS buffer until 15 ml has been collected at a flow rate of 2 to 5 ml/min. Once the column has been set up, buffer should be continuously added, *and at no time should the buffer head at the top of the column go below the resin.* Change the collecting vessel and continue washing through with TRIS buffer until another 15 ml has been collected. Now elute the column with 0.2 M carbonate buffer, pH 10.5, until two additional 15 ml aliquots have been collected. Keep all fractions at 4° C in an ice bath until ready for assay. *In order to save time, proceed to the biuret and enzyme activity determinations as each fraction is collected.*

Part C.   Protein Determination

Biuret Method.   Determine the protein concentration of the filtered and buffered egg white and the four fractions, using samples as shown in Table 25. The filtered and buffered egg white solution is diluted 1:5 for this purpose, and the first TRIS eluate is diluted 1:2 in order for both to be in the appropriate range for biuret assay. The remaining three eluates may be tested without dilution. It must be remembered that the filtered and buffered egg white has been diluted eight fold; the second dilution of five fold brings the total dilution to 40 fold. The concentration of filtered egg white will fall in a concentration range of 1 to 2 mg/ml since the original egg white has a concentration of about 80 mg/ml. In each determination the student should make sure that the absorbance readings are within the limits of the biuret determination. If the sample proves to be too dilute, repeat the assay with undiluted eluate; if it appears too concentrated, dilute it to an appropriate volume, and repeat the biuret determination.

Furthermore, since TRIS buffer interferes with the biuret determination by giving some color of its own, the blank should contain a 1 ml aliquot of a 1:5 or other appropriate dilution of buffer. The blank for TRIS fraction 2 should contain 1 ml of undiluted buffer. The blanks for carbonate fractions 3 and 4 should contain 1 ml of 0.2 M carbonate buffer. The biuret readings are taken at 540 nm.

Direct Spectrophotometric Methods.   If spectrophotometers with U.V. power supply attachments are available, two convenient rapid and direct methods of determining protein concentration are possible. These methods are

Table 25

	Reagents	1 Filtered Egg White	2 Tris Fraction 1	3 Tris Fraction 2	4 Carbonate Fraction 3	5 Carbonate Fraction 4
A.	sample volume, ml	1.0	1.0	1.0	1.0	1.0
B.	H_2O, ml	0.4	0.4	0.4	0.4	0.4
C.	5 per cent Na deoxycholate, ml	0.1	0.1	0.1	0.1	0.1
D.	biuret reagent, ml	1.5	1.5	1.5	1.5	1.5
E.	absorbance					
F.	protein concentration, mg/ml					
G.	total dilution	1/5	1/2	none	none	none
H.	Protein concentration corrected for dilution, mg/ml					

referred to on page 75. Since they are considerably more sensitive methods, very great dilutions must be made, in the order of 1:1000 for the filtered-buffer egg white and 1:500 for the fractions. These may be read against blanks of water.

(a) By measurement of absorbance at 215 and 225 nm, the $\Delta A = A_{215} - A_{225}$ and $\Delta A_{215-225} \times 154 = \mu$g protein/ml.

(b) By measurement of absorbance at 260 and 280 nm, then $1.11 \times A_{280} - 0.755 \times A_{260} =$ mg protein/ml.

Part D. Enzyme Assay

Enzyme activity is also determined by dilution of fractions. The filtered and buffered egg white should be diluted 100 fold and assayed. Take a 0.1 ml aliquot of the filtered and buffered egg white which was saved and dilute it to 10 ml. The best dilutions for the TRIS and carbonate fractions may have to be ascertained. Start with 1:10 dilutions. *These dilutions must be made with 0.1 M phosphate buffer, pH 7.0.*

Lysozyme of egg white will hydrolyze bacterial cell walls. The assay method is based on observing turbidometric changes of a suspension of dried cell walls of *Micrococcus lysodeikticus*. This is a spectrophotometric assay in which the digestion or hydrolysis of the cell wall suspension can be measured at 450 nm. As hydrolysis takes place, the turbidity of the cell wall suspension decreases. This decrease is used to measure enzyme activity. An absorbance change of within 0.020 to 0.040 per minute is considered to be a good rate for measuring activity. One unit of activity is equal to an absorbance change of 0.001 per minute at pH 7.0 and 25° C. The first two minutes of recorded change in absorbance are used to determine the rate of catalysis. The rate during the earliest one minute interval which gives a maximum absorbance change is the rate used as the initial reaction velocity.

$$\text{Specific activity} = \frac{\text{Units } (\Delta A \text{ per minute})}{\text{Total mg protein in actual assay}}$$

Suppose, for example, one of your fractions had a concentration of 2 mg protein/ml. It was diluted 1:100 for the enzyme assay, and 0.1 ml exhibited an initial velocity of $\Delta A = 0.030$ min. Then

Units = 30/min

$$\text{Total protein in the assay} = \frac{2 \text{ mg/ml} \times 0.1 \text{ ml}}{100} = 0.002 \text{ mg}$$

[handwritten: $29.65 \text{ mg/ml} \times 0.1$]

$$\text{Specific activity} = \frac{30}{0.002} = 15,000 \text{ units/mg protein}$$

In order to accurately measure enzyme activity it is very important to establish a dilution of enzyme which will give a reasonable rate of hydrolysis or absorbance change. If the rate is not within the recommended limits cited above, then either a further dilution of enzyme is necessary or the amount of enzyme used must be increased. Undiluted fractions from 0.1 to 1.0 ml may be used in the latter situation. Whatever the case, be sure to record the dilution used in the assay, and note the original protein concentration of the fraction in order to properly calculate activity.

In the actual assay itself, the substrate, a uniform suspension of 0.3 mg/ml of dried *Micrococcus lysodeikticus*, also has been prepared in 0.1 M phosphate buffer, pH 7.0, and kept at room temperature. The original enzyme fractions are kept in ice, but dilutions are made with 0.1 M phosphate buffer at room temperature.

Set the spectrophotometer at 450 nm and zero absorbance with a water blank. In all cases of assay add 0.1 ml of diluted or undiluted enzyme solution to 2.9 ml of substrate previously placed in a cuvette. Record immediately the decrease in absorbance at 450 nm at 15 second intervals for two minutes. Make a few practice runs and attempt to get good duplication. The serious student may do his assays in duplicate or triplicate as time permits.

What would you expect to happen if there were a gravity sedimentation of substrate during the assay?

Table 26

Fraction	1 Total Volume, ml	2 Protein Concentration, mg/ml	3 Total Protein, mg	4 Dilution for Enzyme Assay	5 Specific Activity	6 Total Activity	7 Per cent Recovery	8 Purification
filtered egg white	4						100%	1
TRIS fraction 1	15							
TRIS fraction 2	15							
carbonate fraction 3	15							
carbonate fraction 4	15							

How would you go about checking this out and making corrections for this eventuality?

Fill in Tables 25 and 26 for your report.

Part E. Purification by Gel Filtration—Sephadex Gel Exclusion

This technique involves separating molecules of different size by passing them through a gel column. The polysaccharide, dextran, is carefully cross-linked to give small beads of a hydrophilic but insoluble polymer which swells in water to form a gel. Sephadex gel, as it is known commercially, blocks the passage of large molecules, yet allows the diffusion of small ones. As indicated in Table 27, there is a choice of several types of Sephadex beads.

A low degree of cross-linkage of the dextran produces large pores, a large water region and a higher affinity for larger molecules, rejecting only the very large molecules. A high degree of cross-linkage produces small pores, a low water gain, and the capability of rejecting larger or very large molecules. Thus, molecules larger than the pores of the swollen dextran, above the exclusion range, cannot penetrate the gel and so pass through the bed in the liquid phase outside the particles. These are eluted first. Smaller molecules are able to move in the water or buffer, both within and outside the gel, and are retained longer. The molecules are eventually all eluted in the order of decreasing molecular size. Fractionation of a mixture of molecules occurs when diffusion of the molecules into the gel pores is restricted. The fractionation arises from differences in passage at rates that are inversely related to the total fluid volume accessible within the column to the molecules, including voids within the gel.

The total volume (V_t) of a Sephadex column is:

$$(1) \qquad\qquad V_t = V_0 + V_i + V_g$$

where V_0 is the void volume, or outside volume; V_i is the volume inside the gel matrix and includes bound water of hydration; and V_g is the volume of the gel matrix itself.

The inner gel matrix volume can be calculated in one of two ways:

$$(2) \qquad\qquad V_i = A \cdot W_r$$

where A = the known dry weight of the entire gel column and W_r = water regain; or

$$(3) \qquad\qquad V_i = \frac{W_r \cdot d}{W_r + 1}$$

$$d = \text{wet density}$$

Table 27

Sephadex Type	Limit for Complete Exclusion (Molecular Weight)	Limits of Separation Range (Molecular Weight)	Water Regain (W_r) ml H_2O/gram Dry Sephadex	Wet Density (d) g/ml	Volumes in ml/gram dry Sephadex*		
					V_t	V_0	V_i
G-25	5,000	100–5,000	2.5 ± 0.2	1.13	5	2	2.5
G-50	10,000	500–10,000	5.0 ± 0.3	1.07	10	4	5
G-75	50,000	1,000–50,000	7.5 ± 0.5	1.05	13	5	7
G-100	100,000	5,000–150,000	10.0 ± 1.0	1.04	17	6	10
G-200	200,000	5,000–200,000	20.0 ± 2.0	1.02	30	9	20

* Remember that these figures have to be used with respect to column size.

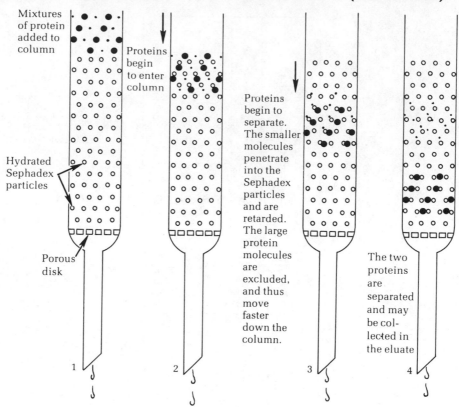

FIGURE 17. Schematic diagram of gel filtration. (From A. L. Lehninger. *Biochemistry*, Worth Publishers, Inc., N.Y. (1970))

The mathematic expression of solute elution is

$$(4) \quad V_e = V_0 + K_d V_i \qquad \text{or} \qquad K_d = \frac{V_e - V_0}{V_i} = \frac{V_e - V_0}{A \cdot W_r}$$

where V_0 may further be referred to as the elution volume of a solute completely excluded from the internal cavities of the gel. Any macromolecule which is completely excluded by a gel can therefore be used to determine V_0. It must be of a molecular weight higher than the exclusion range for each gel. Those used most successfully are thyroglobulin (molecular weight = 670,000) and dextron-blue (molecular weight = 1×10^6). K_d is the partition or distribution coefficient. It is, in fact, the volume fraction of solvent taken up by

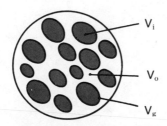

FIGURE 18. Schematic cross section of a gel column.

the gel which is accessible to the solute. Both volumes, V_e and V_0, can be determined from the time between the initial addition of protein to the column and the appearance of the protein peak in the eluate. The precise time of the peak can be defined by triangulation—that is, by extension of the sides of the elution peak to an apex.

For large molecules that cannot enter the stationary phase of the gel particle, $K_d = 0$, and $V_e = V_0$. For small molecules, such as sucrose, for which accessibility is complete, $K_d = 1$. Proteins with K_d between these two extremes will be excluded as a function of molecular size of weight, as is the case for globular proteins. The term V_s is defined as the elution volume between two protein peaks:

$$V_s = V_e' - V_e'' = (V_0 + K_d' V_i) - (V_0 + K_d'' \cdot V_i)$$
$$= (K_d' - K_d'') V_i$$

For a complete separation of two proteins, the sample volume must not be larger than V_s; otherwise overlapping of peaks will occur.

Dialysis or desalting also can be carried out on suitable Sephadex columns. The G-25 is particularly useful for this purpose because all molecules above a molecular weight of 5000 will be excluded from the gel while the smaller molecules will be retained. In this procedure the protein is introduced and elution with buffer is begun. When the void volume V_0 has passed, most of the protein has been eluted, while the smaller molecules will be eluted after $V_0 + K_d V_i$ has passed. The process is very rapid and is very useful when working with labile proteins.

Another application involves using Sephadex to concentrate dilute protein solutions. In this case, the appropriate Sephadex is used in the dry form and is added to the solution to be concentrated. After swelling and water regain take place, the mixture is centrifuged and the supernatant will contain the concentrated protein provided the correct Sephadex is used, that is, one which will exclude the protein of interest.

GENERAL PROCEDURE

Column Packing and Sample Application. In order to get adequate flow rates, Sephadex gels should be soaked in water 2 to 3 days with intermittent stirring. At the end of that time the gel may be agitated once more in order to decant off the fine particles. Follow the instructions given for the ion exchange experiment for setting up the column. Add a slurry of Sephadex G-100 to a column already containing some buffer. Degas to avoid air bubbles in the slurry. Allow the gel to settle by gravity feed and bring it to a height of 40 cm. Place a filter paper disc on top in order not to disturb the gel when adding sample. Evenness of packing of the column can be checked by watching the passage of a colored material such as cytochrome C or dextran blue. By using the latter, it is also possible to determine V_0 simultaneously. Irregularities

Table 28

Reagents	Filtered Egg White	Fraction 1	Fraction 2	Fraction 3	Fraction 4	Fraction 5	Fraction 6
A. sample volume, ml	1.0	1.0	1.0	1.0	1.0	1.0	1.0
B. H$_2$O, ml	0.4	0.4	0.4	0.4	0.4	0.4	0.4
C. 5 per cent Na deoxycholate, ml	0.1	0.1	0.1	0.1	0.1	0.1	0.1
D. biuret reagent, ml	1.5	1.5	1.5	1.5	1.5	1.5	1.5
E. absorbance							
F. protein concentration, mg/ml							
G. total dilution	1 : 5						
H. protein concentration corrected for dilution, mg/ml							

Table 29

Fraction	1 Total Volume, ml	2 Protein Concentration mg/ml	3 Total Protein, mg	4 Dilution for E	5 Specific Activity	6 Total Activity	7 Per Cent Recovery	8 Purification
filtered egg white	2			1:100				
fraction 1	5							
fraction 2	5							
fraction 3	5							
fraction 4	5							
fraction 5	5							
fraction 6	5							

develop at the top of the column after repeated use. Irregularities will be detected by skewness of the colored bands or poor definition of peak maxima in elution diagrams. This may be corrected by replacing 0.5 to 1.0 cm of the top of the gel with fresh gel, and then stirring up the top 2 to 3 cm and allowing the particles to settle again. It is also important to equilibrate the gel with elution buffer (0.05 M TRIS, pH 8.2, in 0.05 M KCl) as long as is practical. Time does not permit this luxury, but in laboratories where the gel column is in constant use, buffer is allowed to pass through continuously between runs.

Preparing Enzymes. Enzyme is prepared according to the instructions on page 179. Carefully pipette 2 ml of the TRIS buffered egg white onto the column just as the buffer reaches the disc. Avoid disturbing the top of the column; put the pipette tip against the side of the column. Save a 1 ml aliquot of the buffered egg white. Collect six 5 ml quantities of eluant. Start your collections when the 2 ml sample has passed into the column. Use the TRIS buffer to elute the column. *Do not let the column go dry.*

Protein Determination. Follow the instructions of the ion exchange experiment. Make a 1:5 dilution of the filtered egg white and a 1:2 dilution of fraction 1. Use the proper dilution of 0.1 M TRIS buffer for a blank in all cases.

Enzyme Assay. Follow instructions of the ion exchange experiment. Use a 1:100 dilution of the filtered egg white and start with 1:10 dilution of the other fractions.

Fill in the appropriate parts of Tables 28 and 29.

When would you expect the lysozyme to come off the Sephadex column? Consider your answer on the basis of the size exclusion principle and the sizes of the proteins in egg white.

CALCULATION OF PURIFICATION DATA

(a) The protein concentration (mg protein ml) of each fraction may be obtained from the data in Table 28 which have been corrected for dilution. Record this in column 2 of Table 29.

(b) Total protein for each fraction = (mg protein/ml) × total volume. Therefore, multiply the figures from column 1 by those from column 2. Record the answers in column 3.

(c) Specific activity $= \dfrac{\text{units } (\Delta A \text{ per min})}{\text{mg protein actually used in assay}}$

See page 181 for instructions. Record in column 5.

(d) Total activity in each fraction $= \dfrac{\text{units}}{\text{mg protein}} \times$ total protein in each fraction or = specific activity × total protein in each fraction or = (units/ml) × total volume of each fraction.

Multiply the figures in column 3 by those in column 5 and record in column 6.

(e) per cent recovery of activity

$$= \frac{\text{total units enzyme activity in each purified fraction}}{\text{total units in the filtered-buffered fraction}} \times 100$$

(f) Degree of purification $= \dfrac{\text{specific activity of purified fraction}}{\text{specific activity of filtered-buffered fraction}}$

REFERENCES

1. *Experimental Biochemistry Laboratory Manual.* Department of Biochemistry, University of Wiconsin, 1967.
2. *Sephadex: Theory and Experimental Technique.* (pamphlet) Pharmacia Fine Chemicals, Inc., Uppsala, Sweden.
3. *Sephadex: Gel Filtration in Theory and Practice.* (pamphlet) Pharmacia Fine Chemicals, Inc., Uppsala, Sweden.
4. Hjerten, S., and Mosbach, R. *Molecular Sieve Chromatography.* Anal. Biochem. **3** 109 (1692).

CHAPTER 16

THE KINETICS OF EGG WHITE LYSOZYME AND HEXOKINASE

FUNDAMENTALS OF ENZYME KINETICS

Kinetics concerns the study of reaction rate and the conditions which affect reaction velocity. It is reasonable to assume that every chemical reaction takes place at a definite rate. Some are extremely slow; others are so fast that they appear to be spontaneous. Between these two extremes, there are reactions for which a rate can be assigned within a reasonable time scale. Kinetic studies are primarily concerned with the measurement and calculation of rate constants, and, where appropriate, of equilibrium constants. Kinetics is also concerned with mechanisms, and is the best tool for predicting or eliminating mechanisms under consideration, since an observed reaction rate must satisfy the proposed mechanism.

In non-biological systems there are a number of considerations when measuring rate. The concentrations of reactants and products, temperature, pressure, solvent effects, stereochemistry, geometry, stoichiometry, effects of electromagnetic radiation, gravitational fields, and magnetic fields are all important variables in kinetic studies.

While these considerations may equally apply to enzyme systems, there are additional variables which have to be taken into account in studies of enzyme catalysis; for example, the capacity and the substrate specificity of the enzyme, the coenzyme, cofactor, and metal ion requirements, the pH, the effects of aging and denaturants, and finally, effects of inhibitors, modifiers, or controlling agents on the enzyme are all important.

Any study of reaction velocity involves measuring the rate either of disappearance of starting material, designated as substrate, or of the appearance of products. These changes take place exponentially in the simple case when $A \rightarrow B$.

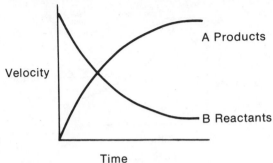

FIGURE 19. First order reaction kinetics.

The initial velocity in this case is a line tangent to the rate curve at $T = 0$. In a unimolecular reaction, when $A \rightarrow B$, the velocity V is proportional to the concentration of A;

$$V \propto C_A$$

and

$$V = kC_A$$

where k is the specific rate constant of the forward reaction. This is designated as a first order reaction. In simple second order reactions, $A + B \rightarrow C$,

$$V \propto C_A C_B$$

and

$$V = kC_A C_B$$

The rate is proportional to the product of the concentrations of the two starting materials. In the special case, which is particularly true in enzyme kinetics,

$$C_B \gg C_A.$$

For all practical purposes, therefore, the concentration of B remains unchanged during the reaction. This applies particularly when water is one of the reactants, and in these circumstances such a reaction is referred to as a pseudo-first order reaction.

In enzyme kinetics there are additional considerations. The rate limit of a given enzyme reaction is dependent upon the catalytic surface of the enzyme itself. This involves the number of active sites on the enzyme surface and the concentration of enzyme, both of which influence rate.

If we regard the enzyme as one of the reactants, then

$$C_A \gg C_{\text{enz}} \quad \text{(where } A \text{ is the substrate).}$$

When substrate concentration is very high, the reaction rate becomes independent of substrate concentration, and we can assume that $C_A = 1$; the enzyme reaction under these conditions is also pseudo-first order, and:

$$V = kC_{\text{enz}}$$

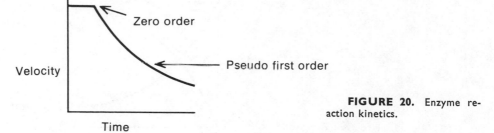

FIGURE 20. Enzyme reaction kinetics.

As a second assumption, we can also state that by definition a catalyst is infinitely dilute, and therefore, $V = k$. An enzyme reaction will be zero order when the rate is independent of either substrate or enzyme concentration. This occurs when substrate concentration greatly exceeds enzyme concentration and all the catalytic sites on the enzyme are occupied. The reaction rate is at a maximum and zero order kinetics are observed. But when these conditions alter with time and substrate concentration becomes limiting, that is, when the enzyme surface is no longer saturated with substrate, a pseudo-first order reaction takes place, and the experimental decay curve for first order reactions is observed.

At the same time, when the rate change in most enzyme catalyzed reactions is measured as a function of increasing substrate concentration, mixed zero order and first order kinetics are observed.

At low substrate concentration not all of the active sites on the enzyme surface are occupied at all times, and reaction rate will be proportional to the concentration of enzyme-substrate complex interactions.

When conditions of saturation are reached, that is, when substrate concentration is very high, and all active centers of the enzyme are occupied maximally, the velocity is independent of substrate concentration; V is at a maximum, and is designated V_m.

A mathematical derivation of these interactions was proffered by Michaelis and Menten in 1913. In this derivation, they formulated the concept of an enzyme-substrate complex as an integral part of an enzyme reaction mechanism. Over the years the Michaelis-Menten derivation of basic enzyme kinetics has undergone modification and refinement. One modification

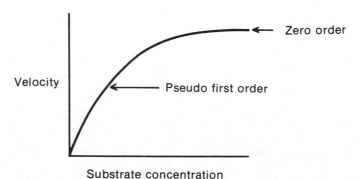

FIGURE 21. Enzyme reaction: velocity as a function of substrate concentration.

involves a derivation of the original Michaelis-Menten equation entirely on equilibrium considerations.

The simplest mechanism of an enzyme induced catalysis may be formulated as follows:

(1)
$$E_0 + S \underset{k_2}{\overset{k_1}{\rightleftharpoons}} ES \underset{k_4}{\overset{k_3}{\rightleftharpoons}} E + P$$

E_0, the free enzyme, reversibly combines with substrate S to form ES, the enzyme-substrate complex; ES breaks down essentially irreversibly to reform enzyme E, and to yield product P.

The reversible formation of ES is subject to mass action laws, and this reaction can be formulated as equilibrium association constant:

(2)
$$K_A = \frac{k_1}{k_2} = \frac{[ES]}{[E_0][S]}$$

or as the dissociation constant in the reverse direction

(3)
$$K_D = \frac{k_2}{k_1} = \frac{[E_0][S]}{[ES]} \quad \text{and} \quad K_A = \frac{1}{K_D}, \quad K_A K_D = 1$$

Certain reasonable assumptions can be made. The rate of the first reaction, the formation of ES, must be much greater than the rate of breakdown of ES; therefore the rate constants k_1 and $k_2 \gg k_3$, and k_4 is considered to be negligible; otherwise no ES would be formed, and apparent zero order kinetics would not be observed. At any given time, the total enzyme concentration $E_t = E_0 + ES$. The overall observable forward velocity will approach a maximum, V_m, when all of E_t is in the form of ES. When the enzyme is saturated with substrate, i.e., all the available active centers on the enzyme surface are occupied by substrate, then

(4)
$$\frac{V}{V_M} = \frac{ES}{E_t} .$$

A plot of velocity versus enzyme concentration therefore will be linear so long as substrate concentration is not limiting, and there are no interferences due to high substrate concentration, product accumulation, or other inhibiting mechanisms.

At steady state conditions, when enzyme is functioning at full capacity,

(5)
$$\frac{\Delta ES}{\Delta t} = 0 .$$

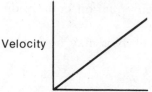

FIGURE 22. Enzyme reaction: velocity as a function of enzyme concentration.

Since only initial velocities are measured, any possible adverse effects of product accumulation on the reaction rate becomes negligible. Since

$$E_t = E_0 + ES, \qquad \text{then} \qquad E_0 = E_t - ES$$

by substitution in Equation (3). Now simply change the designation of K_D to K_m to signify the Michaelis-Menten ES dissociation constant.

$$(6) \qquad K_m = \frac{[E_t - ES][S]}{[ES]}$$

Solving for ES,

$$(7) \qquad [ES] = \frac{[E_t][S]}{K_m + [S]} .$$

Now we look at the second reaction,

$$(8) \qquad ES \xrightarrow{k_3} P + E.$$

This is a first order unimolecular reaction and is expressed mathematically as

$$(9) \qquad V = k_3[ES]; \qquad [ES] = \frac{V}{k_3} .$$

Substituting in Equation (7),

$$(10) \qquad \frac{V}{k_3} = \frac{[E_t][S]}{K_m + [S]} ; \qquad \text{and} \qquad V = \frac{k_3[E_t][S]}{K_m + [S]}$$

V_m or maximum velocity will be reached when S saturates the enzyme and all E_t is ES; then $[E_t] = [ES]$, and $V_m = k_3 E_t$. Therefore

$$(11) \qquad V = \frac{V_m[S]}{K_m + [S]} .$$

When V_m is $2V$, or $V = \frac{V_m}{2}$ (when the observed reaction velocity is half of the maximum velocity),

$$(12) \qquad \frac{V_m}{2} = \frac{V_m[S]}{K_m + [S]} .$$

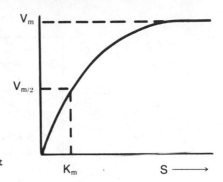

FIGURE 23. Determination of equilibrium constant from kinetic curve.

Divide by V_m and solve:

$$K_m = [S]$$

This means that at $\frac{1}{2}V_m$, the K_m is equal to the substrate concentration. This is the most direct experimental way of determining K_m.

Since K_m, the Michaelis constant, is derived as the equilibrium constant for the dissociation of ES to $E + S$, it is expressed in moles per liter. The smaller K_m, the greater is the affinity of substrate for enzyme surface.

K_m can also be defined experimentally as the concentration at which half maximum velocity is achieved. This is true only if k_1 and $k_2 \ggg k_3$ and k_3 has been eliminated from incorporation into K_m.

If K_m is derived from wholly kinetic considerations, k_3 is part of the constant, and K_m is not an *equilibrium* constant; it is the constant derived at $\frac{1}{2}V_m$ and equals $k_2 + k_3/k_1$.

Most enzymes obey simple Michaelis-Menten equilibrium relationships.

In practical terms, K_m can easily be obtained from a plot of velocity versus substrate concentration.

INHIBITORS OF ENZYME ACTION

There are two major types of inhibitors, competitive and non-competitive. Competitive inhibitors usually resemble the substrate and compete with it at the active center. They do not alter the enzyme by chemically reacting with it. By definition they must inhibit reversibly if they are in competition for an active site. Therefore, in the presence of a competitive inhibitor, V_m remains the same, but K_m is increased by a factor $1 + I/K_i$. The overall effect of a competitive inhibitor is to reduce the affinity of substrate for enzyme.

The equation for a competitive inhibitor is:

(13)
$$V = \frac{V_m[S]}{K_m\left[1 + \dfrac{I}{K_i}\right] + [S]} \quad \text{or} \quad V = \frac{V_m[S]}{K_p + [S]}$$

where

$$K_p = K_m\left[1 + \frac{I}{K_i}\right].$$

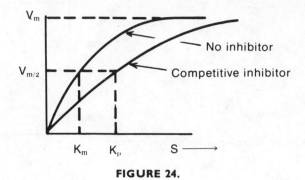

FIGURE 24.

K_p is the apparent K_m in the presence of a competitive inhibitor. K_i can be determined from values for K_p and K_m.

A noncompetitive inhibitor, by contrast, does react with the enzyme, for the most part reversibly but sometimes irreversibly. In either case the effect is to reduce the amount of available enzyme. K_m in this case is not changed but V_m is reduced because the apparent total available enzyme is reduced. The equation for a noncompetitive inhibitor is

(14)
$$V = \frac{\dfrac{V_m}{K_i + I}[S]}{K_i}{K_m + [S]} \ .$$

K_i may be obtained by substitution in the equation

$$V_{m_i} = \frac{\dfrac{V_m}{K_i + I}}{K_i} \ .$$

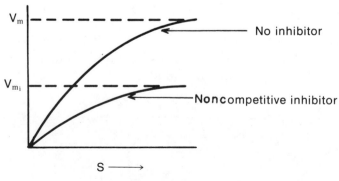

FIGURE 25.

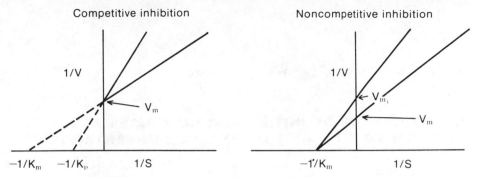

FIGURE 26.   One linear form of the Michaelis-Menten equation.

LINEAR FORMS OF THE MICHAELIS-MENTEN EQUATION

By taking the reciprocal of equation (11),

(15)
$$\frac{1}{V} = \frac{K_m}{V_m[S]} + \frac{1}{V_m}.$$

The original M-M equation becomes a straight line equation in which the two variables are $\dfrac{1}{V}$ and $\dfrac{1}{S}$; $\dfrac{1}{V_m}$ is the y intercept and K_m/V_m is the slope.

A second linear form is obtained by plotting V versus V/S. This form may be obtained by multiplying the reciprocal equations by $V\,V_m$:

(16)
$$V = V_m - K_m \frac{V}{S}.$$

This plot is statistically more reliable than the reciprocal plot because the points are more evenly distributed.

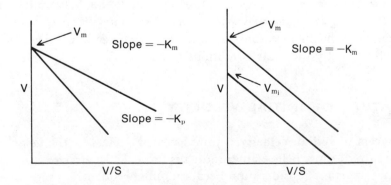

FIGURE 27.   Another linear form of the M-M equation (cf. Fig. 26).

EXPERIMENT 16

Part A. The Kinetics of Egg White Lysozyme

THE ESTIMATION OF INITIAL REACTION RATES AS A FUNCTION OF TIME AND OF ENZYME CONCENTRATION

Work in pairs.

From a stock solution of crystalline egg white lysozyme (50 μg/ml) make up a series of dilutions in 0.15 M NaCl to contain 40, 30, 20, and 10 μg/ml of enzyme.

To a series of standardized spectrophotometer tubes, add the reagents indicated in Table 30.

Table 30

	Tube Number					
Reagents	**1**	**2**	**3**	**4**	**5**	**6**
A. cell wall substrate 0.3 mg/ml in 0.1 M phosphate buffer, pH 7.0	1.9	1.9	1.9	1.9	1.9	1.9
B. 0.15 M NaCl	1.0	1.0	1.0	1.0	1.0	1.0
C. enzyme		50 μg/ml solution 0.1 ml	40 μg/ml solution 0.1 ml	30 μg/ml solution 0.1 ml	20 μg/ml solution 0.1 ml	10 μg/ml solution 0.1 ml

Set the spectrophotometer at 450 mμ, using H_2O as a blank, and determine ΔA at 450 mμ for each tube *immediately* after adding water, in the case of tube 1, and after the addition of enzyme, in tubes 2 to 6. Record your results at 15 second intervals for two minutes, and thereafter at 3, 4, 5, 6, 8, and 10 minutes. *Be sure that you mix by inversion immediately after adding enzyme.* Start reading immediately. The first reading is your zero time reading. Try to do each assay under exactly the same conditions.

ESTIMATION OF INITIAL VELOCITY

Concept of Initial Velocity. Initial velocity refers to the velocity of the reaction at zero time, before any product is formed. In enzyme kinetic studies this is an important consideration because, quite frequently, products exert an inhibitor action on the enzyme. Initial velocity can be determined from the slope of the curve of the change or disappearance of substrate as a function of time. For practical reasons it is difficult to determine initial velocity directly, mainly because the accuracy of this critical measurement is influenced by the interval of time between mixing reactants and the first measurement of the change in substrate concentration. Frequently this can be a very long time in certain enzyme assays, thus subjecting these determinations to considerable error. The most convenient procedures involve measurement of the continuous changes in substrate concentration by spectrophotometric or other instrumental means which allow continuous monitoring of concentration change. The readings from these instruments may be plotted directly to give a measure of velocity since the readings can be converted to concentration of substrate at any time. If the substrate concentration is high and zero order kinetics are being observed, the change in substrate concentration will be linear with respect to time, but if substrate concentration is limiting and first order kinetics are being observed, then an exponential decay curve with respect to substrate concentration will be observed. In the former case it will be relatively easy to determine the initial rates, but in the latter case a line tangent to the curve at zero time will provide the initial rate. It is possible that the linear portion of this curve is so short that only a crude approximation is possible. However, in either case it is possible to arrive mathematically at a reliable initial rate by plotting $\log \dfrac{S}{S_0}$ versus time, where S_0 is the substrate concentration at $t = 0$. For this experiment, use $\dfrac{A_t}{A_0}$. From the slope of this curve and the known value of S_0, it is possible to solve for S at any time. It will probably be useful to treat your data by this means, as there is a rapid drop in observed rates under the experimental conditions.

Procedure. (1) Correct each reading for the amount of apparent change in the blank solution.

(2) Plot the change in absorbance at 450 nm over the 10 minute interval. Determine whether the curve is linear over a sufficiently long time interval. If necessary, replot the points of the first two minute intervals on an expanded scale.

(3) In all cases, arrive at an initial velocity which is expressed as the change in absorbance at 450 nm per unit of time, preferably per minute.

(4) Plot initial velocity versus enzyme concentration and explain the graph.

THE EFFECT OF SUBSTRATE CONCENTRATION ON ENZYME ACTIVITY

Make a series of dilutions in 0.1 M phosphate buffer, pH 7.0, of the dried cell suspension of M. Lysodeikticus from the stock solution of 0.3 mg/ml by diluting it 1:1.5, 1:2, 1:2.5, 1:3, 1:4, 1:5, 1:7, and 1:10 just prior to assay.

Using an enzyme concentration which will give a change of at least 0.040 absorbancy per minute, record the ΔA at 450 nm for one to two minutes at

Table 31

	Reagents	Tube Number						
		1	2	3	4	5	6	7
A.	substrate, ml	1.9	1.9	1.9	1.9	1.9	1.9	1:9
B.	dilution	none	1:1.5	1:2.0	1:3	1:4	1:5	1:10
C.	1.15 M NaCl, ml	1.0	1.0	1.0	1.0	1.0	1.0	1.0
D.	enzyme, 50 μg/ml, ml		0.1	0.1	0.1	0.1	0.1	0.1

15 second intervals. Follow the protocol in Table 31; add reagents A and B, and initiate the reaction by adding enzyme.

Calculate the initial velocities; plot velocity versus substrate concentration, and determine the K_m for the enzyme by using the V versus S, the $1/V$ versus $1/S$, and the V versus V/S plots. Tabulate your results. Make some judgement as to which plot gives the best results.

THE INFLUENCE OF pH ON RATE

As in the previous experiments, determine initial velocities, but in this case, as a function of changes in pH. Follow the protocol in Table 32 and plot V_m versus pH.

How do your results compare with the literature report on the pH effect on this enzyme?

Table 32

	Reagents	Tube Number					
		1	2	3	4	5	6
A.	*cell wall suspension, ml	1.9	1.9	1.9	1.9	1.9	1.9
	phosphate buffer pH	5.8	6.2	6.6	7.0	7.4	7.8
B.	0.15 M NaCl, ml	1.0	1.0	1.0	1.0	1.0	1.0
C.	enzyme, ml		0.1	0.1	0.1	0.1	0.1

* Cell walls are to be suspended as a 0.3 mg/ml solution in 0.1 M phosphate buffer at the pHs indicated.

RATE AS A FUNCTION OF TEMPERATURE

Determine the initial rate under the optimum conditions for 60 seconds at 15 second intervals at 4°, 23°, 30°, 37°, and 45° C. Repeat the assay at each temperature more than once; average the initial rates. Allow a 5 minute temperature equilibration period for reagents A, B, and C prior to assay.

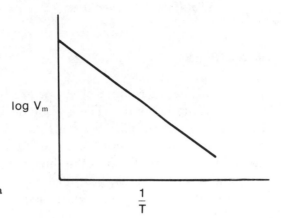

FIGURE 28. Reaction rate as a function of temperature.

Plot $\log V_m$ versus $1/T$. Using the Arrhenius equation

$$\log V_m = \frac{-E^*}{4.576\,T} + C$$

where the slope $= \dfrac{-E^*}{4.576}$, calculate E^*.

Compare the value for E^* obtained above with the average value for E^* obtained by using the integrated form of the Arrhenius equation

$$\log \frac{k_2}{k_1} = \frac{-E^*}{2.303R}\left[\frac{1}{T_2} - \frac{1}{T_1}\right]$$

where k_2 = rate constant at T_2,
$\quad\ \ k_1$ = rate constant at T_1,
and $\quad R = 1.987$ calories/(mole)(°K)

Since

$$V_m = k[E_t],$$

then

$$\log \frac{V_{m_2}}{V_{m_1}} = \log \frac{k'[E_t]}{k[E_t]} = \log \frac{k'}{k} = \frac{-E^*}{4.576}\left[\frac{T_1 - T_2}{T_1 T_2}\right]$$

where the prime indicates conditions at T_2. This equation is the expression for the rate constants at two temperatures derived as a function of V_m. Calculate E^* at several temperature intervals, 4° to 23°, 4° to 30°, 4° to 45°, 23° to 37°, and so on, and obtain an average value for E^*. Express it in terms of a standard deviation of the mean.

Part B. Denaturation and Renaturation of Lysozyme Through Mixed Disulfide Intermediates

PRINCIPLES

The primary structure of egg white lysozyme indicates the positions of the four disulfide bonds. These disulfide bridges serve as anchor points for chain folding and stabilize lysozyme considerably; hence, this enzyme is very resistant to denaturation. These sulfydryl bridges are buried within the molecule, and normally are quite inert; but they can be made accessible by denaturants such as urea, heat, detergents, or organic solvents, which loosen or disrupt the folded chain. Under such conditions the disulfide bonds react smoothly and reversibly with both thiols and other disulfide bonds. The most commonly employed procedures for cleaving these sulfide bonds include oxidation, sulfitolysis, and reduction. The latter reaction is the most readily reversible but is unstable in the presence of oxygen, which leads, through oxidation and disulfide interchange, to the formation of cross-linked intermolecular aggregates. A stable yet easily reversible derivative of the disulfide proteins may be formed by reacting them with excess low molecular weight disulfides such as cystine in the presence of traces of dithiols such as dithiothreitol (DTT). The summary equations are:

$$(1) \quad \text{Protein} \overset{\text{S}}{\underset{\text{S}}{<}} \; + \; \text{RSH} \; \rightleftarrows \; \text{Protein} \overset{\text{SH}}{\underset{\text{S—SR}}{<}}$$

$$(2) \quad \text{Protein} \overset{\text{SH}}{\underset{\text{S—S—R}}{<}} \; + \; \text{R}'\text{—S—S—R}' \; \rightleftarrows \; \text{Protein} \overset{\text{S—SR}'}{\underset{\text{S—SR}}{<}} \; + \; \text{R}'\text{SH}$$

The mixed disulfide enzyme, unlike the reduced form, is stable in the presence of oxygen, and therefore can be manipulated, isolated, and characterized. The reaction may also be reversed by catalytic amounts of low molecular weight thiols at slightly alkaline pH. These principles will be illustrated in this experiment.

PROCEDURE

Work in pairs.

Divide the work. One partner should initiate the viscometry part of the experiment, and the other should set up the electrophoresis part.

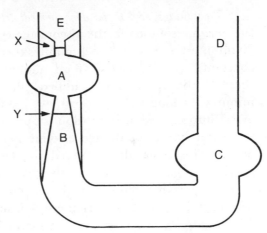

FIGURE 29. The Ostwald viscometer.

Obtain an Ostwald viscometer and a timer (use a watch with a sweep second hand). The viscometer should have been stored in cleaning acid. Drain it carefully. Rinse it thoroughly with water. Set up the viscometer vertically in an ambient temperature water bath at your lab space so that its vertical orientation is not shifted during the experiment. Introduce 3.5 ml of enzyme into D by means of a pipette. When the liquid is in C it can be drawn by gentle suction via rubber tubing into E so that the level is above the mark X. By placing an index finger over the opening at E, the level of liquid can be controlled.

Carefully allow the level of liquid to fall until the reference line marked X is reached by the fluid meniscus. The time is noted and the fluid is allowed to continue its free flow through the capillary B until the meniscus reaches the line marked at Y. Practice timing with water until you can get very reproducible results within ± 0.1 second. The time for water flow is 1.0 minute. Dry the viscometer thoroughly by rinsing with acetone and aspirating.

When you are ready to begin your experiment, set out a series of six clean, dry 13×100 m test tubes labelled 0, 10, 20, 30, 45, and 60 minutes, and add 0.9 ml of 0.05 N formic acid to each. Add 0.5 ml of a freshly prepared 10 per cent solution of crystalline lysozyme to 4.5 ml of a solution containing 4.45 M urea and 0.05 M cystine, pH 9.2. Transfer a 3.5 ml aliquot of this 1 per cent enzyme mixture to the viscometer and 0.85 ml to a sampling test tube set up in the water bath.

Record the outflow time of the enzyme solution in the viscometer until you are satisfied that it is constant within ± 0.05 to 0.1 second.

Remove a 0.1 ml aliquot from the sample tube and transfer it to the 0 time dilution tube containing 0.9 ml of 0.05 N formic acid.

To initiate the disulfide interchange reaction, add to the viscometer 5 μl of 0.5 M DTT, using a micropipet attached to polyethylene tubing. Mix the contents by blowing bubbles gently into the solution from the capillary arm of the viscometer. Record *zero time*.

Your partner should *simultaneously* add 1 μl of 0.5 M DTT to the sampling tube containing 0.75 ml of enzyme solution. Both viscometer tube and sampling tube will contain equal concentrations of DTT.

Two things need to be done immediately. (a) Measure the outflow times of the enzyme solution in the viscometer every 2 to 3 minutes for the first 15 to 20 minutes, and every 3 to 4 minutes thereafter until the outflow time becomes constant. (b) Remove 0.10 ml aliquots from the sampling tube and add them to the appropriate acidified dilution tubes at 10, 20, 30, 45, and 60 minutes (or longer if necessary).

Dilution and acidification stops the disulphide interchange reactions immediately. These diluted enzyme solutions must be assayed for activity at once. The remainder is saved for electrophoretic examination on polyacrylamide gels.

The enzyme solution in the viscometer is saved for renaturation studies (see Part D, page 207). If there is insufficient time these solutions may be frozen, later to be analyzed electrophoretically.

Determine the specific activity of your respective samples, using 0.1 ml aliquots of the diluted, acidified solutions. Remember that the concentration is now 0.1 per cent or 1 mg/ml. In some cases an even further dilution of this enzyme solution may be necessary in order to be in the correct activity range of 0.040 per 15 seconds or 160 units per minute of ΔA at 450 nm.

Plot the log of specific activity versus incubation time. Plot separately viscosity outflow times versus time of reaction. Indicate the time DTT was added but also indicate on this curve the measurement of outflow time prior to adding DTT. Calculate for each time point the *difference* between the *final* or last *outflow* time and the outflow time beginning from before zero time. Now plot the logarithm of this difference versus time. How do you explain the results? What is the half-life of (a) unfolded lysozyme, (b) enzyme activity?

Part C. Polyacrylamide Electrophoresis of Lysozyme and Its Derivatives

Mix 0.20 ml (200 μg of lysozyme) of the formic acid diluted lysozyme products saved at 0, 10, 20, 30, 45, and 60 minutes with an equal volume of 8 M urea containing 0.002 per cent methylene blue. Examine 0.1 ml aliquots of these mixtures electrophoretically.

The separating gels are prepared beforehand, but in this case, the pH is 4.8 and the gel concentration is 7.5 per cent. (See Experiment 5, page 83, and the appendix.) Furthermore, no stacking gel is employed because at most there are only one or two enzyme components.

Fill each tube with electrophoresis buffer, pH 4.9, and then underlay 0.1 ml of the more dense enzyme solution on the gel surface. A melting point capillary attached to a tuberculin syringe through a short piece of narrow hose tubing makes a convenient sample applicator. When all samples have been applied, follow instructions given on page 84 for gel electrophoresis. The alternative staining procedure with coomasie blue dye can be used.

While the electrophoretic separation is proceeding, set up the tubes needed for the method of staining and destaining chosen.

After electrophoresis has been completed, remove the tubes from the upper bath with a twisting action and immerse them in the *proper order* in an ice bath. Follow the instructions on page 86 as required for removing the gels from the tubes and for staining and destaining.

Part D.   Renaturation of Lysozyme and the Mixed Disulfide Derivative

Set up six 13×100 mm test tubes containing 1.0 ml of 0.05 N formic acid and label them 0, 5, 10, 15, 20, and 30 minutes.

Thaw the 1 per cent denatured lysozyme solution saved from the viscometry part of the experiment. Dilute 0.2 ml with 9.8 ml of 0.01 M TRIS buffer, pH 7.4. Withdraw a 1.0 ml aliquot immediately and transfer it to the zero time dilution tube.

Add 0.09 ml of 0.4 M cysteine to the remaining lysozyme derivative in TRIS buffer and incubate this solution at room temperature. Withdraw 1.0 ml aliquots and place them into their respective dilution tubes at 5, 10, 15, 20, and 30 minutes after adding the cysteine. Assay 0.10 ml aliquots of all acidified samples by the standard procedure while the experiment is being carried out.

Compare activity to that of a 10 μg quantity of unmodified lysozyme in 0.1 μl.

Plot specific activity of the enzyme against time of renaturation. What proportion of the specific activity of the native lysozyme did you recover?

Part E.   The Kinetics of Yeast Hexokinase

Hexokinase (trivial name): IUB number 2.7.1.1. Systematic name: D-hexose-6-phosphotransferase. Yeast hexokinase was identified by Otto Meyerhoff in 1927. It was crystallized in 1946. Hexokinases and pentokinases are enzymes that catalyze the transfer of the terminal phosphate group of ATP to any hydroxyl group of 5 or 6 carbon monosaccharides. In actual fact the activity of individual enzymes is much more restricted in terms of substrate utilization. Enzymes of this type are found in all tissues and all cells capable of metabolizing sugars.

Yeast hexokinase is a prototype of this group. It has the broadest specificity, being capable of phosphorylating glucose, mannose, fructose and glucosamine at the six hydroxyl position. Most kinases are much more specific, and only phosphorylate one monosaccharide either at the 1 or 6 position. The turnover number for yeast hexokinase is 13,000 moles per 10^5 gm of protein per minute at pH 7.5 and 30°, or 13,000 molecules per molecule of enzyme per minute.

SUBSTRATE SPECIFICITY OF YEAST HEXOKINASE

Work in pairs.

In this experiment you will determine initial rates with respect to substrate. The assay method for this enzyme is based on the use of an acid-base indicator (cresol red) and a buffer (glycylglycine), each having identical pK_a' values of 8.25 at 25°. The reaction taking place is:

$$ATP + D\text{-hexose} \rightleftarrows ADP + D\text{-hexose-6-phosphate} + H^+ \text{ (Explain)}$$

The amount of acid produced in the reaction is equivalent to the amount of basic buffer neutralized, which in turn is proportional to the amount of basic indicator neutralized. By reading at the wavelength of maximum absorption for the indicator, the decrease in absorbance is directly proportional to Pi transferred or acid produced and becomes a direct measure of enzyme activity.

It can be shown by independent analytical means that the phosphorylation of 1 μmole of glucose produces a decrease in absorption equivalent to that of 1 μmole of added acid. A change of A_{560} of 0.035 normally corresponds to 1 μmole of acid released. The exact value should be determined for each new batch of reagents by adding an equivalent volume of a standard quantity of acid (1 μmole) in the place of enzyme. Solutions of 0.2 M sugars, L-glucose, D-glucose, D-mannose, D-fructose, D-glucosamine, D-galactose, D-sucrose, and D-xylose are to be used for enzyme assay. Add 0.4 ml of the sugars to cuvettes containing 2.5 ml of fresh assay solution of ATP, Mg^{+2}, and glycylglycine buffer, pH 8.25, and cresol red. At zero time add 0.1 ml of enzyme solution containing 20 μg/ml of crystalline yeast hexokinase in 0.2 % albumin. Set all tubes at an absorbance of 0.400 and record the ΔA_{560} at 15 second intervals for 1 to 2 minutes. *Gently invert before reading.*

One unit is defined as the amount of enzyme catalyzing the formation of 1 μmole of acid per minute at 25°. A constant temperature must be maintained during assay since a small rise in temperature can simulate acid formation by the buffer reagent system as temperature increases. To determine the ΔA_{560} for 1 μmole of acid, add 0.1 ml of standard 0.0100 M HCl to the reaction mixture *in the place of enzyme*. You may do this with glucose as part of the standardization procedure.

$$\text{units/mg protein} = \frac{\Delta A/\text{min (Initial Velocity)}}{\Delta A/\mu\text{mole } H^+ \times \text{mg protein in reaction mixture}}$$

For more refined work, to correct activity from the known temperature t to 25° C, multiply by 2.2 to the power $\dfrac{25 - t}{10}$.

Make a table of comparison of the reaction rates for each sugar. What are your conclusions about yeast hexokinase specificity?

Table 33

Reagents	Tube number						
	1	2	3	4	5	6	7
A. MgATP, buffer assay solution, ml	2.5	2.5	2.5	2.5	2.5	2.5	2.5
B. water, ml	0.2	0.2	0.2	0.2	0.2	0.2	0.4
C. dilution	none	1:2.5	1:5	1:10	1:15	1:20	
D. 0.4 M glucose, ml	0.2	0.2	0.2	0.2	0.2	0.2	
E. enzyme 20 μg/0.1 ml, ml	0.1	0.1	0.1	0.1	0.1	0.1	0.1

THE EFFECT OF INHIBITORS ON ENZYME ACTIVITY

Make a series of dilutions of 0.4 M glucose as follows: 1:2.5, 1:5, 1:10, 1:13, 1:20. Follow the protocol given in Table 33, adding enzyme last, and determine the initial rates at 560 nm. Use tube 7 as the blank and set it as well as the other tubes at 0.400 absorbance.

Table 34

Calculation	Tube Number					
	1	2	3	4	5	6
substrate concentration (S)						
1/S						
initial velocity with no inhibitors						
1/V						
V/S						
initial velocity with N-acetylglucosamine or xylose						
1/V						
V/S						
initial velocity with n-ethyl maleimide						
1/V						
V/S						

Repeat the above experiment after substituting 0.2 ml of 0.2 M N-acetyl glucosamine or xylose for the water. Complete the appropriate part of Table 34.

There is some doubt about whether yeast hexokinase is a SH-sensitive enzyme. Some investigators do report that yeast hexokinase is indeed affected by these agents, while other investigators find it difficult to demonstrate. Amino acid analysis of the crystalline enzyme does indicate the presence of four cysteine groups, but in one study it was shown that this enzyme was not affected at 25° by CH_3HgNO_3, p-chloromercuribenzoate, or $HgCl_2$, but at 35° C, 5×10^{-5} M CH_3HgNO_3 abolished activity. This was reversed by cysteine. The reagent n-ethyl maleimide, however, appears to be a more effective sulfydryl group attacking reagent in the case of this enzyme.

If time permits, preincubate 1 ml of enzyme with 2 ml of 0.0015 M N-ethyl maleimide for 10 minutes. Use 0.3 ml aliquots of this mixture in place of enzyme and water and repeat the above experiment. Complete the appropriate part of Table 34.

Complete the following plots for each of the three sets of data: V versus S; $1/V$ versus $1/S$; V versus V/S. Ascertain the values for K_m, K_i, and V_M for all three plots. What is your estimate of the reliability of each type of plot? What type of inhibitors have been employed? What units for K_M and K_i must be used in this case?

REFERENCES

1. Boyer, P. D., Lardy, H., and Myerback, K., eds. *The Enzymes.* Vol. I. Academic Press, New York, 1962.
2. Dixon, M., and Webb, E. C. *Enzymes.* Academic Press, New York, 1958.
3. Segal, I. H. *Biochemical Calculations.* John Wiley and Sons, Inc., New York, 1968.
4. Montgomery, R., and Swenson, C. A. *Quantitative Problems in the Biochemical Sciences.* W. H. Freeman and Co., San Francisco, 1969.
5. Alterton., G., Ward, W. H., and Fevald, H. L. *Isolation of Lysozyme from Egg White.* J. Biol. Chem. **157** 43 (1945).
6. Prasad, A. L. N., and Litwack, G. *Measurement of the Lytic Activity of Lysozymes.* Anal. Biochem. **6** 328 (1963).
7. Blake, C., et al. *The Structure of Crystalline Lysozyme.* Nature **206** 757 (1965).
8. Canfield, R., and Liu, A. K. *The Disulphide Bonds of Egg White Lysozyme.* J. Biol. Chem. **240** 1997 (1965).
9. Cleland, W. *Dithiothreitol, New Protective Reagent for SH Groups.* Biochem. **3** 480 (1964).
10. Smithies, O. *Disulphide Bond Cleavage and Formation in Proteins.* Science **150** 1595 (1965).
11. *Experimental Biochemistry Laboratory Manual.* Department of Biochemistry, University of Wisconsin, 1967.
12. Darrow, R. A., and Colowick, S. P. "Hexokinases from Bakers' Yeast" in *Methods in Enzymology,* Vol. 5, Colowick, S. P., and Kaplan, N. O., eds. Academic Press, New York, 1962.
13. Sols, A., Dela Fuente, G., Villa Palasi, C., and Asensio, C. *Substrate Specificity and Some Other Properties of Bakers' Yeast Hexokinase.* Biochem. et Biophys. Acta 3092 (1958).
14. Webb, J. L. *Enzyme and Metabolic Inhibitors.* Academic Press, New York, 1966.
15. Barnard, E. A., and Amel, A. R. *Studies on the Active Center of Yeast Hexokinase.* Biochem. J. **84** 72 (1962).

CHAPTER 17

THE DETERMINATION OF THE MOLECULAR WEIGHT OF PROTEINS BY GEL FILTRATION

PRINCIPLES

In recent years gel filtration has been successfully used to determine the molecular weights of proteins. Since gel acts as a molecular sieve, protein retention becomes a function of molecular weight. The correlation between elution volume and molecular weight is therefore direct. It is particularly applicable to globular proteins, but proteins with subunits, such as hemoglobin and glutamic dehydrogenases, dissociate in the gel; hence, they will elute at a later time and give lower molecular weight values. Other proteins, such as lysozyme, egg albumin, or proteins containing carbohydrates, tend to form associations with the hydroxyl groups of the Sephadex; in these cases the correlation between molecular weight and V_e is poor. Proteins which aggregate, or which deviate from a globular shape to a considerable extent, will give higher molecular weight values than expected. Other sources of experimental error are: (1) inaccurate estimations of elution volumes, and (2) density differences between solvating protein molecules.

Sephadex G-75 has a useful molecular weight detection range for globular proteins of 5000 to 50,000.

With G-100 Sephadex, the molecular weight can be determined in the range of 10,000 to 150,000. For G-200 the working range is a molecular weight of 40,000 to 200,000. Accuracy of the method is around ± 10 per cent.

Molecular weight determinations are ordinarily carried out by carefully preparing a standard curve of pure proteins with known molecular weights and plotting either V_e or the ratio V_e/V_0 against the logarithm of molecular weight.

211

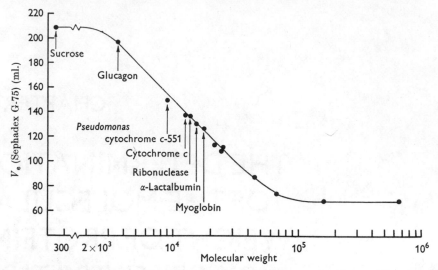

FIGURE 30. Plots of elution volumes, V_e, against log (molecular weight) for proteins on Sephadex G-75 columns [2.4 cm $\times$ 50 cm; equilibrated with 0.05 M tris hydrochloride buffer, pH 7.5, containing KCl (0.1 M)]. (From "Estimations of the Molecular Weights of Proteins by Sephadex Gel Filtration," P. Andrews, *Biochem. J.* **91** 222 (1964).)

In the effective range of the particular gel in use, these plots will give a straight line. The curve tapers off at the extreme end at the point where $V_e = V_0$. The unknown is also run through the same column. Results are plotted on semi-log paper.

Theoretical considerations also predict a linear relationship between (molecular weight)$^{1/2}$ and the cube root of the distribution or partition coefficient, $(K_d)^{1/3}$, of each protein.

An elution diagram of the unknown is plotted; that is, protein or enzyme concentration is measured spectrophotometrically and absorbance is plotted against elution volume (or tube number) in order to get V_e from the peak height. See page 184 for the theoretical treatment of gel filtration.

EXPERIMENT 17

Work in pairs.

There are three major factors affecting resolution in gel filtration: (a) the gel column length, (b) the Sephadex gel particle size, and (c) the sample size. In this experiment some limitations on the column length have to be made because of time, since flow rates become slower as the column length increases. In practice, coarse grade Sephadex has less resolving power than either fine or superfine grades. The sample size can be made as small as 0.3 ml for a 15 cm column height, and 1 ml for a 25 cm column. Where the objective is to obtain as good a separation as possible, the starting zone should be as narrow as possible. This can be accomplished by using the minimum sample volume required to distribute it evenly over the entire bed surface at the top of the column.

The mode of addition of sample to the column can be another factor affecting resolution. This operation can be carried out in several ways: (a) by direct application with a pipette in a manner such that the flow of the sample does not disturb the gel surface, (b) by underlying a more dense sample (made dense by adding glucose or other low molecular weight compound to the sample) in a less dense fluid, usually the eluate buffer which has been placed over the column, or (c) by using a flow adapter. In each of these techniques the object is to prevent disturbance of the top of the gel surface.

Begin your experiment by setting up a coarse grade Sephadex G-100 column at a height of 60 to 70 cm (cf. instructions on page 187). Stop the flow of buffer through the column by means of a screw clamp at the bottom, and leave a 5 cm head of fluid at the top. Cut out a filter paper disc and allow it to pass down to the top of the column. Carefully underlay, just above the paper disc, 0.2 ml of sample containing 2 mg/ml of each of the following: cytochrome C (molecular weight 12,400), myoglobin (molecular weight 17,800), hemoglobin (molecular weight 64,500), and dextran blue (molecular weight 1,000,000) in 10 per cent sucrose. Set up an eluate reservoir consisting of a funnel with a tube extension which can be pinched off with a screw clamp. Now initiate elution with 0.01 M KCl in 0.05 of TRIS buffer, pH 7.5, and allow an inflow of eluate through the funnel at a rate equal to the outflow rate. In this manner, a constant fluid pressure will be maintained at the top. Do not allow fluid from the reservoir to drop from too great a height, thus preventing convection disturbance. Collect the eluate in 0.5 ml amounts in spectrophotometric tubes or other tubes if available in large quantities. Keep track of the tube numbers and eluate volumes. Dilute each tube with 2.5 ml of water and take a reading of absorbance at 660 n m to determine V_0 for the dextran blue, then at 410 n m for the heme proteins as they come off. If the flow rate is too fast you may have difficulty doing all of these operations efficiently. One student should monitor the column and do the collecting while the other student keeps the records and takes readings. While separation is going on the student can take advantage of the fact that it will be possible to visually follow the dextran blue and heme proteins as they pass through the column.

Determine the V_e for each protein as it comes off the column. You should be able to predict the order.

Make a plot of V_e, determined experimentally, versus the log of the molecular weight of each protein. Do they fall on a straight line? If not, explain why.

REFERENCES

1 Andrews, P. *The Gel Filtration Behavior of Proteins Related to Their Molecular Weights Over a Wide Range.* Biochem. J. **96** 595 (1965).

2. Andrews, P. *Estimation of Molecular Weights of Proteins by Sephadex Gel Filtration.* Biochem. J. **95** 222 (1964).

3. Whitaker, J. R. *Determination of Molecular Weights of Proteins by Gel Filtration on Sephadex.* Anal. Chem. **35** 1950 (1963).

4. Porath, S. "Theoretical Treatment of the Gel Filtration Process." In *Proceedings of the International Symposium on Pharmacy and Chemistry, Florence.* Butterworth and Co., Ltd., 1962.

CHAPTER 18

THE TURNOVER NUMBER OF TRYPSIN AND CHYMOTRYPSIN

CONCEPT OF TURNOVER NUMBER (T.O.N.)

The maximum velocity (V_m) of a reaction in the presence of a mole of enzyme is called the turnover number. It is the moles of substrate reacting per minute per mole of enzyme, or the rate at which the active centers of a particular enzyme react. In cases where the molecular weight of an enzyme is not known, it may be expressed as substrate reacting per minute per 10^5 g of protein. When enzymes are composed of subunits and may have more than one active site, it may be better to express the turnover number as rate of catalysis per active center. For instance, catalase decomposes H_2O_2 at a rate of 5×10^6 μmoles of H_2O_2 per minute per mole of catalase. It has, however, four catalytic sites per mole, and it would be better to give the turnover number as 1.25×10^6 min^{-1}, which would measure the reactivity of one active center. Another difficulty arises in reactions in which a polymer is involved and is being degraded or synthesized. Since there is no change in molar concentration in these circumstances, the catalytic activity is expressed as the number of chemical bonds hydrolyzed or synthesized per minute per mole of enzyme.

The Determination of T.O.N. In this experiment you will attempt to ascertain the turnover number of the crystalline proteolytic enzyme trypsin, by quantitatively measuring the number of new amino groups released during hydrolysis. A suitable protein substrate must be chosen that can readily be attacked by them. Proteins used as substrates are modified chemically by converting the primary amine groups (end and epsilon amines) into dimethyl amino groups, a change which does not affect the protein properties but does block reaction with the analytical reactant, in this case trinitrobenzene sulfonic acid. This reagent is used to determine the number of free amino groups released by enzyme action. By methylating protein substrates it is possible to avoid high blanks at zero time and hence give a clear indication of the new terminal amino groups formed as a result of enzyme hydrolysis of the peptide bonds. (Why?)

Trypsin (trivial name): IUB number 3.4.4.4. Molecular weight 23,800.

Chymotrypsin (trivial name): IUB number 3.4.4.5. Molecular weight 24,500.

Systematic name: Not possible in this subgroup, though in general class of catanases, or enzymes which cleave chains.

Trypsin and chymotrypsin belong to a group of digestive enzymes of the gastrointestinal tract that includes pepsin, a digestive enzyme, secreted by the gastric mucosa. Trypsin and chymotrypsin are secretions of the pancreas. Each of these enzymes is synthesized as an inactive precursor or zymogen which is converted just before use to an active form. Pepsinogen (molecular weight 42,000) is cleaved by acid or by pepsin to form a pepsin inhibitor complex (molecular weight 38,000) and five small peptides. The free pepsin has a molecular weight of 35,000. Trypsinogen loses its first six amino acid residues to become trypsin. This conversion is triggered by a special enzyme, enterokinase, and the conversion is subsequently accelerated by trypsin. Chymotrypsinogen is converted by a series of steps involving, in the first step, splitting of the arginine15—isoleucine16 peptide bond by trypsin to form π-chymotrypsin. There are additional cleavages involving the removal of two dipeptides to form δ, α, and γ chymotrypsin, the latter two being the working models. Amino acid analysis of trypsin and chymotrypsin indicates that they are very similar. Forty-one per cent of the chain is identical in the two enzymes. Pepsin has an isoelectric point below 1, a fact which is not easily explained on the basis of its known amino acid composition. Each enzyme acts differently on cleaving the peptide bonds of proteins. Chymotrypsin and pepsin cleave peptide bonds of aromatic amino acids, and to a lesser extent the bonds involving glutamic and aspartic acids. Trypsin hydrolyzes peptide bonds involving the carboxyl groups of arginine and lysine.

EXPERIMENT 18

PROTEOLYTIC ASSAY

Work in pairs:

(a) To suitable test tubes add 0.1 ml of trypsin solution or 0.1 ml of chymotrypsin solution containing 1 μg of enzyme to one ml of 0.1 per cent dimethyl casein, pH 8.1. Incubate the mixture for 30 minutes at 37° C. Do this in triplicate for each enzyme.

(b) Prepare a blank with enzyme omitted. Add 0.1 ml of H_2O to 1 ml of substrate.

(c) Stop the reaction by immersing the samples briefly in a boiling water bath.

(d) Add, consecutively, 1 ml of 0.1 per cent trinitrobenzene sulfonic acid (TBS, or picryl sulphate) and 1 ml of 4 per cent $NaHCO_3$, pH 8.5. Incubate these mixtures in the tubes for 30 minutes at 50° C in the dark. Cover the tubes with aluminum foil for this purpose.

(e) After this treatment add 1 ml of 10 per cent sodium dodecyl sulfate and 0.5 ml of 1 N HCl to each tube.

(f) Read the absorbance at 340 to 380 nm using the blank as a reference. Increase the wave length up to 380 nm if you are not able to zero in the spectrophotometer at 340 nm.

CALCULATION OF THE T.O.N.

The average value of the extinction coefficient for various trinitrophenyl amino acid derivatives is 1.3×10^4 moles^{-1}cm^2 at 340 nm. By using this composite value it is possible to calculate the concentration of liberated amino groups, hence allowing the direct expression of enzyme activity in terms of the number of bonds split. Accordingly, the molar concentration of the total number of peptide bonds cleaved can be calculated from the relationship.

$$[\text{NH}_2\text{-terminal}] = \frac{A_{340}}{1.3 \times 10^4}.$$

The definition of T.O.N. is the moles of substrate reacting per minute per mole of enzyme, or in this case the moles of *peptide bonds* cleaved per minute per mole of enzyme. You need to know the total number of amino groups liberated.

Since a 1 molar solution of released αNH$_2$ groups has an average extinction coefficient of 1.3×10^4 mole^{-1} cm^2 under the conditions of assay, it is the same as saying that a solution containing 0.001 moles of αNH$_2$ *per ml* has an extinction coefficient of 1.3×10^4 mole^{-1} cm^2. Therefore

$$\frac{A_{340}}{1.3 \times 10^4} \times 4.6 = A_{340} \times 3.54 \times 10^{-4} =$$ total moles of αNH$_2$ formed in 30 minutes of reacting time. (The final volume of your experimental assay mixture is 4.6 ml.)

Therefore,

$$A_{340} \times \frac{3.54 \times 10^{-4}}{30} = A_{340} \times 1.18 \times 10^{-5} =$$ moles of αNH$_2$ formed per minute of reacting time.

Likewise,

$$\text{moles of } \textit{trypsin} \text{ used in assay} = \frac{\text{g of enzyme used}}{\text{molecular weight of enzyme}}$$

$$= \frac{0.1 \times 10^{-6}}{2.38 \times 10^4} = 4.20 \times 10^{-12}$$

and,

$$\frac{1 \times 10^{-6}}{2.45 \times 10^4} = 4.25 \times 10^{-11} = \text{moles of } \textit{chymotrypsin} \text{ used in the assay}$$

For trypsin:

$$\text{T.O.N.} = \frac{\text{total moles of amines formed}}{\text{moles of enzyme used in the assay}} = \frac{A_{340} \times 1.18 \times 10^{-5}}{4.20 \times 10^{-12}}$$

$$= A_{340} \times 2.81 \times 10^6 = \text{moles of } \alpha\text{NH}_2 \text{ formed per minute per mole of trypsin}$$

and similarly:

$$\text{T.O.N.} = A_{340} \times 2.77 \times 10^6 = \text{moles of } \alpha\text{NH}_2 \text{ formed per minute per mole of chymotrypsin.}$$

Compare your value for T.O.N. with reported literature values.

REFERENCES

1. Lin, Y., Means, G. E., and Feeney, R. E. *The Action of Proteolytic Enzymes on n,n-Dimethyl Proteins* J. Biol. Chem. **244** 789 (1969).
2. Means, G. E., and Feeney, R. E. *Reductive Alkylation of Amino Groups in Proteins.* Biochem. **7** 2192 (1968).
3. Montgomery, R., and Sorenson, C. A. *Quantitative Problems in the Biochemical Sciences.* W. H. Freeman and Co., San Francisco, 1969.

CHAPTER 19

CATION REQUIREMENTS OF NUCLEOTIDE PHOSPHOHYDROLASES

INTRODUCTION

The enzymes classified as nucleotide phosphohydrolases (ATPase) appear to act as transducers of chemical bond energy into some form of mechanical energy. These enzymes split the terminal phosphate anhydride bond of nucleotide triphosphates, particularly of adenosine triphosphate, and release energy. This compound is the principal energy storage material of the cell, and during catalysis it is split into its corresponding dinucleotide and inorganic phosphate. The reaction is exergonic, and the energy released is utilized to do some form of work in the cell.

As examples of this interrelationship, the hydrolysis of ATP may be coupled to neuronal impulse transmission in nerve tissue, to the contraction in skeletal muscle, and to the excretion of small molecules in the kidney. Though ATPase has been extensively studied in the liver in relation to the oxidative synthesis of ATP, the hydrolytic cleavage of ATP is a catabolic process releasing energy; in fact, is the antithesis of oxidative phosphohydrolation, which leads to ATP synthesis. ATPases are found in a wide variety of tissue, both in the plant and animal kingdoms. In a majority of tissues ATP hydrolysis has been associated with energy-dependent transport of small molecules. This association with ion transport is particularly spectacular in neurons. During impulse transmission, for example, there is an influx and efflux of Na^+ and K^+ through the neural outer membranes. The ATPases of this tissue have been shown to be very sensitive to Na/K ratios, which can either bring about a stimulation or an inhibition of enzyme activity. No other enzymes react quite so uniquely to cationic changes as do neuronal ATPases.

EXPERIMENT 19

The assay for these enzymes will be carried out by spectrophotometrically measuring the amount of inorganic phosphate (Pi) formed after three minutes incubation of ATP with enzyme. Although there are different pH optima for each of these enzymes, for convenience, the assay will be carried out in each case at pH 7.6. Tissue ATPases from two rats will be investigated for cationic requirements and for their modulating effects on enzyme activation. There will be groups of four students, each to be assigned either liver, brain, kidney, or spleen. Two students out of the group of four will set up tubes as indicated in Tables 35, 36, and 37. The other two students are to follow the directions outlined below for removing either the liver, brain, kidneys, or spleen.

PREPARATION OF ENZYMES

Work in groups of four throughout this experiment.

Two animals can serve 16 students or four groups. One group takes two brains, the second group removes the two kidneys and the third group removes the two spleens. The groups assigned the liver need only use one animal. If the liver weighs more than 5 grams only one half needs to be used. In each step care must be exercised to use very clean glassware and equipment. All of it should be rinsed with distilled water; all reagents will be made up with it in order to minimize contamination by metal cations. Tris-ATP, a special ion exchange chromatographic preparation to remove sodium, will be used. As each tissue is removed it is immediately placed into a previously tared beaker containing 20 ml of cold 0.29 M sucrose and 0.001 M EDTA at pH 7.5. After recording the wet weight of the tissue, the sucrose-EDTA solution is decanted from the beaker; the tissue is washed once with fresh cold sucrose-EDTA to remove blood, hair, and other bits of tissue. Three ml of cold sucrose-EDTA are now added, and each tissue is minced into slices with a scissors. The slices are transferred into tissue homogenizer tubes. At this point, the volume of tissue plus sucrose-EDTA should not exceed 10 ml. The tissue is triturated by five to seven complete passes of the homogenizer pestle through the tissue at 1000 rev/min of the drive motor. This operation is performed in the cold room.

After complete trituration, transfer the homogenate quantitatively to a cold 50 ml graduated cylinder. On the basis of the wet tissue weight to volume of sucrose medium, a 10 per cent homogenate (w:v) is made (i.e., if the original tissue weighed two grams then the final volume is twenty ml). This homogenate is thoroughly mixed in the graduated cylinder and then transferred to a pair of polyethylene centrifuge tubes. No more than two tubes should be used; discard the excess if necessary, but under no circumstances should these tubes be placed in the centrifuge unequal in weight. If there is less than 40 ml of homogenate, use one tube for it and balance the mate with water. After properly marking them for purposes of retrieval, place them opposite one another in the rotor. The nuclear, remaining whole cell, and connective tissue are separated by a preliminary centrifugation for one minute at high setting on the Sorvall refrigerated centrifuge; this should include rotor acceleration time as well. The maximum gravitational force attained will be 8200 g at one minute. The supernatant obtained after the first centrifugation is decanted into a second pair of clean plastic centrifuge tubes and balanced once again taking due precaution to leave behind the pellet and the heavy loose layer at the bottom of the tube. Centrifuge the supernatant at 48,000 g force for ten minutes to obtain a crude mitochondrial pellet. The relatively clear supernatant, containing microsomes and soluble protein, is discarded without disturbing the pellet. The crude mitochondrial pellet is uniformly resuspended in a volume of cold sucrose-EDTA equal to the original wet weight of the tissue in the case of liver and brain, but only $\frac{1}{2}$ the original wet weight in the case of spleen and kidney. This step is carried out as follows: add 0.5 ml of cold sucrose-EDTA to the pellet and transfer the suspension to a small tissue homogenizing tube (10 ml capacity). Quantitatively transfer the remainder of the pellet material with 0.3 ml amounts of sucrose-EDTA. Homogenize, and transfer the homogenate to a 10 ml graduated cylinder. Using 0.3 ml volumes, achieve a quantitative transfer once again. Bring the suspension to the correct final volume. Mix thoroughly by inversion. In this way you should have a uniform suspension of tissue homogenate. With the aid of a small homogenizer tube, use 0.5 ml of sucrose-EDTA to resuspend the pellet and use 0.3 ml amounts of medium to quantitatively transfer the homogenate to a 10 ml graduate cylinder. *In order to complete this experiment properly you should have at least 1.0 ml of a crude mitochondrial suspension*, but under no circumstances should it be diluted any more than is necessary to complete resuspension and transfer.

PREPARATION OF REAGENTS

While the enzymes are being prepared, the other two members of the group should set up the series of tubes listed in Tables 35, 36, and 37.

(a) The lettering on the side of the reagents listed in each table indicates the order in which the reagents are added to each tube. Read directions in advance so that you understand what has to be done.

(b) For the preparations in Table 35, *do not add TCA, the enzyme, or the TRIS-ATP until you are ready to start the assay*, that is, until the tissue preparation has been completed. Prepare the tubes in Table 36, except for the biuret reagent and the tissue homogenate. For the tubes designated in Table 37, add reagents *A* and *B* only.

(c) When the homogenate of your particular group is ready, the biuret determination of protein concentration may be performed. Add 0.1 ml of homogenized tissue to tube 2 in Table 36; add 1.5 ml of biuret reagent to *each* tube. After allowing the tubes to sit for thirty minutes, read the absorbance at 540 nm using tube 1 for a blank. Determine the protein concentration from the graph plotted in Experiment 4. *Be sure to use the same spectrophotometer* as

Table 35

Reagents	Tube Number (Use 13 × 100 mm Tubes)							
	1	2	3	4	5	6	7	8
A. 0.5 M TRIS buffer, pH 7.8, ml	0.2	0.2	0.2	0.2	0.2	0.2	0.2	0.2
B. 0.3 M $MgCl_2$, ml		0.2			0.2	0.2	0.2	0.2
C. 0.3 M $CaCl_2$, ml			0.2					
D. 0.75 M NaCl, ml				0.1	0.1		0.1	0.1
E. 0.15 M KCl, ml				0.5		0.5	0.1	0.1
F. 25 μmole Pi/ml, ml								0.1
G. water, ml	1.5	1.3	1.3	0.9	1.2	0.8	1.1	1.0
H. 0.03 M TRIS-ATP, pH 6.0, ml	0.2	0.2	0.2	0.2	0.2	0.2	0.2	0.2
I. tissue preparation, ml	0.1	0.1	0.1	0.1	0.1	0.1	0.1	0.1
J. 25 per cent trichloroacetic acid (TCA), ml	0.5	0.5	0.5	0.5	0.5	0.5	0.5	0.5

used in Experiment 4. Express the concentration of your enzyme in mg/ml, but remember that in the actual assay you used 0.1 ml enzyme preparation.

(d) Set tubes 1 through 7 from Table 35 in a bath at 30° C. Obtain 2.0 ml of TRIS-ATP from your instructor and set it in the bath. Set the homogenized tissue in the bath. After four or five minutes, when temperature equilibrium has been reached, you are ready to carry out the assay. Don't delay once you have started the pre-incubations. Add 0.5 ml of 25 per cent TCA to tube 8, *followed* by 0.1 ml of enzyme. Stir the mixture and then remove it from the bath.

(e) Add 0.2 ml of tris-ATP to each of the other tubes.

Table 36

Reagents	Tube Number (Use Spectrophotometer Tubes)	
	1	2
A. 0.9 per cent NaCl, ml	1.4	1.3
B. *5 per cent Na deoxycholate, ml	0.1	0.1
C. tissue homogenate, ml		0.1
D. biuret reagent, ml	1.5	1.5
E. absorbance		
F. mg protein/ml		

* Protein solubilizing reagent

Table 37

Reagents	Tube Number (Use 18 × 150 ml Tubes)							
	1	**2**	**3**	**4**	**5**	**6**	**7**	**8**
A. 0.1 M NaAc, ml	4.0	4.0	4.0	4.0	4.0	4.0	4.0	4.0
B. 0.1 M HAc buffer, pH 4.0, ml	4.0	4.0	4.0	4.0	4.0	4.0	4.0	4.0
C. aliquot of TCA supernatant, ml	1.0	1.0	1.0	1.0	1.0	1.0	1.0	1.0
D. molybdate reagent, ml	0.5	0.5	0.5	0.5	0.5	0.5	0.5	0.5
E. 1 per cent ascorbic acid, ml	0.5	0.5	0.5	0.5	0.5	0.5	0.5	0.5
F. absorbance for assay at 660 nm								
G. total μmoles Pi formed/minute								
H. specific activity								

(f) At time zero, add 0.1 ml of enzyme to tube 1. Every 30 or 60 seconds (whichever is more convenient), add 0.1 ml of enzyme to the next tube, until all seven tubes contain 0.1 ml each of enzyme. As enzyme is added to each tube, stir the mixture thoroughly. Exactly three minutes after the enzyme was added to tube 1, a second student should add 0.5 ml of 25 per cent TCA. Continue, at the same time intervals as in the enzyme addition, to add TCA to the remaining tubes. *The student adding TCA will have to be ready with the reagent before the first student finishes the enzyme addition.* Each tube, consequently, will have *exactly* three minutes incubation time.

(g) When the assay is complete, remove the tubes from the bath and spin down the precipitate of protein in the clinical centrifuge at three-quarter speed for five minutes. Be sure your tubes are properly marked before you do this.

Table 38

Tissue	Tube Number						
	1	**2**	**3**	**4**	**5**	**6**	**7**
	No Cation	Mg	Ca	Na + K	Mg + Na	Mg + K	Mg + Na + K
liver							
kidney							
muscle							
brain							

(h) With the aid of a long-tipped pipette, carefully remove a 1 ml aliquot from each tube and transfer it to the corresponding tube designated in Table 37. Sequentially, add 0.5 ml of molybdate reagent and 0.5 ml of 1 per cent ascorbic acid to each tube. Mix by swirling and inversion with parawax caps or a vortex mixer after each addition.

(i) Read the absorbance of each tube at 660 nm after 20 minutes using as the reagent blank a tube containing 1.0 ml of 5 per cent TCA and treated as in Table 37.

You are to express enzyme activity in international units or specific activity. In this case the specific activity is expressed as the μmoles of Pi released in one minute by one milligram of protein at 30°. Tube 1 is the endogenous control, representing the amount of Pi present or liberated in the absence of any added cations.

Remember these points in your calculation: Tube 8 is your standard. It contained 1 μmole of Pi per ml. The incubation time was three minutes.

In collaboration with the other groups, complete Table 38. Enter $(+)$ for increased activity; $(++)$ for further increase of activity; and $(-)$ for inhibition of activity. Make comparisons to Tube 1. Make a concise statement of observations and conclusions.

REFERENCES

1. Skou, J. C. *The Influence of Some Cations on an Adenosine Triphosphatase from Peripheral Nerves.* Biochem. Biophys. Acta **23** 394 (1957).
2. Rendina, G. *Enzymic Properties of Subcellular Phosphohydrolases in Bovine Brain Cortex.* J. Neurochem. **13** 683 (1966).
3. Landon, E. J., and Norris, J. L. *Sodium and Potassium Dependent Adenosine Triphosphatase in a Rat Kidney Endoplasmic Reticulum Fraction.* Biochem. Biophys. Acta **71** 266 (1963).
4. Ulrich, F. *Kinetic Studies of the Activation of Mitochondrial ATPase by Mg^{++}.* J. Biol. Chem. **239** 3532 (1964).
5. Lowry, O. H., and Lopez, J. A. *Determination of Inorganic Phosphate in the Presence of Labile Phosphate Esters.* J. Biol. Chem. **162** 421 (1946).

CHAPTER 20

THE RATE OF GLUCOSE ABSORPTION IN RAT INTESTINE (DETERMINATION OF THE CORI COEFFICIENT)

INTRODUCTION

In this experiment the rate of absorption of glucose, expressed as the number of milligrams of glucose absorbed per 100 g of rat per hour, is determined in the presence of phlorizin and iodoacetic acid, which act as inhibitors of sugar absorption.

Iodoacetic acid, in particular, reacts with sulfhydryl groups on an enzyme surface, and it is believed that some key enzymes involved in energy transfer during absorption interact with iodoacetic acid. Phlorizin also inhibits absorption, but its mechanism is less well understood. A particular target organ for phlorizin is the kidney. Phlorizin prevents renal re-absorption of glucose and causes glucosuria and hypoglycemia, both of which partially simulate the symptomatology of diabetis mellitis. Animals treated with phlorizin have been used to study the consequences of prolonged hypoglycemia in animals.

EXPERIMENT 20

(1) Each group of four students will be given four rats deprived of food for 24 hours. Each student member is responsible for one rat of the group of four.

(2) These rats are to be injected intraperitoneally with Nembutal (pentobarbital sodium). Weigh the rats and record these weights in your notebook. Dosage is 25 mg/kilo. Stock solution is 10 mg/ml. A 200 g rat should get 5 mg or 0.5 ml of Nembutal solution.

(3) After the rats show effects of anaesthesia, oral administration is given with the aid of an 8 French catheter. Gently force the catheter into the mouth. Do not press through the bronchial tubes. If you encounter too much resistance, start over, making sure the catheter is lubricated with some saliva or liquid. The catheter should penetrate at least two inches into the stomach before release of sugar.

Table 39

rat #1	2 cc H_2O
rat #2	2 cc 50% glucose solution
rat #3	2 cc 50% glucose in 4×10^{-3} M phlorizin
rat #4	2 cc 50% glucose in 4×10^{-3} M iodoacetic acid

(4) Time of administration should be noted. Exactly one hour later, kill the rat; clamp off the esophagus at the cardiac sphincter and the small intestine at the ileocolic sphincter, and remove the stomach and small intestine. (*Caution:* be very careful when removing the intestine; avoid tearing the walls.)

(5) Insert a funnel into a 200 ml volumetric flask and place the funnel and flask on floor before the desk. Place the stomach end of the digestive tract into the funnel, cut it and connect a 20 cc syringe (without needle) to the opening of the lower end of the ileum. Standing it upright; flush out the contents of the intestine with a syringe containing distilled water until the flask is approximately half full. After 100 ml of washings have been collected, fill the volumetric flask to the 200 ml level. Discard the animal carcass and tissue.

(6) Mix the contents of the 200 ml volumetric flasks by inversion. After allowing any sediment to settle or after filtering, transfer a clear 2.0 ml aliquot to a test tube. Dilute to 5 ml with water, and then transfer a 0.1 ml aliquot to an 18×150 mm test tube for analysis.

(7) Prepare a reagent blank by adding 0.1 ml of water to one test tube and a glucose standard containing 100 μgrams of glucose in 0.1 ml of water to a second tube. Use a 0.1 ml pipette.

(8) Add 5.0 ml of fresh 6 per cent toluidine in glacial acetic acid to all six tubes. Use a propipette for this transfer.

(9) Immerse all tubes in a 100° C water bath for exactly 7.5 minutes; *quickly* remove them and cool to room temperature in a ice-water bath.

The heating period is critical and must be timed exactly. This is a conjugation reaction of an amino phenol with aldoses and ketoses. It will give a blue green color with glucose, galactose, and mannose, and slight color with fructose.

(10) Measure the absorbance at 630 nm in a spectrophotometer. Set the instrument at zero with the reagent blank and record the absorbance of the rest of the tubes relative to the blank.

SAMPLE CALCULATION

$$\text{Weight of the rat} = 200 \text{ g}$$

$$\text{Amount of glucose administered} = 1 \text{ g, } 1000 \text{ mg or } 1{,}000{,}000 \text{ } \mu g$$

$$A \text{ of glucose standard} = 0.070$$

$$A \text{ of unknown} = 0.035$$

$$C_u = \frac{A_u}{A_s} \times C_s \quad \text{where } C_s = 100 \text{ } \mu g/0.1 \text{ ml, the amount of glucose giving the observed absorbance for the standard}$$

$$\frac{0.035}{0.070} \times 100 = 50.0 \text{ } \mu g = \text{amount of glucose in the diluted 0.1 ml aliquot}$$

Since 0.1 ml contains 50.0 μg, then 5 ml contains 2500 μg of glucose. The total amount of glucose remaining is $100 \times 2500 = 250{,}000$ μg, since a 2 ml aliquot was removed from the original cleared wash.

$$\frac{A_u}{A_s} \times 5 \times 10^2 = \text{mg glucose remaining}$$

1000 mg glucose	—	250 mg	=	750 mg
(amount injected)		(amount remaining)		(amount absorbed)

Since the weight of the rat $= 200$ g, the rate of absorption of glucose $= \dfrac{750}{2} =$ 375 mg/100 g of rat per hour, the Cori Coefficient for glucose.

REFERENCES

1. Cori, C. F. *The Fate of Sugar in the Animal Body.* J. Biol. Chem. **66** 691 (1925).
2. Hultman, E. *Rapid Specific Method for the Determination of Aldosaccharides in Body Fluids.* Nature **183** 108 (1959).

CHAPTER 21

GLYCOLYSIS IN CELL FREE EXTRACTS; THE EMBDEN-MEYERHOF CYCLE

INTRODUCTION

Most biological systems utilize glucose or other monosaccharides as an energy source. Glycolysis refers to partial breakdown of these substances and is accompanied by some energy storage in the form of ATP. The end products formed depend on the substrate utilized, the organism or tissue involved, and whether an ample supply of oxygen is present.

Under anaerobic conditions certain micro-organisms, particularly yeast, produce ethanol and CO_2. Mammalian tissue, on the other hand, produces only lactic acid in the absence of oxygen. Anaerobic conditions can be created "artificially" in the mammalian tissue by inducing muscle fatigue. Under these conditions there is an oxygen deficit which leads to a reductive production of lactic acid.

Certain tissues, however, are distinguished metabolically by the fact that the glycolytic rate is very high even in the presence of sufficient oxygen. One way of identifying cancer cells is to measure the rate of aerobic formation of lactic acid. In cancer cells this rate is very high. This observation was made by Warburg many years ago, and has served as a basis for evolving many theories distinguishing cancer cell metabolism from normal cell metabolism. However, this unusual property is not confined to cancer cells alone; some normal tissues, among them brain, also exhibit a high glycolytic rate in the presence of oxygen. Nonetheless, most tissues do not exhibit high glycolytic rates in the presence of oxygen, but only do so when oxygen supply is limited. There are two ways of interpreting this effect: either the absence of oxygen induces or accelerates glycolysis, or that oxygen suppresses it. The suppression of glycolysis in the presence of oxygen is known as the Pasteur effect. It is one of the first cellular control mechanisms to undergo extensive investigations by many prominent biochemists. Actually, the aerobic suppression of glycolysis is not caused by oxygen itself. Any interference in the oxidative phase of carbohydrate metabolism, that is, inhibition at any point beyond the formation of pyruvic acid, will release the glycolytic phase from control by the respiratory phase. The reversal of the Pasteur effect is the enhancement of glycolysis in the presence of oxygen. This can be initiated by many reagents, such as CN^-, azides, amobarbitol, and other inhibitors of respiration. In this respect the Pasteur effect may be regarded as a compensatory mechanism for cellular inadequacy. If respiration, by which we mean the further degradation of these products to CO_2 and H_2O, is restricted in the cell, the major source of energy conservation, in the form of ATP production, is cut off. The cell attempts to compensate for this by accelerating the glycolytic synthesis of ATP. The overall reaction for glycolysis may be represented by the following equation

$$\text{Glucose} + 2\text{ADP} + 2\text{Pi} \rightarrow 2\text{ Lactic acid} + 2\text{ATP}$$

For each mole of glucose broken down to lactic acid, two moles of inorganic phosphate are taken up and two moles of ATP are produced, and there is a conservation of 15 Kcal of energy in the form of phosphate anhydride bonds. There are two enzymes which interfere with this system *in vitro*: (1) NAD^+ases destroy the coenzyme NAD^+, which is required for glycolysis, and (2) ATPases can hydrolyze ATP as it is formed, and a demonstration of inorganic phosphate uptake, which is used as a measure of phosphorylation, will not be possible. The NAD^+ases can be inhibited by nicotinamide. In liver mitochondria, Mg-ATPase is not very active in fresh preparations, and is absent in the soluble fraction. In brain tissue the Mg-Na-K dependent ATPases are partially inhibited by an excess of K^+ in the absence of Na^+. By controlling these factors the interfering effects of ATPases can be minimized.

PROCEDURE

In this experiment you will attempt to do several things. You will measure glucose disappearance by a spectrophotometric procedure using an enzyme as the analytical reagent. You will determine some of the requirements for glycolysis in cell-free systems from rat liver and brain. You will ascertain the stoichiometric relationships of the reaction; you will compare the relative differences in glycolytic activity in these tissues; you will attempt to learn something about the distribution of activity within a given tissue; and lastly, you will attempt to reverse the Pasteur effect.

Work in groups of four. Two members in each group, as in the ATPase experiment, will prepare the tissue and the other two will set up the experiment. This time there are only two tissues of interest, brain and liver, and the fractions desired for study will be mitochondria and the soluble fraction. There will be four groups of four students. Assignment of tissue to each group will be made by the instructor.

Group A: Glycolysis in liver mitochondria
Group B: Glycolysis in liver soluble fraction
Group C: Glycolysis in brain mitochondria
Group D: Glycolysis in brain soluble fraction

Each group of students will set up tubes as indicated in Tables 36, 37, and 38 prior to making the tissue extracts. After the tubes are set up and ready for use, the tissue preparation may proceed.

As in the ATPase experiment, the subcellular fractions will be isolated from a 10 per cent w/v homogenate in 1.1% KCl and 0.0001 M KH_2PO_4, pH 7.2. This time the washed brain mitochondrial pellet is uniformly suspended in a volume of 2 ml. That much is needed to complete the experiment.

PREPARATION OF TISSUE COMPONENTS

All operations are carried out at 4°C.

(a) Tare a 25 ml beaker containing 10 ml of cold KCl-KH_2PO_4 solution (KPS).

(b) Decapitate two adult rats and immediately remove the entire brain and liver. Place in a tared beaker and determine organ weight. Record the wet weight.

(c) Decant the KPS and wash the organ with fresh cold KPS. At this point, those students working with liver need to use only 2 g of tissue.

(d) Add approximately 5 ml of fresh KPS to the beaker and transfer the organ to a 20 ml capacity homogenizer previously set up in the cold room.

(e) Homogenize the tissue by five complete passes of the pestle at 1000 revolutions per minute.

(f) Transfer the homogenate quantitatively to a cold 25 ml graduated cylinder, and on the basis of wet tissue weight to volume of KPS make up a 10 per cent homogenate. For example, if the brain weighed 1.5 g, dilute the homogenate to 15 ml with KPS.

(g) Mix this homogenate in the graduated cylinder; transfer to polyethylene centrifuge tubes, balance them and begin centrifugation. The separation scheme is:

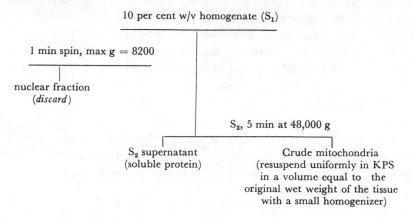

GLYCOLYSIS REACTION

(1) Follow the protocol very carefully and be sure to omit reagents where indicated in Table 40. Do not add tissue, ATP, NAD^+, or $KHCO_3$ until you are ready to start the experiment. You will need at least 1.7 ml of enzyme.

The instructions for Table 41 should be clear. Add only the first three reagents (A, B, and C) as indicated in Table 42.

Table 40

| | | \multicolumn{6}{c}{Flask Number (25 ml flasks)} | | | | | |
	Reagent	1 *−Substrate*	2 *−ATP*	3 *Complete*	4 *+Amytal*	5 *Blank*	6 *Zero Time*
A.	0.05 M glucose in 0.0250 M phosphate buffer, pH 7.2, ml		0.2	0.2	0.2		0.2
B.	0.08 M $MgCl_2$, ml	0.1	0.1	0.1	0.1	0.1	0.1
C.	0.005 M NAD^+ in 0.25 M nicotinamide, ml	0.1	0.1	0.1	0.1	0.1	0.1
D.	0.01 M ATP, ml	0.1		0.1	0.1	0.1	0.1
E.	0.01 M amobarbitol, pH 8.0, ml				0.1		
F.	0.25 M $KHCO_3$, ml	0.1	0.1	0.1	0.1	0.1	0.1
G.	H_2O, ml	0.3	0.2	0.1		0.6	0.1
H.	tissue, ml	0.3	0.3	0.3	0.3		0.3

Table 41

	Reagents	Tube Number 1	2
A.	0.9 per cent NaCl, ml	1.4	1.35
B.	5.0 per cent Na deoxycholate, ml	0.1	0.1
C.	tissue, ml		0.05
D.	biuret reagent, ml	1.5	1.5

(2) Number two sets of heavy-walled conical centrifuge tubes 1 through 6. To one set add 0.5 ml of 2 per cent $ZnSO_4$. Reserve the second set for the TCA precipitation of protein (see Table 40).

(3) Now number six 25 ml flasks 1 through 6 and place them in position on the shaking platform of a Dubnoff shaker bath set at 37° C.

(4) Obtain 0.5 ml each of ATP, NAD^+ and $KHCO_3$, and add them to the appropriate flasks. Don't delay the experiment when you reach this point.

(5) Initiate the enzyme reaction in tubes 1 to 4 *only* by adding your particular tissue extract at 30 to 60 second intervals.

Gently swirl the flasks at 5 minute intervals. Note the time and allow the incubation in each tube to proceed for exactly 30 minutes.

(6) While the incubation is proceeding you may prepare the blank (Tube 5) and the zero time control (Tube 6) as follows:

Add 0.3 ml of the enzyme to flask six; mix the contents well, and immediately transfer a 0.5 ml aliquot to the #6 centrifuge tube containing $ZnSO_4$.

Table 42

	Reagents	Tube Number (Use 18 × 150 mm tubes)					
		1 −Substrate	2 −ATP	3 Complete	4 +Amytal	5 Reagent Blank	6 Zero Time
A.	0.1 M NaAc, ml	4.0	4.0	4.0	4.0	4.0	4.0
B.	0.1 M HAc buffer, pH 4.0, ml	4.0	4.0	4.0	4.0	4.0	4.0
C.	molydate reagent, ml	0.5	0.5	0.5	0.5	0.5	0.5
D.	TCA supernatant, ml	1.0	1.0	1.0	1.0	1.0	1.0
E.	1 per cent fresh ascorbic acid, ml	0.5	0.5	0.5	0.5	0.5	0.5
	A at 660 nm						
	μmoles Pi/ml of aliquot						

Have your partner immediately add 2.0 ml of 6.25 per cent TCA to the remaining incubation mixture in the flask.

Add 0.5 ml of 1.8 per cent $Ba(OH)_2$ to the Zn precipitated solution. Add 3.5 ml of H_2O to give a final volume of 5.0 ml.

Transfer the TCA precipitated solution in the flask to a clean centrifuge tube labeled #6.

Store these centrifuge tubes till the incubation period is over.

(7) Repeat this procedure for flask #5 *but omit the enzyme solution.*

(8) At the end of the 30 minute incubation period, treat flasks 1 to 4 as you did flask #6 in step 6. That is, transfer 0.5 ml aliquots to the appropriate $ZnSO_4$ tubes, immediately followed by addition of 2.0 ml of 6.25 per cent TCA to the flasks; then add $Ba(OH)_2$ and H_2O as above. Transfer the remaining TCA precipitated solution to centrifuge tubes.

(9) Spin all tubes for 3 minutes in the table model centrifuge. You are now ready to proceed with the analysis.

ANALYSIS

One member of the group is responsible for one of the four analyses.

1. Inorganic Phosphate Analysis. After centrifugation to sediment the precipitated protein, transfer 1.0 ml aliquots of the clear TCA-precipitated extract to the appropriate tubes designated in Table 42. Mix contents well, adding the ascorbic acid last, and read the absorbance at 660 nm after 20 minutes. Use the reagent blank to set the spectrophotometer at zero.

2. Glucose Analysis. The glucostat reagent mixture is a commercial preparation of two enzymes, glucose oxidase and peroxidase, and a chromogen used for the specific determination of glucose based on the following reaction sequence:

$$\text{Glucose} + O_2 + H_2O \xrightarrow[\text{oxidase}]{\text{glucose}} H_2O_2 + \text{gluconic acid}$$

$$H_2O_2 + \text{reduced chromogen} \xrightarrow{\text{peroxidase}} \text{oxidized chromogen}$$

Add a clear 1.0 ml aliquot of the Zn-Ba supernatant to 9.0 ml of glucostat reagent. After exactly 10 minutes at room temperature, add one drop of 4 M HCl to stop the reaction and stabilize the color. Let all tubes stand another five minutes; read at 400 mμ after setting the reagent blank (tube 5) at zero absorbance.

3. Biuret Determination. Analysis of protein concentration is carried out simply by adding 0.05 ml of your tissue extract to tube 2 of Table 41. Set the tubes aside for 30 minutes, and read at 540 nm, using tube 1 as the blank. The protein concentration in mg/ml of extract is determined from the graph obtained in Experiment 4.

4. Lactic Acid Analysis. A semi-quantitative estimation of the amount of lactic acid formed can be made by the following simple procedure: When lactic acid is heated in concentrated sulfuric acid, acetaldehyde is formed. Acetaldehyde condenses with p-hydroxydiphenyl to form a colored complex that may be used to estimate lactic acid.

Transfer 1.0 ml of each of the TCA supernatants to glass conical centrifuge tubes. Add 3 ml of H_2O, 1 ml of 20 per cent $CuSO_4$ and a spoonful of powdered $Ca(OH)_2$.

Stir the mixture well with a clean stirring rod. Centrifuge, and transfer a clear aliquot of the supernatant to 20×150 mm test tubes.

Slowly add 6 ml of concentrated H_2SO_4 to each tube; use a propipette. Mix the contents of each tube well. Place the tubes in a boiling water bath for 10 minutes. Cool quickly to room temperature and add 3 drops of p-hydroxydiphenyl reagent to each tube. Mix well with the same stirring rods. Let the tubes stand for 30 minutes at room temperature. Mix at 10 minute intervals. A violet color indicates the presence of lactic acid. Record the color intensity as $-$, $+$ or $++$.

CALCULATIONS

The disappearance of Pi and glucose is calculated as follows:

(1) At the start of the experiment ($t = 0$), 10 μmoles of glucose and 5 μmoles of Pi were contained in a volume of 1.0 ml of reaction mixture.

(2) At time $t = 30$ minutes, 0.5 ml of the mixture was diluted to 2.5 ml for the Pi analysis and to 5.0 ml for the glucose analysis. This means, particularly for the zero time control, that there should be 1 μmole/ml of Pi and glucose in each of these mixtures, provided no metabolic action has taken place. This is our point of reference. The absorbance measured in this tube is directly proportional to the concentration of starting material. An absorbance value less than that of the zero time control is an indication of glucose or Pi utilization or disappearance.

(3) In order to properly assess the net amount of glucose and Pi disappearing, we must calculate *the total amount remaining*.

Table 43

Tissue	Micromoles of Glucose Disappearing in 60 Minutes at 37°/g Protein.
liver mitochondria	
liver supernatant	
brain mitochondria	
brain supernatant	

Subtract the absorbance obtained in tube 1 from those of 2 to 4, since the amount of glucose and Pi found in tube 1 is a control representing the amounts of the substances present in the tissue preparation and reagent at the beginning.

$$C_u = \frac{C_s \times A_u}{A_s}$$ where A_s = absorbance found with tube 6.

The total Pi remaining in tubes 2 to 4 $= \dfrac{1 \times A_u}{A_s} \times 2.5 \times 2$

The total glucose remaining in tubes 2 to 4 $= \dfrac{1 \times A_u}{A_s} \times 5 \times 2$

Why do you have to multiply by 5 and by 10?

(4) The amounts of glucose and Pi taken up are found by subtracting the μmoles of each found at the end from the μmoles of glucose and Pi at the start, which are 10 μmoles and 5 μmoles respectively.

(5) For every μmole of glucose metabolized, 2 μmoles of Pi should be taken up and

$$\frac{Pi}{glucose} = 2$$

What is your experimental ratio?

(6) In collaboration with your fellow students, complete Table 43 and compare fractions from both tissues by expressing glycolytic activity as the rate of disappearance of glucose in 60 minutes at 37° per gram of tissue for tube 4 only.

REFERENCES

1. *Laboratory Manual of Physiological Chemistry*. Department of Physiological Chemistry, The Johns Hopkins University, Baltimore 1957.
2. Lowry, O. H., and Lopez, J. A. *The Determination of Inorganic Phosphates in the Presence of Labile Phosphate Esters*. J. Biol. Chem. **162** 421 (1946).
3. Keston, A. S. *Specific Colorimetric Enzymatic Reagents for Glucose*. Abstract of Papers, 29th Meeting of the ACS, p. 31C (1956).
4. Seifer, A., and Gerstenfeld, S. *The Photometric Microdetermination of Blood Glucose with Glucose Oxidase*. J. Lab. Chem. Med. **51** 448 (1958).
5. Barker, S. B., and Summerson, W. H. *The Colorimetric Determination of Lactic Acid in Biological Material*. J. Biol. Chem. **138** 535 (1941).
6. Umbreit, W. W., Burris, R., and Stauffer, J. F. *Manometric Techniques*. Burgess Publishing Co., Minneapolis, 1963.

CHAPTER 22

THE ISOLATION OF LIPIDS AND THE ISOTOPIC DETERMINATION OF LIPID TURNOVER IN BRAIN

INTRODUCTION

Since lipids are found in all living tissue as complex and heterogenous mixtures, the isolation, purification and identification of lipids becomes both fascinating and challenging. Consequently, the objectives of this experiment are manifold and deliberately designed to introduce students to many useful techniques employed by the contemporary lipid biochemist.

The experiment is divided into four parts, each sequentially dependent on the first. They are also arranged in order of increasing complexity, requiring greater skills and more sophisticated equipment at each level of operation. This design also affords the advantage of selectivity. The experiment can be carried out either with or without isotopes, and it can be stopped at any point in the sequence. By taking the entire sequence the students will experience the use of isotopes and counting techniques; he will also learn various lipid isolation techniques, including classical solvent extraction methods (part 1) and further purification either by column (part 2), thin layer (part 3), or gas-liquid chromatography (part 4). As a general outline of the experiment, students will isolate the cholesterol, phospholipid, and glycolipid fractions from rat brain by solvent extraction methods. They can stop at this point, or proceed further and attempt to purify each of these fractions by any or all of the chromatographic methods indicated. In addition, if the animals have been injected with isotope prior to sacrifice, then determination of turnover numbers of the various lipid fractions become possible at any stopping point in the experiment.

THE USE OF ISOTOPES IN BIOCHEMISTRY RESEARCH

Isotopes of Biological Interest. The tracer technique, while relatively new in comparison with balanced metabolite methods, has yielded information which has literally revolutionized ideas about many metabolic processes. It is now known that living organisms are capable of synthesizing large and complex molecules from simple precursors, and of carrying out stepwise enzymatic degradations of various metabolites in well-ordered sequences. The availability of isotopically labeled compounds has provided a powerful tool for establishing the identity of intermediate compounds involved in many rather complicated sequences, such as the biological synthesis of fatty acids, pyrimidines, purines, and biopolymers, i.e., the polysaccharides, nucleic acids, and proteins.

Many isotopes, or more properly, radionuclides, spontaneously and continuously emit characteristic types of radiation. The isotopes of most common importance in biochemistry are deuterium, tritium, carbon 14, nitrogen 14, oxygen 18, phosphorus 32, and sulfur 35. Three of these, deuterium, oxygen 18, and nitrogen 14, are stable isotopes which are measured by mass spectrographic techniques. There are additional stable isotopes used less frequently in biological research. These are O^{19}, S^{34}, C^{13}, Fe^{54}, and Zn^{66}. Carbon 14, phosphorus 32, and hydrogen 3 are unstable radioactive isotopes which emit only beta particles, and are used mostly in biochemical research.

The term "isotope" means the species of atoms having the same atomic number but a different mass number. The atomic number refers to the number of protons in the nucleus of the atom, hence also to the number of orbital electrons. The mass number refers to the sum of the protons and neutrons. Thus an atom of hydrogen with a mass number of 1 must have no neutrons in its nucleus. This atom is represented as $_1H^1$. The H identifies the element as hydrogen. The subscript number refers to the atomic number and the superscript number refers to the mass number of this species of atom. The two isotopes of hydrogen have mass numbers of 2 and 3, and they are designated $_1H^2$ and $_1H^3$. The former is deuterium, a stable non-radioactive nuclide, and the latter is tritium, an unstable but very weak beta particle emitter; it disintegrates as follows:

$$_1H^3 \rightarrow {}_2He^3 + {}_{-1}\beta^0$$

This reaction represents a simple beta particle type of decay. The subscript -1 indicates the charge on the β particle (electron). This equation indicates that tritium decays to helium with a mass number of 3 and an atomic number of 2, a consequence of the disintegration of one neutron to an electron, emitted as a β particle, and a proton left behind in the nucleus.

The nuclide C^{14} also undergoes a simple β decay:

$$_6C^{14} \rightarrow {}_7N^{14} + {}_{-1}\beta^0$$

Other isotopes of carbon are $_6C^{12}$ and $_6C^{13}$. The natural abundances of these two isotopes are 98.9 per cent for C^{12}, and 1.11 per cent for C^{13}, thus giving an average atomic weight for natural carbon of 12.011.

Since our concern is mainly with beta emitters, the discussion will be limited to their properties. Beta particles are negatively charged electrons which are emitted with varying initial velocities from the nucleus of an atom. The average velocity of emission is less than 1/10 the velocity of light. When beta radiation is given off, the particles exhibit a spectrum of energy resembling an unsymmetrical Gaussian distribution. By contrast, gamma radiation is essentially monoenergetic.

The energy of radiation is measured in electron volts. An electron volt is defined as the energy acquired by a particle with a unit charge as it travels through a field potential of 1 volt. It is equal to 1.6×10^{-12} *ergs;* hence, 1 mev $= 1.6 \times 10^{-6}$ ergs. An erg is a unit of energy which may be expressed as 1 g cm^2/sec^2, that is, as a force times a distance. The *dyne* is the unit of force equal to 1 g cm/sec^2, that is, the force which gives a mass of 1 gram an acceleration of 1 cm/sec^2. The greater the energy of radiation, the greater is its penetrating power. The penetration power of beta particles is slight as compared to gamma rays. Hard radiation has an energy greater than 1 mev; soft radiation is considered to be less than 0.2 mev, and medium radiation is assigned energies between these two extremes. With the single exception of phosphorous 32, which has a maximum energy of 1.71 mev and a mean of 0.76 mev, all of the radioactive isotopes of major biochemical interest are soft or weak beta emitters. Tritium is the weakest with a maximum energy of 0.018 mev and a mean of 0.0056 mev; the carbon 14 figures are 0.156 and 0.050 mev respectively.

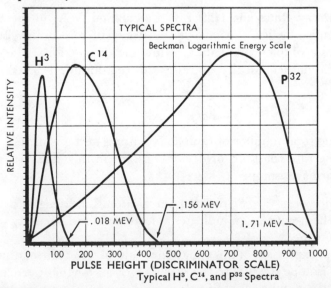

Typical H³, C¹⁴, and P³² Spectra

FIGURE 31. Beta emitter spectra. From "Liquid Scintillation Systems", Beckman Instruments, Inc., 1967, pages 1–5.

Table 44. Nuclides Commonly Used in Biological Research

Isotope	$t_{1/2}$	Maximum Decay Energy	
		β	γ
calcium 45	163 days	0.254	
carbon 14	5700 yrs	0.154	
chlorine 36	440,000 yrs	0.714	
cobalt 60	5.3 yrs	0.31	1.17
			1.33
hydrogen 3 (tritium)	12.3 yrs	0.0179	
iodine 131	8.1 days	0.250	0.080
		0.31	0.284
		0.608	0.364
			0.638
iron 55	2.9 yrs	K capture 0.231	
iron 59	45.1 days	0.29	0.19
manganese 54	314 days	1.0	0.84
phosphorus 32	14.3 days	1.718	
potassium 42	12.4 hrs	1.98	1.51
		3.58	
sodium 22	2.6 yrs	0.58	0.510
			1.28
sodium 24	15.06 hrs	1.390	1.38
			2.758
strontium 90	28 yrs	0.54	
sulfur 35	87.1 days	0.167	
zinc 65	245 days		0.201
			1.11

Isotope Decay Rates and Half Life Calculations. As in spectrophotometry and enzyme kinetics, the law of the decay of radioactive isotopes can be expressed as a simple exponential, or first order, reaction. The differential equation which mathematically expresses this relationship is given as:

(1)
$$-\frac{dN}{dt} = \lambda N$$

where $N =$ the total number of radioactive atoms at a given time

$\lambda =$ the rate constant, or decay constant in this case. It is different for each isotope.

$-\dfrac{dN}{dt} =$ the count rate or the number of atoms decaying in a given time. The negative sign indicates a decrease in radioactivity with respect to time.

The verbal expression of this equation would be: the rate of disappearance of radioactive isotope is equal to the decay constant for that particular isotope multiplied by the number of atoms present at any given point in time.

By rearranging the equation, we obtain

(2)
$$\frac{dN}{N} = -\lambda \, dt$$

Integrating between the limits of N_0, the number of radioactive atoms at time $t_0 = 0$, and N atoms at time t we have

(3)
$$\int_{N_0}^{N} \frac{dN}{N} = -\lambda \int_{t_0}^{t} dt$$

(4)
$$\ln \frac{N}{N_0} = -\lambda t.$$

Eliminating the negative sign,

(5)
$$\ln \frac{N_0}{N} = \lambda t$$

and

(6)
$$2.303 \log \frac{N_0}{N} = \lambda t.$$

The exponential form is

(7)
$$N = N_0 e^{-\lambda t}.$$

The half life $(t_{1/2})$ of a nuclide is the time required for half of the original number of atoms to decay. By substituting $t_{1/2}$ for t in Equation 6, which means that $N_0 = 1$ and N at $t_{1/2} = 0.5$, then

(8)
$$2.303 \log \frac{1}{0.5} = \lambda t_{1/2}$$

$$2.303 \log 2 = \lambda t_{1/2}$$

$$(2.303)(0.301) = \lambda t_{1/2}$$

(9)
$$0.693 = \lambda t_{1/2}$$

and

$$\lambda = \frac{0.693}{t_{1/2}} \quad \text{or} \quad t_{1/2} = \frac{0.693}{\lambda}.$$

This form of the basic equation allows the calculation of either the half life of any isotope, knowing the decay instant, or the calculation of the decay constant from experimental data. Plotting the log of the decay rate $\frac{dN}{dt}$ versus time, using the straight line form of Equation 6,

$$\log N_0 - \log N = \frac{\lambda t}{2.303} ,$$

the slope is equal to $\frac{\lambda}{2.303}$, from which the decay constant is obtained.

The average life of a nuclide is called the mean life ($\bar{T}$), and is the reciprocal of the disintegration constant λ:

$$\bar{T} = \frac{1}{\lambda} .$$

Thus, from Equation 9,

$$\bar{T} = \frac{t_{1/2}}{0.693} = 1.443 t_{1/2}$$

C^{14} is an artificial isotope produced by nuclear bombardment of nitrogen with neutrons or other accelerated particles, as for example:

$$_7N^{14} + {}_0n^1 \rightarrow {}_6C^{14} + {}_1H^1$$

Carbon 14 is also produced continuously in the upper atmosphere by cosmic neutron bombardment of N^{14}. For this reason, all carbon compounds currently being biosynthesized in nature have sufficient C^{14} to yield 14 disintegrations per minute per gram of substance. By using this information and the rate equations given above, it is possible to calculate the age of biological samples. For instance, it can be shown by calculation that one out of every 8230 atoms of carbon initially present in a given sample decays per year. It can also be shown that a sample of material of organic origin now giving 3 disintegrations per minute per gram is 12,650 years old. This type of analysis is called the carbon-dating technique, and has been used very successfully in geological and anthropological studies.

Methods of Measurement. All radiations emanating from unstable isotopes are capable of ionizing gases as they pass through. This ionization must not be confused with ionization of electrolytes in solution. The ionization process results from bombardment by charged particles, and serves as the basic means of detecting radioactive decay. Gamma rays do not directly ionize gases but cause ionization by secondary beta particle production resulting from their absorption. In the course of the ionization of a gas, beta particles are slowed down by the collisions with the gaseous molecules. They also are scattered and bounced about in their collisions, and therefore can produce secondary collisions.

The Geiger tube, which was developed as the first and most successful radiation detection device, is simply a gas-filled chamber containing oppositely

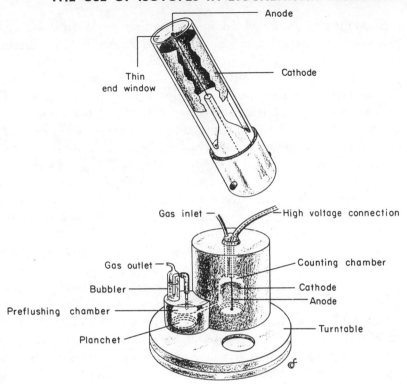

FIGURE 32. Radiation detectors. (From *Experimental Biochemistry*, John M. Clark, Jr., ed. W. H. Freeman and Company. Copyright © 1964.)

charged electrodes. When an ion pair is produced by collision with a beta particle, the ion pair will separate and drift toward the electrodes of opposite sign. Because of the difference in mass between electrons and their positively charged counterparts, the atomic nuclei, the two species will have different rates of travel to the electrodes. In fact, very few positively charged particles will have moved significantly by the time electrons have collected at the anode. In the Geiger tube the travel of electrons is further increased by application of high voltages to the electrodes. Accelerated electrons in the atmosphere within the tube produce secondary and tertiary ion pairs, giving rise to a cascade of electrons. Their arrival at the electrode will result in a potential drop, or pulse, which can be measured and is used to determine the amount of radiation being produced. This pulse is used to make a record of these ionization events. In fast counting tubes, a heavy polyatomic vapor such as EtOH or EtCOOH is included with the tube gas of argon or helium to prevent excessive cascading of electrons and tube burn-out. This is called quenching.

The measurement of soft beta emitters such as tritium and carbon 14 present special problems. They have very limited penetrating power, and even the end window of a Geiger tube can severely restrict counting efficiency. Some extremely thin windows have been developed to overcome this disadvantage. Another solution has been to utilize a windowless gas-flow detector that improves counting efficiency considerably. A more important development has been the use of liquid scintillation spectrometers to count soft beta emission.

These instruments offer many advantages, including very high counting efficiencies, 57 per cent for tritium and 92 per cent for carbon 14. In addition, they can count more than one isotope simultaneously.

Liquid scintillation counting depends on a different set of principles than is involved in Geiger measurement of isotope decay. In liquid scintillation detecting systems the radioactive material is dissolved in a solvent containing fluorescent substance, which, when exposed to radiation, produces a light burst lasting no more than a fraction of a second. When a beta particle is released in a solution containing a scintillator, the energy is largely absorbed by the solvent. If the solvent is one which facilitates the liquid scintillation process, such as toluene or p-xylene, the energy is transferred by molecular interactions to activate the scintillator. As activated scintillator molecules return to their ground state, light is emitted.

Because different scintillators produce different light emissions, the choice of scintillator is dependent upon the type of photomultiplier used in the detection unit. The most common types of scintillators used are standard mixtures of 2,5 diphenyl oxazole (PPO) and 1,4 bis 2,5 (phenyl oxazolyl) benzene (POPOP). The photomultiplier is the most critical part of the system. It consists of a quartz-faced tube containing a number of amplification stages or dynodes. Deposited in the quartz face or window of the photocathode is a light sensitive material. When light strikes the cathode, photoelectrons are liberated. They are accelerated to the dynodes where each bombardment causes a cascading effect. Amplification between each dynode is 3- to 6-fold, and a combination of dynodes can ultimately provide gains of 3^{10} to 6^{13} from a single photoelectron. This is transmitted to an analyzer and is counted as an event of radiation decay.

Nomenclature and Units of Measurement. The standard unit measure of radioactivity is the curie, abbreviated as c or C. It is historically defined as the amount of radioactivity equivalent to 1 g of radium, which has been measured as 3.7×10^{10} disintegrations per second or 2.22×10^{12} disintegrations per minute (DPM). A millicurie (mC) is therefore 2.22×10^{9} DPM, and a microcurie (μC) is 2.220×10^{6} DPM.

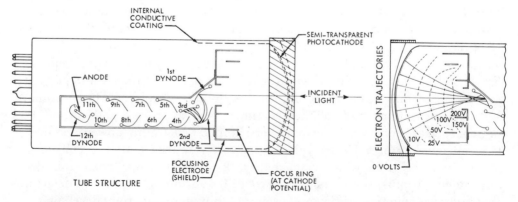

FIGURE 33. Beckman-RCA-Developed Bialkali 12-Stage Head-on Type of Photomultiplier Tube Liquid Scintillation System, Beckman Instruments, Inc., 1967, pages 2–3.

In biochemical experiments it is also important to know something about the distribution of the radioactive label in a compound. For instance, if we are studying the metabolic fate of glyceraldehyde, an intermediate of both carbohydrate and fat metabolism, it is important to know something about where the label is, that is, whether it is on carbons 1, 2, 3, or some combination of the positions. By using very selective means of synthesis it is possible to label such compounds in a variety of ways. When the label is on carbon 1 it is designated as $1C^{14}$ glyceraldehyde; when it is on carbon 2, it is designated as $2C^{14}$ glyceraldehyde, and similarly for every possible combination; and finally when it is on all carbon atoms of a given molecule it is designated as a UC^{14} compound.

When a compound such as glyceraldehyde is designated as being uniformly labeled, it does not necessarily follow that all of the carbon present is C^{14}. Some molecules could contain no C^{14}, others could contain C^{14} at one or more positions or at all three positions. The important point is that, on degradation analysis, the C^{14} distribution is equal in the three positions. It follows also that the more unlabeled molecules there are in a population of radioactive molecular species of the same compound, the lower will be the total radio-activity. The extent of radioactive labeling is expressed as the specific activity of that compound.

$$\text{S.A.} = \frac{\text{number of disintegrations per unit of time}}{\text{mg or moles of compound}}$$

The specific activity can also be expressed as curies per unit mass. High specific radioactivity indicates high radioactive content or purity. In this case, more and more of the atoms in the molecule are radioactive, ultimately achieving 100 per cent radioactivity as the extreme limit (all of the atoms under consideration are radioactive). In this case, there are no carrier or non-radioactive molecules present. (cf. Isotope Dilution Techniques.)

The common practice in biochemistry is to report results in units such as counts per minute (CPM) and specific activity as CPM/mg or mole of sample. The term CPM now refers to an empirical observation, that is, the actual or observed counts as measured by a particular detecting instrument under a specified set of conditions. This figure is subject to a number of corrections which can ultimately lead to a calculation of DPM. Presenting data in the form of DPM, rather than in CPM, is becoming an increasing practice in the literature.

There is another expression frequently encountered in isotope research, the atom per cent excess of isotope; that is, the degree or extent to which a compound is labeled, given in terms of specific activity. The degree of labeling of a compound with *stable* isotope, such as H^2 or O^{18}, is expressed in terms of atoms per cent excess. These terms represent the proportion of stable isotopes above that normally present in nature.

Sources of Error and Data Treatment

BACKGROUND. When assaying for radioactive isotopes certain corrections must be made on the raw data. In the first place, the counter tube itself is continuously being bombarded from without by cosmic rays and other ionizing radiations which are recorded by the counter. Other contributions toward background are made by electronic noise and radioactive containments in the detection or analytical units of the measuring equipment, in the glassware and other containers, and from prior isotope spills in the laboratory. The sum of these conditions is referred to as "background." When a sample is placed in the counter, the observed counts are therefore the sum of the background activity and the sample activity. In order to determine the activity of the sample alone it is thus necessary to subtract the background activity (counts per minute) from the observed counts per minute. Therefore, corrected counts/min. = observed counts/min. − background counts/min. This correction is best accomplished by counting a non-radioactive sample which most closely resembles the radioactive sample in point of origin and method of isolation, or to use a set of experimental conditions which omits the radioisotope.

SELF-ABSORPTION. When assaying for weak beta emitters such as C^{14}, a correction for "self-absorption" must be made. The necessity for this arises from the fact that the average energy of the electrons emitted by C^{14} is quite low, and some of the electrons arising from the lower layers of the sample will be absorbed by layers of material above them, very much as photons are absorbed by colored materials. Because of this self-absorption, if increasing amounts of a material of constant C^{14} concentration are placed on planchets and the total number of counts per minute are measured, a plot of CPM versus the mg of material added will show a curve of the form indicated by the lower line of Figure 34.

It can be seen from this curve that as the sample gets thicker and thicker a point is reached at which all of the radiation from the bottom layer will be absorbed. Now, regardless of the amount of sample added, the observed counts will not increase. In the region, where the sample becomes infinitely thin or weightless, i.e., <1.0 mg/cm$_2$ of surface area, the CPM is directly proportional to the sample weight. This region determines the shape and position of the theoretical curve. As the thickness increases in this region, the deviation from linearity is real, and increases markedly until a plateau is

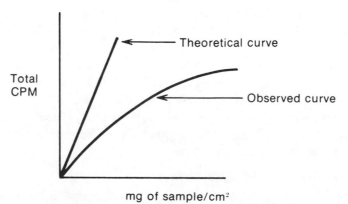

FIGURE 34. Effect of self-absorption of beta emissions.

reached. This is the region of infinite thickness which begins at >10 mg/cm². If there is sufficient sample, a good relative count can be made of samples having different specific activities, since in this region samples with different specific activities will reach the limiting values at counting rates which are directly proportional to the radioactivity of the different samples. The major limitations in using this method include the sample size required and the considerable reduction in observed counts resulting from excessive self absorption.

It can be seen that the total counts observed theoretically should be directly proportional to the amount of radio-active material added if there were no self-absorption. However, because of self-absorption this sample proportionality does not exist. Instead, the total number of counts observed becomes progressively less than the theoretical number as the sample size is increased, and eventually approaches a limiting value. In order to correct for this, a correction factor is used which converts the counts per minute of a sample of any weight to the counts per minute which that sample would have if one mg of it were assayed.

Another more useful method is to correct observed counts to infinitesimal thickness. If the curve is examined again it becomes obvious that, by comparing the actual counts obtained to the counts expected from the theoretical curve as the size of the sample increases, the discrepancy between the two curves becomes progressively greater. By extending the theoretical curve as far as is necessary, it is possible to plot the data in another way. By plotting the difference between the theoretical count and the observed count as a percentage of the theoretical figures, we achieve a correction factor for a specific type of sample. This curve will now allow correcting for self-absorption.

A third correction for self-absorption is achieved as follows. Increasing amounts of radioactive sample are added to planchets under experimental conditions and the counting rates are plotted. The thickness of sample applied should never exceed the average penetrating range of the beta particle or $\frac{1}{3}$ the maximum penetrating range; otherwise, only the most energetic particles

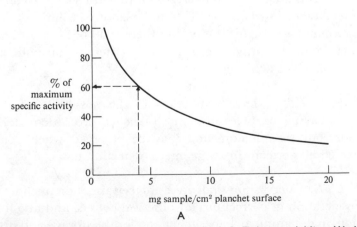

A

FIGURE 35A. From *Basic Biochemical Calculations*, J. S. Finlayson, Addison-Wesley Publishing Co., Reading, Mass., 1969, page 249.

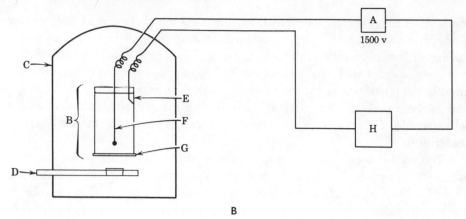

FIGURE 35B. Schematic diagram of a Geiger-Müller tube and scaler. *A* is the source of potential: *B,* the Geiger-Müller tube: *C,* lead shield; *D,* sample pan holder; *E,* cathode shield; *F,* anode wire; *G,* mica window (1.5 to 2.0 mg/cm²); *H,* scaler. (From R. Cowgill and A. Pardee, *Experiments in Biochemical Research Techniques,* John Wiley and Sons, New York, 1957.)

will pass through. For P^{32} the average penetrating range is 200 mg/cm², and for C^{14} and S^{35} it is 10 mg/cm². The plot of the log of the count rate per mg versus the weight or thickness of the sample should give a straight line. The linearity of this graph constitutes a good test of the reliability of the sample spreading technique. The graph may be conveniently drawn on semi-log paper. A parallel straight line drawn through log 100 at zero thickness gives a correction curve from which the per cent of theoretical counts at zero thickness can be extrapolated, corresponding to any required sample weight or thickness.

A fourth method is pragmatic: simply accurately weigh out samples weighing up to 2.00 mg, apply them uniformly thinly on wide planchets, and count them.

GEOMETRY. It is apparent from the design of a Geiger tube counter that only about 50 per cent of the total number of beta particles emitted by the sample will travel in a direction which will allow them to enter the "sensitive volume" of the counting tube, namely those that travel up. The remaining 50 per cent will travel down and away from the sensitive volume of the counting tube. The terminology used is to say that there is 50 per cent or 2π geometry for decay detection. By providing a means of detecting on both sides of the sample, 4π geometry is achieved.

BACK SCATTER. This term refers to the possibility of the upward reflection of decaying particles from the planchet. No adequate correction can be made, but the phenomenon may be observed when efficiencies in counting exceed 50 per cent in 2π geometry detection units.

HALF LIFE. Very short half lives for certain radionuclides is an important consideration in corrections for observed counts, and also for practical considerations in designing an experiment and for the storage and discarding of isotopes. As can be seen from Table 44, tritium has a half life of 12.3 years and its disintegration rate is roughly 0.5 per cent every thirty days. Phosphorus 32 has a half life of 14.3 days, so that every two weeks it will have 50 per cent less radioactivity; one microcurie, 2.22×10^6 DPM, will become essentially non-radioactive in six months, and one millicurie, 2.22×10^9 DPM, will vanish in a year.

RESOLVING TIME OR DEAD TIME. Each type of detection instrument has a certain resolving time. This simply means that the instrument cannot record or detect two decay events which occur in a time interval which is less than the resolving time of the instrument. As technology improves, the resolving time of instruments becomes less and less. The simple type of windowless, gas flow counter you may use in the laboratory may have a resolving time of 100 to 200 microseconds, and a sophisticated liquid scintillation spectrometer may have an overall resolving time of 1 to 2 microseconds.

$$N_0 = \frac{N}{1 - NT}$$

where T = resolving time in seconds

N = observed counts per second

N_0 = corrected or true counts.

The error involved is usually not more than 0.5 per cent if the resolving time is less than 300 microseconds. This correction, however, should be applied if high counting rates are encountered.

COINCIDENCE LOSSES. This is due to disintegrations occurring at the same instant. Since the detecting instrument records only one count, there will be a loss due to coincident decay. This loss will be significant at high counting rates.

EFFICIENCY. Only those particles which enter the sensitive volume will be counted; hence, geometry is an important consideration in efficiency calculations. In addition, some of the beta particles which enter the sensitive volume will have insufficient energy to "fire" the tube, and will not be counted. Dead time and coincidence losses will contribute to loss of counting efficiency. In addition, absorption by the window, if present, will also cut down detection rate. Thus only a fraction of the nuclei which disintegrate will be registered. However, corrections for these factors can be made and lumped together in a single constant which we will denote as K_e.

After correcting observed counts for background and self-absorption or quenching where required, the CPM obtained can be corrected for efficiency losses arising from the instrument itself. The final computation hinges on the notion as to whether the recorded count is equivalent to disintegrations on a one to one basis (100 per cent efficiency) or whether one observed count stands for something less. This is what is implied by the factor K_e and the term efficiency, which take these additional errors into account. By efficiency we mean, therefore, the fraction of the total disintegrations counted.

$$\text{per cent efficiency} = K_e = \frac{\text{number of disintegrations counted}}{\text{number of disintegrations}} \times 100$$

and therefore:

$$\text{DPM} = \frac{\text{CPM}}{K_e} \times 100$$

The value of K_e for most counters varies somewhat over a period of time. K_e actually indicates the efficiency of the counter, and with 2π geometry, efficiency has to be less than 50 per cent. *However, this value will be determined for you just before the experiment and given to you at that time.* K_e is determined by counting a standard sample with calibrated DPM. The instrument therefore is evaluated on its ability to count a percentage of the total known disintegrations occurring per minute in the standard sample.

QUENCHING. Quenching in liquid scintillation counting arises from two factors: from color quenching by colored materials within the sample which interfere with light transmission, and from chemical quenching by colorless materials which may inhibit the energy transfer in the scintillation process. Both of these can reduce counting efficiency. Quenching can be corrected for by double counting techniques, either by adding a precise amount of radioactive material to each sample after counting, or by the method of external standardization. The latter method is now the most common. In this case a radioactive source of fixed radioactivity is mechanically moved in the vicinity of the sample to a precise locality, either at its side or at the bottom, and the sample is counted before and after exposure to the external standard. Quenching in the sample can be determined by evaluating the amount of quenching of the external standard and applying the proper correction.

STATISTICAL TREATMENT. No counting measurement is complete until a statistical evaluation of the counts can be made.

$$\text{Error } (e) = K\sqrt{x + 1}$$

where $x =$ the value of the observation, or the variable, and $K =$ probability constant. When x is very large then $e = K\sqrt{x}$.

If we observe a count of 552 CPM, the *standard* error would be $\sqrt{552} = \pm 23.5$. This is the standard deviation, σ.

This means that we can say that this count is reliable to the extent that we can expect it to be 552 ± 23.5 CPM 68 per cent of the time upon repeated

Table 45

Probability Constant (K)	Error	Confidence Level
0.6745	probable error	50 per cent
1.000	standard error	68 per cent
1.6449	reliable error	90 per cent
1.96	95 per cent error	95 per cent
2.5758	99 per cent error	99 per cent

counting. This also means that if a series of measurements are made, 68 per cent of them should fall within $\pm \sigma$.

The *reliable* error is $1.6449\sqrt{552}$ and the confidence level is 90 per cent. It may also be noted that the larger the initially observed count, the greater our confidence will be, as a long count is equivalent to repeated measurement. By counting longer the statistical error can be reduced.

If a sample giving 100 CPM were counted until 100 counts were recorded, the standard deviation $= \sigma = \sqrt{100} = 10$. This is $\frac{1}{10}$ of the total count or 10 per cent. If the sample were counted until 10,000 counts were recorded,

$$\sigma = \sqrt{10,000} = 100.$$

This is now $\frac{1}{100}$ of the total count or 1 per cent. The statistical error has been reduced to $\frac{1}{10}$ of the error in counting only 100 counts, because a 10,000-count sample is equivalent to 100 samples of 100 counts each.

At the 95 per cent confidence level it should be expected that repeated measurements of this sample should give values within $\pm 2\sigma$ of the mean: $\pm 2 \times 100 = \pm 200$. Since in this case σ is 1 per cent of the total, any single count will be within ± 2 per cent of the mean in 95 per cent of the cases.

Since it is not practically feasible to count a given sample many times, there is a need to evaluate the reliability of a single measurement. In this case if 10,000 counts are recorded there is only a 5 per cent chance, or 1 in 20, that the measured count will be greater than 10,200 or less than 9800.

The Isotope Tracer Method for Determining the Turnover of Tissue Metabolites. One unique advantage of tracer methods is that they allow us to study the "turnover" or regeneration of tissue constituents in the intact animal while the animal is in a "dynamic steady state," a state in which the rates of synthesis and breakdown of tissue components are approximately equal. That is, they allow us to study the *rates* of synthesis and degradation of a given tissue constituent under consideration. For example, UC^{14} glucose can be incorporated into the tissue lipids and therefore may serve as a *precursor* of these tissue constituents. Although under certain experimental conditions there would be little or no net change in the quantity of tissue lipids if these were measured analytically, it will be quite apparent *from the experimental results using isotopes* that these lipids are nevertheless turning over at more or less rapid rates. Furthermore, it will be found that the rates of turnover will vary from tissue to tissue in a characteristic manner.

The incorporation of an isotopic precursor into a labeled product has, in most cases, been found to proceed as illustrated in the following graph:

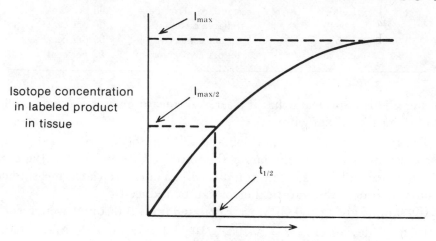

Isotope concentration
in labeled product
in tissue

Time of continuous administration of
isotope precursor

FIGURE 36

It can be seen that with prolonged administration of the precursor, the isotope concentration (I) in the labeled product approaches a limiting value, depicted on the graph as I_{max}.

When I reaches I_{max}, every molecule of the product which was present initially has been replaced by a new molecule. But for I to reach I_{max} takes infinite time, and hence is impractical to measure. However, it is possible to make a series of measurements of the isotope concentration at known time intervals and to calculate I_{max} from these. From this the biological "half life" (designated "$t_{1/2}$") of the tissue constituent can then be calculated. The "half life" is defined as the time required for the isotope concentration in the tissue constituent to reach $\frac{1}{2}I_{max}$, and corresponds to the time required for replacement of half of the molecules of the tissue constituent by new molecules derived from radioactive metabolic precursors.

There is another method for determining biological half lives which is more convenient and more frequently used. In this method an animal is injected with a single small dose of an isotopically labeled precursor and the isotope concentration in a labeled product is then followed.

When a radioactively labeled isotope enters a metabolizing system in tissue, it does several things: (a) It mixes with non-radioactive molecules of the same kind, giving it a certain specific activity. This molecule has become, in this sense, an equal part of the total metabolic pool of metabolites in that particular tissue. (b) At $t = 0$ after the isotope has been administered, the isotope concentration (or specific activity) of the metabolite in the pool increases rapidly to a maximum, and then declines. The decline is a result of

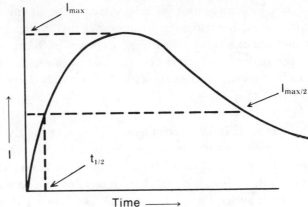

FIGURE 37. Determination of biological half life.

two processes, the metabolic alteration of the radioactive substance and its replacement by non-radioactive molecules in the tissue. (c) The curve of specific activity after I_{max} is reached most resembles the exponential first order decay of the isotope itself and can be expressed mathematically by the same form of a first order equation.

$$\log \frac{I_0}{I_x} = k(t)$$

where $I_0 = \text{CPM}$ at $t = 0$, and

$$I_x = \text{CPM at } t_x = x$$

$$t = t_x - t_0$$

The $t_{1/2}$ or biological half life can be taken from the plot as shown, or by mathematical calculation. From it λ_b, the biological rate constant, may be determined.

The biological rate constant, λ_b, is related to the biological half life, T_b, by

$$T_b = \frac{0.693}{\lambda_b}$$

The overall or effective half life (T_0) of any given radioactive metabolite is a function of both the biological half life, T_b, and the physical nuclide half life, T_a.

$$\frac{1}{T_0} = \frac{1}{T_a} + \frac{1}{T_b}$$

If the time span of the experiment is short as compared to T_a or T_b then this correction is not significant.

The biological half life has proven to be the most useful way of expressing the rate at which tissue constituents are metabolized and replaced. In general, the half life of a tissue constituent depends both on the nature of the compound and the tissue in which it occurs. For example, liver total protein has a *longer* half life than the liver fatty acids, but a much *shorter* half life than muscle total proteins.

Isotope Dilution Technique. This is an analytical method which makes measurement of very small samples possible. It can involve the determination of an inactive compound by dilution with a radioactive species of the same compound, or it can be used to determine the radioactive species by dilution with the inactive form. If a compound has a known CPM or DPM, for example, and is mixed with a precise weight of the same inactive compound, the specific activity of the compound will be diluted. The SA' of the diluted compound will be:

$$SA' = SA \times \frac{W^*}{W^* + W}$$

where W^* is the weight of the radioactive form, W is the weight of the non-radioactive form, and SA is the specific activity of the undiluted radioactive compound.

From this relationship either W or W^* can be calculated, provided the weight of one of the two species is known.

Methods of Handling and Discarding Isotopes. In all laboratory experiments involving the use of the soft and medium β emitters, only simple precautions need to be taken, essentially those similar to handling a poisonous or caustic substance. Laboratory benches, particularly in classroom work, may be covered by hard commercial wrapping paper, and students should wear disposable plastic gloves and aprons. Harder isotopes and gamma emitters require more careful handling, and all spills should be marked off and thoroughly cleaned up, followed by monitoring of the contaminated area with a survey meter to assure removal. Personnel should wear safety film badges and have periodic medical check-ups for excessive radiation exposure. Laboratories and storage areas should display conspicuous signs and labels indicating isotope usage.

Glassware contaminated with radioactive material should be rinsed several times in special containers set aside for that purpose, prior to putting them in with the pool of non-radioactive glassware for further washing. In cases where sample counting involves very low specific activities, it may be best to discard the containers, such as liquid scintillation vials and planchets, rather than wash them.

Solutions from pre-rinse treatment and other solutions containing carbon 14 and tritium may be discarded in a designated sink of running water. The water should be allowed to run for five or more minutes. This sink should also be connected to the main sewage effluent system. If it is at all convenient, all solutions including rinse solutions of the hard isotopes with short half lives, such as P^{32}, ought to be stored for six or more months before disposal in the discard sink.

Solid wastes may be handled in two ways, either by incineration or by burial. Radioactive contaminated animal carcasses should be incinerated. Such animals may be sealed in plastic bags and frozen for temporary storage prior to incineration. Other solid wastes such as glassware and incinerator ashes may be stored in sealed labeled containers and buried or be taken away for burial by commercial firms performing this service.

THE SOLUBILITY PROPERTIES OF LIPIDS

The isolation of lipid fractions from tissue is based on the differential solubility properties of the simple lipids and the complex lipids.

1. Simple Lipids

(a) *FATS AND OILS:* The glycerides of C_4 fatty acids are soluble in water. Glycerides of C_6 fatty acids and above are insoluble in water and soluble in ether, chloroform, petroleum ether, and carbon tetrachloride. All fats and oils are slightly soluble in cold methanol, ethanol, and acetone, and very soluble in hot solvent. After saponification of fats and oils, the fatty acids liberated can be classified as being either volatile or non-volatile. These fatty acids can also be water soluble (10 carbon atoms or less) or water insoluble (10 or more carbon atoms).

(b) *CHOLESTEROLS AND STEROLS:* Cholesterol is insoluble in water and soluble in ether, chloroform, and acetone; it is also soluble in hot methanol and ethanol. The sterols in general have similar solubility properties.

2. Complex Lipids

(a) *THE PHOSPHOLIPIDS:* Lecithins, plasmologens, cephalins, cardiolipids, and sphingolipids, such as sphingomyelin, are soluble in ether, chloroform, and cold ethanol, but insoluble in acetone, which is a major distinction between cholesterol and phospholipid.

(b) *THE GLYCOLIPIDS:* Cerebrosides, gangliosides, and sulpholipids are soluble in ether, chloroform, and hot methanol and ethanol, but only slightly soluble in ether and insoluble in acetone.

General Extraction Procedures for Lipids. Hot methanol or hot 95 per cent ethanol are excellent solvents for the total extraction of lipids, as are some combinations of solvents, e.g., 2:1 methanol–chloroform. A procedure using this mixture in a ratio of 0.8 g of tissue to 3 ml of solvent is extremely useful in semimicro-extractions of tissue components like mitochondria. In this extraction procedure, insoluble protein is removed by filtration or centrifugation, and a biphasic system is produced by adding an equal volume of 2 N KCl. The phases are separated by centrifugation. The lipids will be found in the lower phase. The lipid phase can be washed with water to remove traces of upper phase. This procedure may be used on any scale as long as the proportions of solvent to tissue are maintained. For samples rich in lipids, the amount of solvent should be increased. This procedure, however, is less attractive than one involving as a first step a cold extraction of tissue with 95 per cent ethanol followed by an ether extraction of the residues. In this manner, fats, steroids, and glycolipids are left in the tissue residue and the phospholipids are extracted into the ethanol. The phospholipids can be precipitated out by addition of acetone to the cold ethanol extract. Ether extraction of the residue removes the fats and sterols, and some of the glycolipids. The glycolipids can be further extracted by treating the residue with hot ethanol. There are several alternatives to this approach: the one outlined below is a variant.

EXPERIMENT 22

THE ISOLATION AND TURNOVER OF CHOLESTEROL, PHOSPHOLIPID, AND GLYCOPROTEIN OF BRAIN

General Instructions. If the isotope incorporation study is possible, rats should be injected with 5 μcuries of UC^{14} glucose, and the incorporation rate and as well as turnover can be calculated for each lipid fraction. If the isotope studies are not required, then the experiment can be treated as an exercise in lipid isolation, in which case the amount of tissue used for extraction can be scaled upward as desired. However the experiment is performed, one pair of students will be responsible for extracting one or more rat brains.

If the isotope incorporation studies are to be performed, seven 150 g rats are injected intraperitoneally with 0.5 ml of 0.055 M UC^{14} glucose containing 10 μC/ml. The animals are sacrificed by decapitation at 0, 15, 30, 60, 90, 120, and 180 to 240 minutes after injection. Some of these animals should be injected prior to the beginning of the laboratory period so that all students can start at about the same time. During this experiment all table tops will be covered with brown wrapping paper. Each student will wear plastic gloves while handling radioactive tissue or lipid fractions. All radioactive tissue should be carefully wrapped and disposed of by incineration. Small quantities of radioactive carbon 14 compounds can be discarded in the sink. At the end of the day, table tops and areas where there might have been spillage should be monitored with a Geiger-Müller survey meter for radioactive contamination.

Preliminary Extraction and Separation of Cholesterol Fraction from Phospholipids, Glycolipids and Sphingolipids. Both for the isotope labeling experiment and the non-isotope experiment, the same extraction procedures are to be followed for each of the four main experimental parts.

(a) The brain of a young adult rat will weigh between 1.0 and 1.5 g. Decapitate the rat assigned to you; remove and weigh the brain and place it in a heavy-walled glass conical centrifuge tube. Thoroughly mash the brain with a glass stirring rod. Exercise caution; do not break through the bottom of the tube. Extract the ground-up brain with 3 ml of acetone. Allow the mixture to stand for 10 minutes, but agitate with the stirring rod continuously. Centrifuge on a clinical model centrifuge at $\frac{3}{4}$ speed for 1 minute; then decant the acetone supernatant into 25 ml flask and stopper it. Re-extract the residue with 1.5 ml of acetone. Combine the acetone extracts, and *save the residue*. Store the acetone extract in the cold. This is fraction I; it consists principally of cholesterol, some neutral fat, and phospholipids. Cholesterol is approximately 3.6 per cent of brain weight.

(b) Allow the residue to air dry partially by breaking it up in the centrifuge tube. If necessary, warm in a steam bath for a few minutes to speed up the drying. Extract this dry residue with 2 ml of ether. Decant the ether into a 10 ml flask after centrifugation for 1 minute. Repeat and combine the ether extracts. Stopper the flask, and label it as fraction II. It contains the phospholipids, some glycolipids and cholesterol and fats. There is 5.7 per cent phospholipid and 2.6 per cent glycolipid in brain tissue.

(c) In two consecutive steps, extract the brain residue with 1 ml quantities of hot 95 per cent ethanol, centrifuge, and decant into a 10 ml flask. Label this as the glycolipid-sphingomyelin fraction III. It contains glycolipid and other lipids not previously extracted. Discard the extracted brain residue.

Separation of Phospholipids from Glycolipids and Sphingomyelins in Fraction II.

(a) To the combined ether extract (fraction II) add two volumes of acetone (up to 9 ml). Centrifuge the resulting precipitate of phospholipid and glycolipid. Add the supernatant to cholesterol fraction I. Wash the acetone precipitate three times with 1 ml portions of acetone and combine the washings with the cholesterol fraction I in cold storage.

(b) Add 1 ml of ether dropwise to the acetone extracted precipitate from part (a). Stir 3 to 5 minutes. Allow the precipitate to settle. Transfer the supernatant with a disposable pipette into a *previously tared* 10 ml flask. Repeat this step twice more. Combine the ether extracts. *Do not discard the residue.* Label as the cephalin-lecithin fraction (phospholipid fraction). Bring it to dryness very gently by flushing with N_2 under mild heat. Determine the yield on a wet tissue weight basis. As soon as this is completed, reconstitute this fraction in 0.5 to 1.0 ml of chloroform. Stopper and store in the refrigerator until ready for further fractionation if this is to take place (cf. silicic column or TLC experimental section). If further fractionation is not to be done and the sample is radioactive, transfer it quantitatively to a planchet; dry the sample and measure radioactivity after all the fractions have been isolated. If the liquid scintillation counting method is to be used, transfer the sample quantitatively with chloroform to a counting vessel, add 14.5 ml of a toluene cocktail containing scintillators, and proceed to count after all samples have been collected.

Proper plating technique is important. Place your planchets on a warm hot-plate. Drop a chloroform solution of lipid into the planchet one drop at a time. The hot-plate should be hot enough to evaporate the chloroform almost on contact but not hot enough to cause rounding up of the solvent or to cause sputtering and splattering of radioactive sample in every direction. Try a practice plating with unlabeled solvent; use this plate for background counting.

(c) Extract the residue from part (b) with 2 ml of hot ethanol. Centrifuge and collect the filtrate in a *tared* 10 ml flask. Repeat extraction with 1 ml amounts of hot ethanol and combine this extract with fraction III. Bring the combined fraction III, containing mostly glycolipid and sphingomyelin, to dryness and determine the yield. Reconstitute this fraction in 0.5 to 1 ml of chloroform. Stopper and store until ready for further fractionation (cf. silicic column experimental section, or proceed with measuring of radioactivity by either Geiger-Müller or liquid scintillation methods of analysis. Follow the instructions given above for plating samples.

FLOW DIAGRAM FOR LIPID EXTRACTION

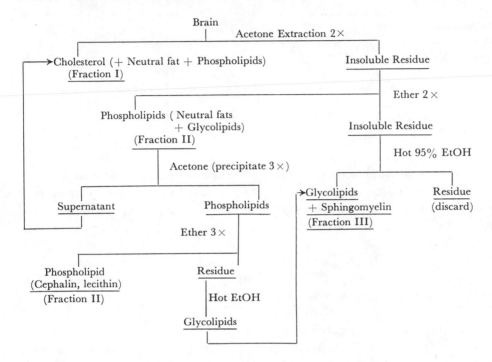

Purification of Cholesterol. Concentrate the combined acetone extracts of fraction I, containing the cholesterol, to about $\frac{1}{5}$ of its original volume on a steam bath. Set the flask in the cold overnight until the precipitation of cholesterol is complete.

After storage, collect the precipitate. Dissolve it in 1 to 2 ml of acetone. At this point not all of the precipitate will dissolve. This is probably some phospholipid which can be added to fraction II. Meanwhile, transfer the cholesterol, which did dissolve in the acetone, to a watch glass. Dry it and determine the yield. Save the cholesterol residue.

If the sample is radioactive and the biological half life is to be determined, dissolve in 0.5 ml amounts of acetone and transfer to a previously tared planchet and measure its radioactivity.

If liquid scintillation counting methods are to be used, there may be a quenching problem associated with acetone. Attempt to dissolve the cholesterol in the scintillator cocktail or in chloroform for transfer to scintillation vials.

Complete only Table 46 if no radioactive determinations are to be made. If radioactive measurements are to be made, complete Table 47 as well.

At this point there are two choices: (a) measurement of radioactivity in each fraction, or (b) further fractionation by column chromatography, TLC, or GLC.

Table 46

Tissue	Tare of Planchet or Vessel Alone	Tare Plus Sample	Weight of Sample	Per Cent Yield
whole brain				100
cholesterol fraction				
cephalin-lecithin fraction				
glycolipid– sphingomyelin fraction				

Measure of Radioactivity. Follow the instructions placed on the instruments. Turn the instrument on and set the voltage at the Geiger plateau lever indicated for your instrument. Never turn the counting gas pressure above the level marked on the instrument. Insert your samples in the changer. Gas the first sample at least one minute before counting. Reset the counter register and timer. Turn the stop count switch to count. Count the sample to 500 counts and record the time. If the counting rate is prohibitively low, count for 10 minutes and record your results.

Calculations. Calculations for columns 2, 3, 4, and 5 of Table 47 are obvious. The correction for self-absorption of lipids under conditions of assay used for planchet counting is:

$$\text{CPM/mg lipid} = \text{CPM corr for BG} = \left[\frac{\dfrac{1}{\text{mg sample on planchet}} + 0.1235}{1.1235} \right]$$

In this case, the correction is derived from an experimental curve described on page 252. However, correction for self-absorption does not apply when liquid scintillation counting is employed.

The value is entered in column 6. To convert this from CPM to DPM, you must correct for efficiency:

$$\text{DPM} = \frac{\text{corrected CPM/mg lipid}}{K_e} \times 100$$

where K_e is given as a percentage figure. This value is entered in column 7. In order to determine the total number of disintegrations per minute incorporated into the lipids of each fraction, multiply the value in column 7 by the value in column 1 and enter this new figure in column 8.

In order to determine the per cent isotope recovery in each fraction, the DPM of total lipid is divided by dosages of DPM administered. One microcurie

Table 47

Fraction	1 Total Lipid mg	2 mg Lipid Used for Sample Counting	3 Observed Sample CPM	4 Background CPM	5 CPM Corrected for Background	6 CPM Corrected for Self-Absorption, CPM per mg Lipid	7 Correction for Efficiency Expressed as DPM/mg Lipid	8 Total DPM in Recovered Lipid	9 Per Cent of Administered Isotope Recovered	10 Microatoms of Administered Carbon Incorporated
brain cholesterol										
brain phospholipid										
brain glyolipid and sphingo-myelin										

Per cent total counts recovered

(μC) is defined as 2.22×10^6 DPM; the dosage for each animal was 5 μC or $5 \times 2.22 \times 10^6 = 1.11 \times 10^7$ DPM.

Therefore:

$$\text{per cent radioactivity recovered} = \frac{\text{Total DPM per fraction}}{1.11 \times 10^7} \times 100$$

Enter this figure in column 9.

Determine the total DPM recovered by adding up the figures in column 7. Divide this by 1.11×10^7 to calculate the fraction recovered.

Calculate the microatoms of administered C^{14} incorporated into each tissue lipid fraction.

Since C^{14}-glucose was administered as a 0.055 M solution containing 1.11×10^7 DPM, it can also be expressed as containing 55 μmoles per ml (0.055 $\times$ 1000, the conversion factor from moles per liter to μmoles per ml). Consequently, the total number of μmoles of labeled carbon administered is

$$0.5 \times 55 = 27.5 \ \mu\text{moles}.$$

The total number of microatoms of labeled carbon is $27.5 \times 6 = 165$, since UC^{14}-glucose, containing six labeled carbons per molecule, was injected. Thus, the DPM per μatom of carbon administered is

$$\frac{1.11 \times 10^7}{165} = 6.73 \times 10^4.$$

Therefore, the microatoms of administered C^{14} incorporated into each lipid fraction is

$$\frac{\text{DPM in total lipid (column 8)}}{6.73 \times 10^4}.$$

Enter these figures in column 10.

In collaboration with your class, an attempt to determine the half life of each of these fractions will be made by plotting DPM per mg of lipid versus time for each fraction.

What fraction has the fastest turnover? What fraction had the greatest incorporation of lipid?

How do you account for the incorporation of glucose into lipids?

EXPANSION ON PRINCIPLES OF CHROMATOGRAPHY

General Comments. All chromatographic processes have two basic parts: a mobile phase and an immobile phase. The immobile phase may be either a liquid or solid. In column chromatography (CC) the stationary phase is contained in tubing. In thin layer chromatography (TLC), of which paper chromatography is a variant, the immobile phase is supported on a glass plate on resistant plastic backings, or impregnated fiber glass backings. For paper chromatography the immobile phase is self-supporting. In CC, a gas or a liquid percolates through a granular immobile phase, which can be either a solid with absorbant properties or a liquid supported on an inert solid. In TLC the mobile liquid phase moves by capillary action through the stationary phase. In CC the mixture of unknown substances is added to the moving phase as a gas or a liquid, whereas in TLC, the sample is spotted in the thin layer of absorbant. Consequently, there are four basic types of chromatography; gas-solid, gas-liquid, liquid-liquid, and liquid-solid.

Two additional generalizations can be made about chromatography. In the first place, all types of chromatography are based on interactions of phases on surfaces. In column chromatography the surface of absorbants are vastly increased with finer and finer particle sizes. Secondly, all chromatography involves the differential distribution or partitioning of one or more components between the phases, whether the active phases are gases, liquids, or solids. Partitioning can be based on solvation, adsorption, absorption, ion exchange, or the size exclusion properties of one of the phases.

Thin Layer and Column Chromatography. See other sections for discussions of these techniques. For TLC, cf. pages 94 and 142. For CC, cf. page 175 on ion exchange chromatography and page 184 on gel permeation chromatography. For absorption chromatography, consult pages 168 and 268.

Gas-Liquid Chromatography. Gas-liquid chromatography is a technique for separating volatile substances by percolating a gas stream over a stationary phase. If the stationary phase is a solid, this is referred to as gas-solid chromatography (GSC). The adsorptive and absorptive properties of the column packing separates vapors from one another. If the stationary phase is a liquid, this is referred to as gas-liquid chromatography (GLC). The liquid is spread as a thin film over an inert solid. Separation is based on partitioning of sample vapor in the liquid coating the large surface of the solid inert column packing particles.

In GLC a mixture of two or more compounds, or one believed to be pure, is vaporized into gases and separated according to the volatility and solubility of the compounds with respect to the stationary liquid phase. An inert carrier gas which does not dissolve or adsorb on the liquid-solid mixture, such as He, N_2, H_2, or CO_2, is used to sweep the samples through.

Interestingly, GLC is not yet 20 years old, but in that time it has dramatically altered separation techniques for the analysis of complex mixtures of gases, liquids, and solids which can be volatilized readily. It has the advantage of automatic detection of volatilized material as it emerges from the column, and it can be used to separate very minute amounts of sample, in the order of $\mu\mu$g quantities, by the most sensitive capillary columns. The main limitation is that samples must be volatilized without thermal decomposition at the temperature of column operation. For this reason its greatest use is to the organic and lipid biochemist.

THEORY. The interpretation of chromatograms is based on two characteristics of the operation of a chromatograph: (1) On the same column, at the same temperature and carrier gas flow rate, a given compound will always elute in the same length of time after injection, this *elution time* being characteristic of the compound. (2) The ratio of the area under any one peak in a chromatogram to the sum of the areas under all the peaks in that chromatogram is equal to the concentration in the injected mixture of the compound causing the peak, if the heat capacities of the various compounds are similar. That is, given an injected mixture of n compounds, the concentration of the ith compound is given by

$$C_i = \frac{A_i}{A_T},$$

where

$$A_T = \sum_{j=1}^{n} A_j.$$

The air peak is not included in the total.

There are a number of methods for obtaining these areas. In applications where great accuracy is required, an electronic integrator is attached to the detector of the chromatograph, and the peak areas are automatically recorded along with the peaks themselves. For less exacting work, the following procedure will suffice.

Allow the injected mixture to elute long enough to make certain that all compounds have eluted completely. Notice that the peaks are not exactly triangular, most of them having curves at the "corners." With a ruler, draw a base line for all peaks (they should all be on the same line, but may not be if the detector temperature varied during elution). Then draw lines along the sides of the peaks, projecting them to form triangles which approximate the peaks themselves. Under ideal conditions, the two small areas at the bottom of each peak that are not included in the triangle will exactly compensate for the area at the top of the triangle, which is not included in the peak. The formula for the area of a triangle, $A = \frac{1}{2}hb$, can now be applied, using the height and base of the triangle for h and b, respectively. Using the areas of the triangles as approximations to the peak areas, the concentrations of the compounds can now be determined.

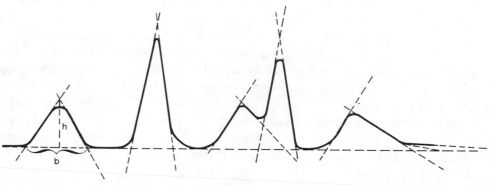

FIGURE 38

There are several complications which may arise. Two or more peaks may overlap because of closely spaced elution times. The peaks may be separated by lowering the gas flow rate in the column and injecting another aliquot of sample; if, however, this is impractical and there is enough of each peak showing to allow estimation of the slopes of the intersecting sides, the triangles may be drawn as usual and the area of each calculated as though the other were not there. If the heat capacities of the compounds in the mixture vary widely, a series of pure standards must be run to obtain a volume = area standard curve.

A common difficulty in chromatography is that of *tailing*, or distortion of a peak such that the leading edge of the peak is very steep while the back has a very gentle slope and is slightly curved. This will cause the area under the peak to be underestimated if the triangle method is used. The problem can be solved by (a) drawing the main triangle as usual, and then drawing smaller triangles under the tail to include more of the peak area; (b) by counting the graph-paper blocks under the peak; or, (c) cutting the shapes of the peaks in heavy paper or cardboard and weighing them to within 0.01 g.

Figure 38 illustrates each of the manual methods by which the peak areas can be measured. Gas chromatography is comparable to distillation. The distillation column can be considered as a stack of units or theoretical plates. In each plate equilibrium is established between ascending vapor and descending liquid; the larger the number of theoretical plates, the more efficient the distillation column. In GLC the column can also be treated as a collection of plates where each plate is defined as a section of the column in which there is equilibrium between the mobile and stationary phases. While the two plates are not strictly comparable, there has been established a rough equivalence between them: ten distillation plates equals 30 GLC plates. Thus, to achieve a separation power of 1000 distillation plates a gas chromatographic column must have up to 30,000 plates. This can be accomplished in GLC rather easily. Furthermore, a distillation column may require a long time to reach maximum efficiency, but a GLC column may separate in minutes or hours.

The HETP value, or height equivalent of a theoretical plate, is a measure of the performance of a chromatographic column. This value is determined from the time a material spends in a column, i.e., its passage or retention time through the column, and the extent of peak broadening during this passage. At a given temperature and for a given compound, the flow rate μ of the carrier gas determines the HETP value. This is defined mathematically by the Van Deemter equation:

$$\text{HETP} = A + \frac{B}{\mu} + C(\mu)$$

where A = the eddy diffusion (independent of flow rate)
$\quad B$ = the molecular diffusion (greatest at low flow rates)
$\quad C$ = the resistance to mass transfer in the column (increases as flow rate increases)
$\quad \mu$ = the linear gas velocity (flow rate)

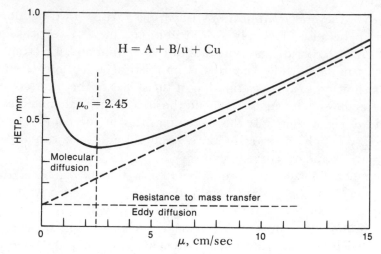

FIGURE 39. Height equivalent of theoretical plate (HETP) vs. linear gas velocity (μ). (M. L. Hair and A. M. Filbert, Gas Chromatography after Research/Development: Vol. 20, No. 12, p. 31, Dec. 1969.)

and

$$\mu = \frac{\text{column length (cm)}}{\text{retention time of air (sec)}} .$$

For any given substance, a minimum value for HETP occurs at a given carrier gas velocity, which is an indication that the column gives a maximum performance at this velocity.

If a given substance is swept through the column at too fast a rate, partitioning into the liquid phase is incomplete, equilibrium between the gas phase and the liquid phase is not established, and the sample will emerge as a broad diffuse band. If, on the other hand, the sample passes through too slowly, considerable molecular diffusion takes place in the column, and a broadening of bands will be seen again. For every substance, therefore, there is an optimum elution flow rate that must be established.

The contribution due to eddy diffusion is a property of the support medium holding the liquid phase. If it is porous, then the path of the gas vapors through the column is very tortuous, and A becomes very large. An efficient column should be filled with a support material which is not porous, and it should have a uniform structure and size. The ideal column consists of a single capillary tube about 300 feet long, coated with the liquid phase. The eddy diffusion in this case becomes negligible, and high efficiencies are obtained. Capillary columns may have as many as 500,000 theoretical plates as compared to a norm of 10,000 for ordinary columns.

The last factor which effects the elution peak pattern is in the nature of the packing material. If the liquid phase is supported on a substance which adsorbs the solute or sample vapor, then this activity causes the sample molecules to be retained in the column. Consequently, the elution peak becomes skewed. Polar molecules adsorb strongly as compared to non-polar molecules and the process of separation becomes less efficient for polar substances.

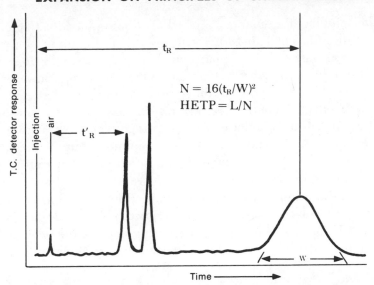

$$N = 16(t_R/W)^2$$
$$HETP = L/N$$

FIGURE 40. Chromatogram showing quantities for calculation of HETP and N. (M. L. Hair and A. M. Filbert, Gas Chromatography after Research/Development: Vol. 20, No. 12, p. 31, Dec. 1969.)

Skewing also prevents reliable quantitation since the amount of material passing through the column is proportional to the area under the peak.

In any GLC run, if a substance does not absorb at all it will come out in a (short) finite time after injection into the column. One such substance is air, and it is used as a convenient reference point since the detector will record its passage. The theoretical plates for a given substance can easily be measured from the chromatogram of elution peaks. Tangents are drawn to the peak at the point of inflection and down to the base line. The number of theoretical plates is given by

$$N = 16 \left(\frac{t_R}{w}\right)^2$$

where w is the length of the base line cut by the two tangents and t_R is the distance from the injection to the peak maximum. t_R = retention time of air = μ, the column flow rate, and $HETP = \dfrac{\text{length of column}}{N}$.

The lower the value for HETP, the more the number of theoretical plates for a given column length.

In summary, column efficiency is improved by the following factors. (1) By use of small particles of uniform size for the immobile phase. Supports which give the highest efficiency are the diatomaceous earths, with a mesh range of 100 to 120. Some success has also been achieved with glass particles etched by acid to increase the surface area. (2) By operating the column at optimum flow rates. This is found by plotting HETP against flow rate. (3) Using a high molecular weight carrier gas. However, where rapid analysis is desired, a low molecular weight carrier gas such as He, H_2, N_2, or CO_2 is preferred. (4) The liquid phase should consist of a low viscosity, low vapor pressure fluid with good solubility for the sample. It also should exhibit differential solubility for separation. (5) Low liquid phase loading of the solid support, as thin films of 1 to 10 per cent provide fast analyses, lower temperature operation, but also lower capacities. (6) Ideally, the pressure of the carrier gas should be the same at both ends of the column. However, packing does not allow this and a pressure differential does occur. Best efficiencies are obtained when the inlet/outlet pressure ratio is low. The pressure should be very well regulated at the inlet end by two-stage regulators, with a tolerance of 1 per cent. The outlet end will fluctuate even more, but the pressure should be 1 atmosphere. (7) Lowering column operating temperature improves resolution, but too low a temperature will increase the analysis time. It may also increase adsorption and decrease decomposition. If temperatures are too high, the more volatile components emerge too fast and bunch up, thus reducing resolution. (8) Finally, by decreasing column diameter to capillary size, the highest efficiencies are achieved because eddy diffusion is eliminated.

SILICIC ACID COLUMN CHROMATOGRAPHY OF GLYCOLIPIDS AND PHOSPHOLIPIDS

Beyond this point the experiment is designed to continue purification of lipid fractions obtained by the procedures on pages 257 to 259. Further separation of these fractions may be attempted by column, thin layer, or gas-liquid chromatography, or by any combination as is desired with the limitations of time and equipment. This part can also be integrated with the isotope studies.

Chromatographic Properties of Silicic Acid. Silicic acid is a versatile absorbant for lipids of widely varying composition. As contrasted to alumina and some basic absorbants, losses due to irreversible adsorption or adsorbant-catalyzed reactions are slight. The affinity of lipids for silicic acid increases as lipid polarity increases.

Groups contributing to polarity are double bonds, carboxyl ester, amino, hydroxyl, carboxyl and ionic groups (listed in order of increasing contribution). The least strongly absorbed compound will be washed off the column first, and the most strongly absorbed, last.

The general procedure for elution of the adsorbed lipids is to wash the column with gradients of increasingly polar solvent. The more polar the eluent, the faster are the absorbed compounds washed through. A gradient of petroleum ether containing increasing concentrations of ethyl ether is useful for eluting non-ionic lipids. Chloroform–methanol gradients are useful for eluting ionic lipids. Solvent flow rates of less than one ml per minute per cm^3 of column give optimum resolution, but limited laboratory time forces us to use more rapid flow rates. Solvents used should be reagent grade, anhydrous or redistilled. Greased joints and rubber tubing should be avoided.

For lipid separations, silicic acid is selected to be uniformly and finely granulated. The extreme fines are removed since they markedly decrease the flow rate by clogging the interstices. The silicic acid is heated at 105° to 115° to dehydrate it. Higher heat or heating in a vacuum removes more water and increases the affinity for all lipids. For this reason also, the sample and solvents must be made free of water.

Efficient separations are obtained with a weight ratio of absorbant to added compounds from 50:1 up to 1000:1.

Batch Adsorption and Elution of Sphingomyelin and Cerebrosides (Fraction III).

Pack a 1.0×50 cm column with 5 cm of dry activated silicic acid. Use ordinary glass tubing which has been stretched and constricted at one end. Plug with glass wool at the constricted end. The column is efficiently and uniformly packed by making a paper funnel containing absorbant, and allowing a small but continuous stream of absorbant to pour into the column. As this is being done your partner should apply an electric vibrator to the column. The column may be freed of air pockets and bubbles by clamping it off at the top and applying suction. Allow 5 to 10 ml of 3 per cent methanol in $CHCl_3$ to wash through under suction, but once solvent is added, do not let the column run dry. Clamp it off at the bottom and keep a head of solvent at the top until you are ready to apply the sample.

Make sure your sample is completely dissolved so as not to clog the column and apply it quantitatively to the column using several 0.5 ml $CHCl_3$ aliquots to complete the transfer. Progressively and gently pull the sample through the column by suction if necessary. Rinse down the column with 3 ml of 3 per cent v/v methanol in $CHCl_3$. Discard the eluate. Now pass through 5 ml of 3 per cent methanol in $CHCl_3$ and collect the elute in a tube; label as fraction 1. Add 3 ml of 10 per cent v/v methanol in $CHCl_3$. Collect the eluate; repeat 3 times and label these as fractions 2, 3, 4. Repeat with 3 ml of 50 per cent v/v methanol in $CHCl_3$ and label these as fractions 5, 6, 7. Take each sample to dryness under an air stream and moderate heat.

At this point there are two alternatives: (a) C^{14} counting, and (b) further fractionation by TLC or GLC.

For counting, dissolve the samples in 0.5 ml amounts of chloroform, and transfer quantitatively to tared planchets. The lipid weight is determined, and radioactivity is measured in a gas-flow windowless counter. If liquid scintillation counting is to be used, then add the lipid sample to 14.5 ml of a toluene scintillation cocktail and count. (See page 258.)

For TLC and GLC chromatography concentrate the sample and apply as needed for each technique.

Column Chromatographic Separation of Lecithin and Cephalin (Fraction II).

For this separation follow the directions for packing the column as given in the preceding section with two exceptions: (a) Pack a 1×10 cm column in a 50 cm tube, and (b) use chloroform instead of 3 per cent methanol in chloroform under vacuum as part of the procedure for getting rid of air pockets and wetting the column.

As soon as the column is ready, draw the sample (fraction II) in at the top of the column. Rinse the column with 3 ml of $CHCl_3$. Discard the eluate, and pass through successively 5 ml portions of 10, 20, 30, 40, and 50 per cent methanol in $CHCl_3$. Use the full amount of eluate to create and maintain a fluid head at the top of the column. Collect each fraction and label as 1, 2, 3, 4, and 5.

Follow instructions as in previous experiment either for counting or for further analysis by TLC or GLC.

TLC SEPARATION OF SOLVENT EXTRACTED CHOLESTEROL, PHOSPHOLIPID, AND GLYCOLIPID FRACTIONS OR SILICIC ACID COLUMN SUBFRACTIONS

TLC of Fraction III or the CC Subfraction of Fraction III. Precoated sheets 20×20 cm layered with silica Gel G (J. T. Baker) or Adsorbosil-5 (Applied Science Laboratories) are spotted with 0.1 per cent solutions of pure cerebrosides and sphingomyelin, or a standard TLC-2 mixture containing cholesterol, cerebrosides, sulphatides and sphingomyelin (Applied Science Laboratories), as well as with the combined fraction III. Apply spots with capillary tubes 0.5 cm in diameter.

If the fractions from the silicic acid column are to be further analyzed by TLC, then these can be spotted on plates along with 0.1 per cent pure solution or with standard known commercial mixtures.

Place the precoats in a TLC chamber containing a $\frac{1}{2}$ inch layer of chloroform, methanol, and water (v/v 65:25:4). Quickly reseal the chamber. Let the chromatogram run until the solvent rises within 5 to 10 cm of the top edge (30 to 60 minutes). Remove the precoats and let them air dry.

The lipids may be located by spraying with Rhodamine 6G, a universal lipid spray reagent. Immediately after spraying, examine the precoats under ultraviolet light while still wet and outline the spots with a sharp pencil.

An alternate means for locating lipids is to expose the precoats to iodine vapor. The iodine is soluble in the lipids and produces brown spots. These spots must be marked quickly with a sharp pencil since the iodine will evaporate.

TLC of Fraction II or the CC Subfractions of Fraction II. As in the previous experiment, spot a 0.1 per cent solution of lecithin, phosphatydyl serine or phosphatydyl ethanol amine with TLC mixtures 2 and 3 (Applied Science Lab) containing cholesterol, phosphatydyl ethanol amine, lecithin, and lysolecithin. There can be many choices. Place the precoated plates in the same solvent as in the preceding section and develop the chromatogram.

For detection of phospholipids, the technique is as follows: First spray the precoats with Rhodamine 6G and pencil in the spots under U.V. while still wet. Then allow the precoats to dry and spray with ninhydrin. Note the spots that give a characteristic purple-red color for amino acids. Finally, treat the precoats with Molybdate Blue spray reagent. Compounds containing phosphate ester immediately show up as blue spots.

TLC of Fraction I.   It may be useful to examine the cholesterol fraction for identification of impurities.

Finally, radioautographs of the thin layer plates can be made as a means of determining the distribution of radioactivity in the lipids spots.

GLC ANALYSIS OF FRACTIONS I, II, AND III AND SILICIC ACID COLUMN SUBFRACTIONS

General Description of a Gas Chromatograph.   A gas chromatograph unit consists of three main parts: a simple injection section, the separation column, and a detector with recorder. The sample injection part, or *septum*, is usually made of silicone rubber and has long injection life and high temperature stability to prevent bleeding. The separation column can be custom manufactured or homemade. It is usually made of an inert diatomaceous earth consisting of silicone dioxide. It is coated with silicones having the following basic structure:

$$
\begin{array}{ccccc}
& CH_3 & \left[CH_3 \right] & CH_3 \\
& | & | & | \\
CH_3-Si-O- & SiO & -Si-CH_3 \\
& | & | & | \\
& CH_3 & \left[CH_3 \right]_n & CH_3
\end{array}
$$

The quantity n can range from zero to many thousands. Silicone oils and gums contain mixtures of chains of various molecular weights. They can be cross-linked by using appropriate catalysts, but the resulting compound is a silicone rubber, not used as the liquid phase. The temperature stability and volatility of a silicone liquid depend on (a) impurities in the gum or oil, (b) the activity of the solid support, and (c) molecular length of the chain. Long chain methyl silicones can be produced which have negligible vapor pressures close to their thermal decomposition temperature (which approaches 400° C). Methyl groups along the chain can be replaced by other organic substituents. If all the methyl groups are replaced by phenyl groups, this is called 100 per cent phenyl substitution. A substitution with less than that is some per cent of the total. Some of the most common substitutions are vinyl, phenyl, and nitrile. Polarity or selectivity can be achieved by replacing methyl groups with other substitutions. As a general rule of thumb, polar liquid phases will retain polar substances and allow the non-polar substances to come off first, and non-polar columns will retain non-polar molecules, letting polar molecules come off first. Some of the more useful liquid phases are OV-I, dimethyl silicone, and GESE-30 (methyl silicone) for separating fatty acid methyl esters, cholesterol, trimethylsilyl derivatives, and estrogens and for general screening; OV-17 (50 per cent phenyl methyl silicone) for cholesterol and estrogens; and XE-60 for ketosteroids, some vitamins, pregnanes and estrogens. There are many more types of liquid phases to choose from. Commercial firms offer not only silicones, but also esters, ethers, alcohols, nitriles, amides, animides, halocarbons,

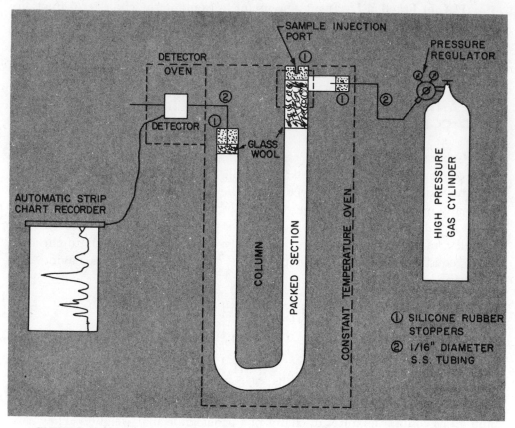

FIGURE 41. A gas chromatograph consists of three main parts: sample injection section, separation column, and a detector which monitors the gas mixture leaving the column. The detector indicates relative amounts of components and time they leave the column. S.S. = stainless steel. (From W. R. Supina and R. S. Henly, "Gas Chromatography Snooper Par Excellence." Chemistry: Vol. 37, No. 11, p. 13, Nov., 1964.)

hydrocarbons, polyglycols, polyhydric alcohols, and polyesters for use as liquid phases.

The columns themselves are made from glass or metal tubes which can be straight, U-shaped, or coiled. Diameters range from $\frac{1}{8}$ to $\frac{1}{4}$ of an inch, and lengths from 1 to 20 feet. The chromatographic columns are maintained at constant temperatures, depending on the volatility of the sample being analyzed. The detector oven is separate and is maintained at a temperature higher than the column oven to ensure that no condensation of sample components takes place in the detector. The vaporization chamber on the other side of the injection plug is considered as part of the column.

The detector is a thermal conductivity device consisting of a chamber containing coiled wire with a known resistance. This wire is connected to a Wheatstone bridge and to a recorder. When only carrier gas such as helium is passing through the detecting chamber, the effect is to cool the wire. The helium atoms pick up thermal energy and dissipates it along the walls of the detector. This cools the wire, changing its resistance, and a base current is recorded. When a sample molecule or air passes through, containing molecules heavier and larger than He, the thermal energy is not taken up so rapidly from the wire and it is not cooled as rapidly. Its resistance increases, and the change in resistance is recorded as a change in current in the recorder.

Making Packing Material. The support is slurried with an excess of solution containing a volatile solvent for the liquid phase. The excess solution is vacuum filtered off the support in a Büchner funnel. The wet support is dried either by spreading it out in a tray, drying in a vented oven, under infrared lights in a hood, or in a rotary vacuum evaporator.

Column Packing. In order to clean the column, attach an aspirator to one end of the column and draw 300 ml of chloroform and acetone through it. The column is flushed for a second time with acetone, and dried under vacuum by blocking the inlet end with the thumb. Do not draw air through the column. Alternatively, pass N_2 through to dry it.

Silane treatment of glass columns is recommended for analysis of pesticides, drugs, and steroids. A fresh solution of 10 per cent (v/v) of dimethyldichlorosilane in toluene is poured into the dry tube so that both legs are filled. After 10 minutes the solution is drained, and the column is rinsed with 300 ml of toluene under vacuum. The column is filled with dry methanol, drained, and rinsed again with methanol under vacuum till the pH of the effluent is neutral. Pack the column immediately.

Attach a small plastic funnel to the column with 1 to 2 inches of clean rubber tubing. Keep the column vertical during packing. Pour packing into the column through the funnel until it is filled to 3 inches from the bottom. Make a mark of the level on the outside of the column and attempt to achieve closer packing with the aid of a vibrator. Remove the etching tip. Hold the vibrator in the palm of the hand and curl the index finger around the tubing. Now add packing in 6 inch increments, followed by vibration. The process is repeated till the leg is filled. Pack the other leg in the same way. Do not pack the inlet leg in the vicinity of the heater zone. Place glass wool plugs in the inlet and outlet legs. They should be firmly installed but not so tight as to stop gas flow. Silanize the glass wool plugs as needed.

Operating the Apparatus. Follow instructions given in the manual for your particular instrument. Call on your instructor for guidance.

Be sure that before you heat up the instrument, you have initiated gas flow. If this precaution is not taken the detector thermocouple wire will oxidize.

Ideally the sample should be vaporized as quickly as possible, but the column should not be so hot that thermal decomposition of the sample occurs. At the same time a temperature of operation must be found which keeps the sample vaporized but is not hot enough to vaporize the liquid phase of the column. As a rule of thumb, the temperature of the injection port should be 50° above the boiling point of the highest boiler in the sample, but the column should be at a temperature 50° below the highest boiler. This is necessary because partitioning depends on the solubility of the sample in the liquid carrier. If the column temperature is too high then the sample vapor will not dissolve in the liquid since the solubility of a gas in a liquid is inversely proportional to temperature.

Sample Injection by the Solvent Flush Technique.   This technique ensures quantitative transfer of the entire sample

Thoroughly dry and clean the syringe. Wash the plunger with solvent by repeatedly pulling it through a piece of tissue paper. Do not touch the plunger at any time.

Fill the syringe with solvent to wet down the barrel and plunger. A rule of thumb is to use 1 μliter of solvent flush for a μliter of sample. For a 2 μliter flush, dip the needle in solvent and draw 1.2 μliters into the barrel. The needle volume must be known; assume it to be 0.8 ml. The needle is removed from the solvent and air is let in as shown in Figure 42.

Preparation of Fraction I, II, and III or Their Subfractions for GLC.   The direct chromatographic separation of glycerophospholipids, cerebrosides, and sphingolipids has not been very successully employed. Indirect methods are necessary. Usually a derivative is made of some portion of these molecules which can be analyzed by GLC. For example, after preliminary fractionation by TLC or CC or both, the phospholipids isolated are converted to diglyceride acetates or to dimethyl phosphatides. This is followed by fractionation by argentation TLC. Analysis of the fatty acids of the phospholipids is now possible by GLC. Sphingolipids which may also have many different fatty acids can be directly analyzed as trimethylsilyl derivatives. Cerebrosides are analyzed

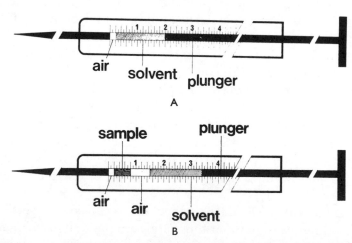

FIGURE 42A.   The needle is now immersed in the sample and the desired amount drawn into the syringe. In this example we draw up 0.6 microliter by moving the plunger six scale divisions outward. The needle is removed from the solution, and the plunger pulled back until the sample size can be read in the syringe.   The configuration for reading and injection then appears as below. (See reference 21 at the end of this chapter.)

FIGURE 42B.   The sample may now be injected. If air is undesirable during injection because of sample problems, minor changes which decrease the air pockets can be made with a little experimentation. (See reference 21 at the end of this chapter.)

first by separation in Silica Gel G thin layers. They are separated into two main spots in chloroform-methanol-water (65:25:4 v/v), the two groups are sub-fractionated on silicic acid columns into four fractions and followed by analysis of fatty acids by GLC.

A quicker method, which is more practical for student work, is to prepare derivatives of the fractions and subfractions of interest in the following manner. Fatty acids can be released from all complex lipids by chemical or enzymatic hydrolysis, and then analyzed by GLC as methyl esters. Methanolysis, however, does both simultaneously.

Anhydrous methanolysis is ideally suited for analysis of carbohydrates and for fatty acids of any lipid fraction. This process hydrolyzes lipid fractions to individual constituents and simultaneously converts all carbohydrates to methylglycosides or ketols, and the free carboxyl groups of fatty acids to methyl esters which can be analyzed by GLC.

PREPARATION OF METHYL ESTERS OF FATTY ACIDS FOR GLC ANALYSIS. Transfer your fractions containing 5 to 10 mg of lipid to a screw cap vial fitted with a Teflon liner in the cap. Evaporate to dryness if necessary with gentle heat in a stream of N_2. Add 0.5 ml of dry 0.5 N methanolic HCl per mg of lipid to extract the methyl esters of the fatty acid residues. *Do not discard the methanolic layer.* The hexane layer is saved for fatty acid analysis. These fractions are compared to standard mixtures of known fatty acid methyl esters for both qualitative and quantitative analysis. Standard mixtures of fatty acid methyl esters are available commercially. Since cerebrosides contain large quantities of C_{24} nervonic, cerebronic, and lignoceric acids, selection of standards with these fatty acids might be very useful for analysis of the glycolipids.

There are widely divergent specifications for the column to be used. One recommendation is: Gas-Chrom P, 100/120 mesh, using 10 per cent EGSS-X for the liquid phase, and a column made from 6 ft × 4 mm I.D. glass U-tube and maintained at 190° C.

THE PREPARATION OF TRIMETHYLSILYL DERIVATIVES (TMS) OF THE GLYCOLIPID FRACTION. After the hexane layer is drawn off for the fatty acid analysis, the methanolic layer is evaporated to dryness as previously with a stream of N_2. The TMS derivatives are prepared by addition of hexamethyl disilazane, trimethyl chlorosilane, and dry pyridine (v/v 3:1:9) in amounts of 0.5 ml per 5 mg of original lipid. After 15 minutes a sample of the reaction mixture is injected into a chromatographic column of 3 per cent GCSE-30 on 80/100 Supelcoport or on 1 per cent EGSS-X on 100/120 Gas-Chrom Q. The unknown can be compared to TMS derivatives of methyl α-glucopyranoside and methyl α-galactopyranoside.

REFERENCES

1. Kamen, M. D. *Isotopic Tracers in Biology.* (3rd ed.) Academic Press, New York, 1957.
2. Wang, C. H., and Willis, D. L. *Radiotracer Methodology in Biological Science.* Prentice-Hall, Englewood Cliffs, N.J., 1965.

3. Mann, W. B., and Garfunkel, S. B. *Radioactivity and Its Measurement.* D. Van Nostrand and Co., Princeton, N.J., 1966.
4. Birks, J. B. *The Theory and Practice of Scintillation Counting.* Pergamon Press, 1964.
5. Finlayson, J. S. *Basic Biochemical Calculations.* Addison-Wesley Publishing Co., Reading, Mass., 1969.
6. Montgomery, R., and Swenson, C. A. *Quantitative Problems in the Biochemical Sciences.* W. H. Freeman and Co., San Francisco, 1969.
7. Segal, I. H. *Biochemical Calculations.* John Wiley and Sons, New York, 1968.
8. Bligh, E. O., and Dyer, W. J. *A Rapid Method of Total Lipide Extraction and Purification.* Can. J. Biochem. Physiol. **37** 911 (1959).
9. Müldner, H. G., Wherett, J. R., and Cummings, J. N. *Some Applications of Thin-Layer Chromatography in the Study of Cerebral Lipids.* J. Neurochem. **9** 607 (1962).
10. Marini-Bettolo, G. B. (ed.) *Thin-Layer Chromatography. International Symposium on Thin-Layer Chromatography, Rome, 1963.* American Elsevier Publishing Co., New York, 1963.
11. Dittner, J. C., and Lester, R. L. *A Simple Specific Spray for the Detection of Phospholipids in Thin-Layer Chromatograms.* J. Lipid Res. **5** 126 (1964).
12. Arvidson, G. A. E. *Fractionation of Naturally Occurring Lecithins According to Degree of Unsaturation by Thin-Layer Chromatograph.* J. Lipid Res. **6** 574 (1965).
13. Purnell, H. *Gas Chromatography.* John Wiley and Sons, New York, 1962.
14. Marinetti, G. V. *Lipid Chromatographic Analysis.* (vol. 1) Marcel Dekker, Inc., New York, 1967.
15. Viswanathan, C. V. "Chromatographic Analysis of Molecular Species of Lipids," in *Chromatographic Reviews, Vol. II.* M. Lederer, ed. Elsevier Publishing Co., Amsterdam, 1969.
16. Burchfield, H. P., and Storrs, E. E. *Biochemical Applications of Gas Chromatography.* Academic Press, New York, 1962.
17. James, A. T. "Qualitative and Quantitative Determination of Fatty Acids by Gas-Liquid Chromatography," in *Methods of Biochemical Analysis.* D. Glick, ed. John Wiley and Sons, New York, 1960.
18. Sweeley, C. C., and Walker, B. *Determination of Carbohydrates in Glycolipids and Glycosides by Gas Chromatography.* Anal. Chem. **36** 1401 (1964).
19. Sweeley, C. C., Bentley, R., Makita, M., and Wells, A. W. *Gas-Liquid Chromatography of Trimethylsilyl Derivatives of Sugars and Related Substances.* J. Am. Chem. Soc. **85** 2497 (1963).
20. Folchi-Pi, J., et al. *Preparation of Lipid Extracts From Brain Tissue.* J. Biol. Chem. **191** 833 (1951).
21. Kruppa, R.F., and Henly, R.S. In *GAS-CHROM Newsletter* (Applied Science Laboratories, Inc.) *10*, No. 1 (1969).

CHAPTER 23

ENZYMES OF OXIDATION-REDUCTION REACTIONS IN MAMMALIAN TISSUE

OXIDATION-REDUCTION EQUATIONS AND FREE ENERGY CHANGE

The subject of biological oxidation has interested some of the greatest minds in chemistry and biology in the last 100 years. The enzymes and enzyme systems involved in biological oxidation deserve some special attention not only because of the relatively complex mechanisms and reaction sequences involved, but also because biological oxidations are so important in the energy economy of a cell. The large amounts of energy yielded in biological oxidations are in fact the main source of energy for work in respiring cells. Adenosine triphosphate (ATP) serves as the intermediate energy carrier between the energy-yielding biological oxidations and the energy-requiring systems. The experiments in the following section have been devised to illustrate some of the important aspects of biological oxidations.

Keep in mind the general pattern of mammalian oxidation and electron transport system as illustrated by Figure 43.

There are three ways in which oxidation takes place: by addition of O_2, by removal of hydride (H·), and by removal of electrons.

Actually, all oxidation-reduction reactions involve a transfer of electrons, from compounds having a low affinity to compounds having a high affinity for electrons.

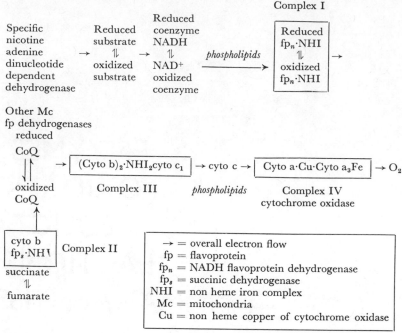

FIGURE 43.   (H. R. Mahler and E. H. Cordes, *Biological Chemisty*. Harper and Row, N.Y., 1966.)

Any O—R reaction can be written as equations representing two half reactions, one of which involves the reduction of oxidant, the other the oxidation of reductant.

For example:

(1) $Fe^{2+} \xrightleftharpoons{\text{oxidation}} Fe^{3+} + 1e$ (a loss of electrons)

(2) $H^+ + 1e \xrightleftharpoons{\text{reduction}} \frac{1}{2}H_2$ (a gain of electrons)

(3) $Fe^{2+} + H^+ \rightleftharpoons Fe^{3+} + \frac{1}{2}H_2$ (electron balance)

Then:

(4) $$K = \frac{[Fe^{3+}][H_2]^{1/2}}{[Fe^{2+}][H^+]}$$

We also know that the change in standard free energy, the maximum energy available for work of a reaction (ΔG^0), can be expressed as a function of the equilibrium constant K by the equation:

(5) $$\Delta G^0 = -RT \ln K,$$

where R = gas constant, and T = absolute temperature.

The conditions for determining the standard free energy ΔG^0 require that all reactants to be at unit concentration at a temperature of 25° in aqueous solution. The standard free energy of formation of a pure substance, ΔG_f^0, can be obtained from standard enthalpies of formation H^0 from the relationship $\Delta G_f^0 = \Delta H_f^0 - T \Delta S_f^0$. However, the values for substances in aqueous solutions at unit *activity* are more useful and can be converted from ΔG_f^0 to ΔG^0. In this equation, $T = 298°$; $R = 1.98$ cal/mole/degree Kelvin, and 2.3 = conversion of natural logs to the base 10; then:

(6)   the standard free energy, $\Delta G^0 = -1.98 \times 298 \times 2.3 \log K$

$$= -1360 \log K$$

(7)   and $\Delta G^0 = -1360 \log \dfrac{\text{Products } (P)}{\text{Reactants } (R)}$

If the concentration of any one reactant is changed, the free energy change, ΔG, can be calculated at any concentrations of reactants and products in the following equation, provided ΔG^0 is known.

(8) $$\Delta G = \Delta G^0 + 1360 \log \frac{(P)}{(R)}$$

and where $\dfrac{R}{P} = 1$ (or $R = P$), under standard conditions at unit concentration reactants, this equation reduces to $\Delta G = \Delta G^0$, which is the definition of the standard free energy. If, on the other hand, the equilibrium concentrations of reactants were substituted into this equation, then calculations would show that $\Delta G = 0$, since at equilibrium no work can be done. Of course, ΔG^0 also can be calculated from the equilibrium constant itself by substituting in Equation 4.

(9) $\Delta G = \Delta G^0 + 1360 \log \dfrac{[Fe^{+3}][H_2]^{1/2}}{[Fe^{+2}][H^+]}$.

This equation can also be written in a form for two half reactions:

(10) $\Delta G = \Delta G^0 + 1360 \log \dfrac{[Fe^{+3}]}{[Fe^{+2}]} + 1360 \log \dfrac{[H_2]^{1/2}}{[H^+]}$.

Since free energy is equal to the maximum energy available to do work in electrochemical cells, free energy change is related electrical work by:

(11) $$\Delta G = -n\mathscr{F}E$$

This is a work term involving a capacity factor $\mathscr{F}$ (the faraday) and an intensity factor E (volts), where n = the number of electrons involved. Under standard conditions, therefore,

(12) $$\Delta G^0 = -n\mathscr{F}E^0.$$

By substitution into (10),

$$(13) \quad -n\mathscr{F}E = -n\mathscr{F}E^0 + 1360 \log \frac{[Fe^{+3}]}{[Fe^{+2}]} + 1360 \log \frac{[H_2]^{1/2}}{[H^+]}$$

and:

$$(14) \quad E = E^0 + \frac{1360}{n\mathscr{F}} \log \frac{[Fe^{+3}]}{[Fe^{+2}]} + \frac{1360}{n\mathscr{F}} \log \frac{[H_2]^{1/2}}{[H^+]}$$

Under standard conditions, when H_2 = 1 atmosphere, H^+ activity = 1 M, and pH = 0 at 0° for the hydrogen electrode, the last term drops out and $E = E_h$. Since $\mathscr{F}$ = 23,000 cal/mole and $n = 1$, then for the half-cell:

$$(15) \quad E_h = E^0 + 0.06 \log \frac{\text{oxidant}}{\text{reductant}} \, .$$

When $n = 2$:

$$(16) \quad E = E^0 + 0.03 \log \frac{\text{oxidant}}{\text{reductant}} \, .$$

Since biological reactions are not run at H^+ = 1 (pH = 0) but at pH 7, and are usually two electron transfers, this equation is written as:

$$(17) \quad E_h = E_h' = E_0' + 0.03 \log \frac{\text{oxidizing species}}{\text{reducing species}}$$

This is a Henderson-Hasselbalch type of equation, and it is a straight line function:

$$y = ax + b$$

where $y = E_h'$, $x = \log \dfrac{O}{R}$ and $E_0' = b$, the y intercept, when $\dfrac{O}{R} = 1$ and $E_h' = E_0'$.

By plotting E_h' versus $\log \dfrac{O}{R}$ we get a plot as in Figure 44. The value for E_0', the

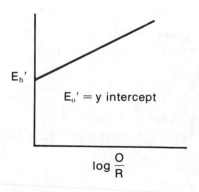

FIGURE 44. Determination of E_0'

standard change in electrical potential of a half-cell at pH_7, like the pK_a value, is a measure of strength. In the case of pK_a, it is a measure of acid strength; in the case of E_0' it is a measure of the ability of an $O-R$ system to create an "electron pressure."

The convention for biochemical potentials is that a flow of electrons proceeds from a lower potential $(-)$ to a higher potential, i.e., to less negative or to $(+)$. Therefore, in any table listing E_0' of $O-R$ systems, the reduced form of a chemical with a negative or less positive potential will reduce the oxidized form of any chemical with a more positive potential. The opposite sign convention is frequently used by chemists, and this can lead to confusion.

	E_0' (pH = 7.0)
NAD $\leftrightarrows$ NADH	-0.32
FMN $\leftrightarrows$ FMNH$_2$	-0.12
NADH will reduce FMN.	
Methylene Blue dye	$+0.01$
NADH and FMNH$_2$ both will reduce MB	

THE INTRACELLULAR LOCALIZATION OF DEHYDROGENASES AND RESPIRATORY ENZYMES

Principle. One purpose of this experiment and the next is to demonstrate the activity of dehydrogenases using methylene blue as the terminal electron acceptor. A second purpose is to localize respiratory enzymes in subcellular organelles which have been isolated from one another by differential centrifugation. This is an active field of biochemical and cytological research leading to information on subcellular function.

All components of an enzyme system, substrate, activator, and dye, will be placed in an reaction tube. At zero time the enzyme or tissue source is pipetted into the tube containing the other components, and the time for the formation of the white leuco-base of methylene blue (MB), or the color change of other dyes, is noted.

Certain dyes are colored in the oxidized form and colorless in the reduced (leuco) form. Some of these dyes can react readily with biological oxidation-reduction systems, and the reaction rate is followed by observing the rate of disappearance of color of the oxidized form either visually or with a spectrophotometer.

The dye MB has played an important part in establishing the properties and function of dehydrogenases. The pioneering work of Thunberg of Sweden in the period 1905–15 was mainly responsible for initiating many subsequent investigations.

MB$^+$ (*blue*)

MBH   (*leuco or colorless*)

The reduced or leuco form is spontaneously and very rapidly reoxidized by atmospheric oxygen. For this reason it mimics the respiratory chain in tissue very well, since oxygen is the terminal electron acceptor in respiring cells. But for this reason also, in order to observe dehydrogenase activity, oxygen has to be excluded from the system so as to prevent MB reoxidation. Thunberg designed a special tube for this purpose which allows evacuation of O_2 by aspiration. For expediency, however, it is more convenient to overlay mineral oil in the reaction tube after initiation of enzyme action.

Certain dehydrogenases catalyze the transfer of electrons from substrate either to NAD$^+$ or to NADP$^+$, e.g., malic, lactic, glutamic, pyruvic, α-keto-glutaric, and triose phosphate dehydrogenases transfer electrons specifically to NAD$^+$, whereas others, like isocitric dehydrogenase, transfer electrons specifically to NADP$^+$ from the appropriate substrates. Some dehydrogenases, like the flavoproteins, e.g., succinate and choline dehydrogenase, do not require a nicotinamide adenine phosphate coenzyme as the first acceptor in the respiratory chain, and the flavin coenzyme of these dehydrogenases act as the primary acceptors. However, whether the electrons are transferred to NAD$^+$, NADP$^+$ or to FP, the next step is a transfer to MB, as indicated in the equation:

$$\begin{array}{ccccc} & \text{NAD}^+ (\text{NADP}^+) & \text{FMNH}_2 & \text{MBH}+\text{H}^+ & \tfrac{1}{2}\text{O}_2 \\ \textit{substrate} \rightarrow & \Updownarrow & \to \quad \Updownarrow \quad \to & \Updownarrow \quad \to & \Updownarrow \\ \textit{dehydrogenase} & \text{NADH (NADPH)} & \text{FMN} & \text{MB} & \text{H}_2\text{O} \\ & & \uparrow & & \\ & & \textit{succinic} & & \\ & & \textit{dehydrogenase} & & \end{array}$$

EXPERIMENT 23

Part A. Preparation of Liver Mitochondria and a Soluble Enzyme Fraction

Work in pairs:

This experiment will take two laboratory periods. Make up a fresh preparation at the beginning of each laboratory period.

As in previous experiments, obtain one fresh liver from a decapitated rat. Make a 10 per cent homogenate in isotonic 1.1 per cent KCl after washing and mincing the liver. Be sure to resuspend the pellet uniformly by passing it through the homogenizer tube 3 to 4 times.

After a minute preliminary spin at maximum g of 8250, discard the pellet, and recentrifuge the supernatant for 5 minutes at maximum speed (48,000 × g). Save the supernatant and resuspend the mitochondrial pellet in 1.1 per cent KCl in a volume twice the original wet weight of the liver; i.e., from 5 grams of liver you should get 10 ml of resuspended liver mitochondria; you will need at least this much mitochondria in order to complete the experiment up to Section 7. Another preparation should be made for Section 8 during a second laboratory period.

I. THE LOCATION AND INHIBITION OF SUCCINIC DEHYDROGENASE

Test the two fractions for the presence of succinic dehydrogenase by following the protocol in Table 48.

Table 48

	Reagents	Tube Number			
		1	2	3	4
A.	0.1 M phosphate buffer, ml	0.5	0.5	0.5	0.5
B.	0.1 M Na succinate, pH 7.4, ml	0.5	0.5	0.5	0.5
C.	0.02 per cent methylene blue, ml	0.5	0.5	0.5	0.5
D.	0.2 M Na malonate, ml		0.2		0.2
E.	water, ml	0.2		0.2	
F.	liver mitochondria, ml			0.4	0.4
G.	liver supernatant, ml	0.4	0.4		

Add enzyme last; mix contents well; and immediately add 0.5 ml of mineral oil. Allow the tubes to stand without disturbance for at least 30

283

minutes. Record tubes showing decolorization and time of decolorization. Stopper the tubes with your finger and shake vigorously.

Does the dye color return? Is this proof that MB is an electron donor? Which fraction has the bulk if not all of the activity? What might be one flaw in interpreting your observations? Why would you expect malonate to inhibit this enzyme?

2. NAD⁺ DEPENDENT DEHYDROGENASES IN MITOCHONDRIA

Follow the procedure outlined in Table 49. Add the mitochondria fraction last; mix contents well and immediately layer with 0.5 ml mineral oil if MB is used. If the dye 2,6-dichlorophenolindolphenol (DICPIP) is used, addition of mineral oil is not necessary.

Meaningful results in this experiment are dependent upon the fact that, although liver contains much NAD^+ after homogenization, it is destroyed fairly rapidly by NADase in the liver, which hydrolyzes the molecule at the β glycosidic linkage to liberate nicotinamide and ribosyl-5'-ADP. By the time you perform this experiment, virtually no NAD^+ should be present. Addition of NAD^+ should, however, restore the action of the various dehydrogenases. Nicotinamide is added in large quantity to act as a competitive inhibitor of the enzyme which hydrolyzes NAD^+. (Why?)

Explain your results in tubes 4 and 5. Why does tube 5 appear to go even faster than 4?

3. THE COENZYME REQUIREMENT AND THE LOCATION OF LACTIC DEHYDROGENASE

L-lactic dehydrogenase is the terminal enzyme in the glycolytic phase of carbohydrate metabolism. Like all members of the glycolytic cycle in liver, it is found in the soluble fraction of liver. It is an NAD^+ dependent enzyme.

Table 49

	Reagents	Tube Number								
		1	2	3	4	5	6	7	8	9
A.	0.1 M phosphate buffer, pH 7.4, ml	0.5	0.5	0.5	0.5	0.5	0.5	0.5	0.5	0.5
B.	0.1 M substrates, pH 7.4, ml		glutamate 0.2	0.2	succinate 0.2	0.2	α-keto-glutarate 0.2	0.2	malonate 0.2	0.2
C.	0.02 per cent MB, ml; or 0.02 per cent DICPIP	0.5	0.5	0.5	0.5	0.5	0.5	0.5	0.5	0.5
D.	water, ml	0.2	0.2		0.2		0.2		0.2	
E.	0.01 M NAD^+ in 0.25 M nicotinamide, ml	0.2		0.2		0.2		0.2		0.2
F.	liver mitochondria, ml	0.5	0.5	0.5	0.5	0.5	0.5	0.5	0.5	0.5

Table 50

Reagent		Tube Number					
		1	2	3	4	5	6
A.	0.1 M phosphate buffer, ml	0.5	0.5	0.5	0.5	0.5	0.5
B.	0.1 M sodium lactate, pH 7.4, ml		0.2	0.2		0.2	0.2
C.	0.01 M NAD$^+$ in 0.25 M nicotinamide, ml	0.2		0.2	0.2		0.2
D.	0.02 per cent methylene blue, ml	0.5	0.5	0.5	0.5	0.5	0.5
E.	water, ml	0.2	0.2		0.2	0.2	
F.	liver mitochondria, ml	0.4	0.4	0.4			
G.	liver supernatant, ml				0.4	0.4	0.4

In certain other tissues, lactic dehydrogenase is found mainly in the particulate fraction. Yeast cells contain an L-lactic dehydrogenase, which is a hemoflavo-protein containing cytochrome b_2, and also a D-lactic dehydrogenase which is a separate enzyme. Follow the procedure outlined in Table 50. Add liver fractions last; mix contents well, and immediately layer with 0.5 ml of mineral oil.

Record the time it takes for each tube to decolorize. What do the results from tubes 1 and 4 tell you? What can you conclude?

4. COENZYME DEPENDENCE AND DISTRIBUTION OF ISOCITRIC DEHYDROGENASE

Isocitric dehydrogenase is found in most animal and plant cells in two forms, one specific for NADP$^+$ and the other for NAD$^+$. Each is located in a different area of the cell. The NAD$^+$ isocitric dehydrogenase of mitochondria is an enzyme exhibiting kinetics suggesting a regulatory role in the Krebs cycle. It is activated by ADP and inhibited by NADH and NADPH. The NADP$^+$ isocitric dehydrogenases do not exhibit these properties and they are located in the cytoplasm of the cell. Proceed as previously, but follow instructions in Table 51 in this case.

Have the facts been borne out by your observations in this experiment? What additional information is provided by tubes 1 and 4?

5. DETERMINATION OF THE DISTRIBUTION OF THE CYTOCHROME-CYTOCHROME OXIDASE SYSTEM

The principle of this test is as follows. P-phenylene diamine is a substance which is very slowly attacked by O_2. In the presence of a cytochrome system and O_2 it is very rapidly oxidized.

$$H_2N-\langle\ \rangle-NH_2 + 2 \text{ Cytochrome c} \xrightarrow{Fe^{+3}} HN=\langle\ \rangle=NH$$

reduced form (beige)
p-phenylene diamine (pPDH$_2$)

oxidized form (brown-black)
p-phenylene diimine (p-PD)

$$+ 2 \text{ Cytochrome c} + 2H^+$$
$$Fe^{+2}$$

Table 51

| | | Mitochondria | | | Supernatant | | |
		1	2	3	4	5	6
A.	0.1 M phosphate, pH 7.4, ml	0.5	0.5	0.5	0.5	0.5	0.5
B.	0.1 M isocitrate, pH 7.4, ml	0.2	0.2	0.2	0.2	0.2	0.2
C.	0.01 M NAD^+ in 0.25 M nicotinamide, ml			0.2	0.2		
D.*	0.01 M $NADP^+$ in 0.25 M nicotinamide, ml	0.2					0.2
E.	0.02 per cent methylene blue, ml	0.5	0.5	0.5	0.5	0.5	0.5
F.	water, ml		0.2			0.2	
G.	liver mitochondria, ml	0.4	0.4	0.4			
H.	liver supernatant, ml				0.4	0.4	0.4

* Do not waste the $NADP^+$. It is very expensive.

The diamine is a very reactive substance which spontaneously condenses and polymerizes to form a mixture of dark pigments.

In tissues containing the respiratory chain, p-PDH_2 reduces cytochrome c; reduced cytochrome c then transfers an electron to cytochrome a, in turn reducing it. Reduced cytochrome a transfers its electrons to oxygen as it picks up a proton from solution. This results in the formation of water, and thus terminates the O–R chain of electron transfers. Oxidized cytochrome c is regenerated and many react with another molecule of p-PDH_2 and so on, until sufficient p-PD has formed to cause an accumulation of black pigment.

The formation of such pigments from p-phenylene diamine by tissues and tissue extracts was discovered many years ago, and was an important link in the identification of the cytochrome system by the English investigator Keilin.

In order to determine the distribution of the cytochrome system, add 1.0 ml of 1 per cent fresh p-PDH_2 to each of two test tubes. To one add 1.0 ml of liver supernatant and to the other 1.0 ml of the mitochondria. *Do not add oil.* Shake occasionally to permit O_2 to diffuse into the solution. After 5 minutes note where a deep brown or black color appears. Do not discard the mitochondria.

6. DEMONSTRATION OF THE CYTOCHROME SYSTEM

You have already used p-phenylene diamine as an indicator for following the activity of the cytochrome system. In the following experiment you will use it again to determine the role of cytochrome c in electron transport and to establish the effect of cyanide on the cytochrome system. Review the principles underlying the p-phenylene diamine test and prepare tubes according to Table 52.

Table 52

Reagents	Tube Number					
	1	**2**	**3**	**4**	**5**	**6**
A. 0.1 M phosphate buffer, pH 7.4, ml	1.0	1.0	1.0	1.0	1.0	1.0
B. 1 per cent *p*-phenylene diamine, ml	0.5	0.5	0.5	0.5	0.5	
C. 4×10^{-5} M cytochrome *c**, ml		0.5		0.5	0.5	0.5
D. 0.01 M NaCN†, ml					0.5	
E. H_2O, ml	1.5	1.0	1.2	0.7	0.2	1.2
F. liver mitochondria, ml			0.3	0.3	0.3	0.3

* *Do not waste*—this is *very* expensive.

† Use a propipette. Do not pipette if you value the cytochrome system in your own tissues!

Add the liver fraction last and mix contents well. *Do not cover with mineral oil.* Shake occasionally to allow diffusion of oxygen into the solution. Observe appearance of brown pigment, and record changes after 10 to 15 minutes. Pay special attention to time difference between tubes 3 and 4. Interpret your findings on the basis of the explanation of action of *p*-phenylene diamine already given.

The requirement of cytochrome *c* for maximum activity in the liver preparation you are using derives from the following facts. All the cytochromes are closely bound to particulate elements of the cell. One of these, cytochrome *c*, is relatively loosely bound, and can be "leached out" of the mitochondria during tissue preparation. In fact, cytochrome *c* is the only cytochrome which has been isolated in very pure form to date; the others, being so closely bound to mitochondrial structure, have not been successfully purified by simple procedures. As a result of "leaching out" of cytochrome *c*, the liver fraction you are using contains such a low concentration of cytochrome *c* that this component is the "bottle-neck" or rate limiting step in the action of the cytochrome system in this liver preparation. By adding authentic cytochrome *c* to the liver fraction the maximal rate of *p*-phenylene diamine oxidation can be restored.

Note the low concentration of cyanide required to inhibit the system completely. Which cytochrome is affected by cyanide?

7. MEASUREMENT OF REDUCED PYRIDINE NUCLEOTIDE OXIDATION

The measurement of the oxidation of a substrate can be made directly as well as by the indirect method involving dyes. One of the most useful procedures for this purpose is based on the fact that some substrates, when oxidized, transfer electrons to the coenzymes NAD+ or NADP+ to form a reduced coenzyme counterpart NADH or NADPH. Unlike the oxidized form,

Table 53

Reagents		Tube Number	
		1	**2**
		Blank	*Experimental Tube*
A.	0.1 M phosphate, pH 7.4, ml	2.7	2.5
B.	0.01 M NAD$^+$ in 0.25 M nicotinamide, ml	0.2	0.2
C.	0.1 M glutamate, pH 7.4, ml		0.2
D.	mitochondria, ml	0.1	0.1

the reduced forms of the coenzymes have an absorption maximum at 340 nm. As a result we can follow the oxidation of substrate by the rate of formation of reduced coenzyme at 340 nm. At the same time it is possible to measure the rate of disappearance of reduced coenzyme as it transfers its electrons to the next member in the respiratory chain, i.e., the flavoproteins.

In this experiment you will attempt to demonstrate both possibilities. Since most spectrophotometers in the student laboratories are not adequate for measurement at 340 nm, the measurement may be performed at 370 nm instead.

Use a fresh preparation of liver mitochondria. Keep it stored in ice for subsequent experiments. Just prior to performing the experiment, take a 0.1 ml aliquot and dilute it to 1.0 ml with 0.1 M phosphate buffer, pH 7.4.

Follow the protocols outlined in Tables 53 and 54. Initiate the reaction by adding enzyme and mixing thoroughly. Follow the rate of reduction of NAD$^+$ at 370 nm in the spectrophotometer for 2 to 3 minutes and record the change in absorbance. Set the instrument at zero absorbance with the blank prior to each reading.

The reaction may be written as glutamate + NAD$^+$ ⇌ α-ketoglutamate + NADH + H$^+$.

Follow the instructions given for the previous experiment. In this reaction there is an oxidation of NADH by flavoprotein. Set the blank at 0.5 absorbance

Table 54

Reagents		Tube Number	
		1	**2**
		Blank	*Experimental Tube*
A.	0.1 M phosphate, pH 7.4, ml	2.8	2.6
B.	1 × 10^{-3} M NADH, ml		0.2
C.	mitochondria, ml	0.2	0.2

and measure the decrease of absorbance of the experimental tubes.

$$NADH + H^+ + FP \rightleftharpoons NAD^+ + FPH_2$$

The overall reaction in the presence of O_2 is $NADH + H^+ + \frac{1}{2}O_2 \rightleftharpoons H_2O + NAD^+$.

8. MEASUREMENT OF SUCCINIC DEHYDROGENASE ACTIVITY WITH TRIPHENYL TETRAZOLIUM CHLORIDE AND 2,6-DICHLOROPHENOL INDOLPHENOL

Unlike MB, certain dyes can be reduced under aerobic conditions, and the reduced form is stable in the presence of air or O_2. Two such dyes will be used in succeeding experiments: (a) 2,6-dichlorophenol indolphenol (DICPIP) ($E_0' = +0.217$ volts), which is blue when oxidized and colorless when reduced; and (b) triphenyl tetrazolium chloride (TTC) ($E_0' = +0.04$ volts), which is colorless and water soluble when oxidized, and forms a red, water insoluble formazone when reduced.

The substrate being studied in this experiment is succinic acid. The standard potential of succinate-fumarate (E_0') is $+0.03$ volts.

Various ratios of fumarate to succinate will be used in the presence of each dye to show that the extent of reduction of the dye at equilibrium depends on this ratio, as well as on the relative E_0' of the systems involved. Succinate and fumarate are present in many times the concentration of the dyes, so that the concentration of the succinate and fumarate will change very little during the reaction; consequently, the final ratio fumarate/succinate will be known. The amount of dye reduction will adjust itself to the O/R potential set up by the fumarate/succinate ratio, according to the equation:

$$E = E_0' + 0.03 \log \frac{[\text{oxid}]}{[\text{red}]}$$

This adjustment of the ratio of oxidized form to reduced form to the existing O/R potential occurs many times in the body. One common example is the ratio of pyruvate/lactate: when the O/R potential in the tissues decreases as a result of oxygen debt, the ratio pyruvate/lactate decreases as the production of lactate increases sharply.

After making a fresh mitochondria preparation from one rat liver, proceed as directed in Table 55.

Let all tubes stand for 30 minutes at 37° and shake from time to time. At the end of that time, add a few crystals of sodium dithionite to tubes 5, 6, and 7 only, in order to completely reduce the dye for a standard curve. To all tubes then add 10 ml of acetone, shake, and centrifuge at $\frac{3}{4}$ speed for 5 minutes using glass tubes. Pour the clear layer into a spectrophotometer tube and read at 420 mμ. Use 10 ml of acetone as a blank.

Table 55

Reagents	Tube Numbers						
	12 × 100 mm Experimental Tubes				Standard Curve		
	1	2	3	4	5	6	7
A. 0.1 M phosphate buffer, pH 7.4, ml	1.0	1.0	1.0	2.0	3.90	3.80	3.70
B. 0.2 M succinate, pH 7.4, ml	1.0	0.5	0.1				
C. 0.2 M fumarate, pH 7.4, ml		0.5	0.9				
D. liver mitochondria, ml	1.0	1.0	1.0	1.0			
E. 3×10^{-3} M TTC, ml	1.0	1.0	1.0	1.0	0.10	0.20	0.3

Determine the amounts of oxidized and reduced dye in each tube by extrapolating from your standard curve.

Explain what is happening in tubes 1 through 4.

Calculation of O/R Potentials.   The oxidation potential of the succinate-fumarate system may be approximated by a simple experiment. This system will react with any dye having a more positive potential than its own, e.g., $E'_{0 \text{ succ-fum}} = 0.03$, and for MB, TCC, and DICPIP, $E'_0 = +0.04$, $+0.05$, and $+0.217$, respectively.

The equation for DICPIP becomes

$$E = +0.217 + 0.03 \log \frac{[\text{DICPIP}]}{[\text{DICPIPH}_2]}$$

Under the same conditions of pH and temperature, the succinate-fumarate system can be represented by

$$E = E'_0 + 0.03 \log \frac{[\text{fum}]}{[\text{succ}]}$$

When both systems are allowed to react in the presence of the appropriate enzymes, an equilibrium will be established, at which time

$$E_{\text{DICPIP}} = E_{\text{succ}}$$

Therefore, $0.217 + 0.03 \log \dfrac{[\text{DICPIP}]}{[\text{DICPIPH}_2]} = E'_0 + 0.03 \log \dfrac{[\text{fum}]}{[\text{succ}]}$, and solving,

$$E'_{0 \text{ succ}} = 0.217 + 0.03 \log \frac{[\text{DICPIP}]}{[\text{DICPIPH}_2]} - 0.03 \log \frac{[\text{fum}]}{[\text{succ}]}$$

and

$$E_0' = 0.217 + 0.03 \log \frac{[\text{succ}][\text{DICPIP}]}{[\text{fum}][\text{DICPIPH}_2]}.$$

The amount of oxidized DICPIP and the ratio of DICPIP/DICPIPH$_2$ in experimental tubes containing different succ/fum ratios may be approximated by comparison with a series of tubes in which fumarate alone is added to maintain the DICPIP in the *oxidized* form and to which graded amounts of DICPIP have been added.

PROCEDURE: Once again in this experiment the amounts of succinate and fumarate used will be high in relation to the amount of dye present. Thus the amount of these two substances undergoing oxidation and reduction will be negligible. At equilibrium, therefore, the concentration of succinate and fumarate may be taken as the concentration originally added. One correction is necessary, because the enzyme preparation also contains fumarase, which converts fumarate to malate via addition of water. At equilibrium for this conversion, the ratio of fumarate/malate = 3:1. Therefore, a sufficiently high concentration of fumarate must be added (0.4 M) to compensate for this conversion.

Again set up two series of tubes as indicated in Tables 56 and 57. Add the mitochondria suspension last. Place these tubes in a 37° bath for 10 minutes in order to establish equilibrium.

Transfer the contents of the tubes to plastic centrifuge tubes and centrifuge for 5 minutes at 40,000 × g. Determine the standard curve for different concentrations of oxidized DICPIP using the contents of Tube 8 of Table 56 as the reference blank. Measure DICPIP concentration at 600 nm. Determine the concentration of oxidized DICPIP remaining in experimental tubes 1 to 6. By comparison to tube 6, the ratio of DICPIP/DICPIPH$_2$ can be determined, and the equation for E_0' can be solved.

Table 56

Comparison Tubes or Standard Curve for Oxidized DICPIP

Tube Number	DICPIP 1.7×10^{-4} M (ml)	Fumarate, pH 0.4 0.4 M (ml)	H$_2$O (ml)	Mitochondria (ml)
1	2.0	0.4	1.6	0.2
2	1.6	0.4	2.0	0.2
3	1.2	0.4	2.4	0.2
4	1.0	0.4	2.6	0.2
5	0.8	0.4	2.8	0.2
6	0.4	0.4	3.2	0.2
7	0.2	0.4	3.8	0.2
8	0.0	0.4	4.0	0.2

Table 57

Experimental Tubes

Tube Number	DICPIP 1.7×10^{-4} M (ml)	0.1 M succinate, pH 7.4 (ml)	0.4 M fumarate, pH 7.4 (ml)	Mitochondria (ml)
1	2.0	2.0	0.0	0.2
2	2.0	1.6	0.4	0.2
3	2.0	1.2	0.8	0.2
4	2.0	0.8	1.2	0.2
5	2.0	0.4	1.6	0.2
6	2.0	0.0	2.0	0.2

Calculate the values for E_0' obtained under each set of conditions for succinate/fumarate.

Average the values for E_0' for succ/fum. Does it agree with the accepted values? *Do not use any tubes which indicate there was complete reduction of dye.*

PROBLEM

The *O–R* potential E_0' of malate-oxalacetate is -0.102, and for MB it is $+0.04$. Under equilibrium conditions, what would be the ratio of MBH$_2$/MB and malate/oxalacetate?

At equilibrium,

$$-0.102 + 0.03 \log \frac{[\text{oxalacetate}]}{[\text{malate}]} = 0.04 + 0.03 \log \frac{[\text{MB}]}{[\text{MBH}_2]},$$

$$0.142 = 0.03 \log \frac{[\text{oxal}]}{[\text{malate}]} - 0.03 \log \frac{[\text{MB}]}{[\text{MBH}_2]}$$

$$0.142 = 0.03 \log \frac{[\text{oxal}]}{[\text{malate}]} + 0.03 \log \frac{[\text{MBH}_2]}{[\text{MB}]}$$

$$0.142 = 0.03 \left[\log \frac{[\text{oxal}]}{[\text{malate}]} + \log \frac{[\text{MBH}_2]}{[\text{MB}]} \right]$$

For each molecule of MBH$_2$ formed, one molecule of malate is oxidized. Therefore, we can write the ratio:

$$\frac{\text{oxal}}{\text{malate}} = \frac{\text{MBH}_2}{\text{MB}}.$$

Consequently:

$$0.142 = 2 \times 0.03 \log \frac{[\text{oxal}]}{[\text{malate}]} \quad \text{and} \quad 2.3666 = \log \frac{[\text{oxal}]}{[\text{malate}]}.$$

Solving for the ratio

$$\frac{MBH_2}{MB} = \frac{234}{1} = \frac{oxal}{malate}$$

$\frac{234}{235} \times 100 = 99.5 =$ per cent of MBH_2 99.5 per cent oxalacetate

             or

            $0.5 =$ per cent of MB   0.5 per cent malate.

REFERENCES

1. *Laboratory Manual of Physiological Chemistry*, Department of Physiological Chemistry, Johns Hopkins University, Baltimore, 1957.
2. *Experimental Biochemistry Laboratory Manual*, Department of Biochemistry, University of Wisconsin, 1967.

CHAPTER 24

OXIDATIVE PHOSPHORYLATION IN RAT LIVER

PRINCIPLES

The formation of ATP, which occurs as electrons pass from substrate to oxygen along the chain of cytochrome carriers, is called oxidative or aerobic phosphorylation. During oxidation, the chemical bond energy stored in carbohydrates, lipids, and proteins is ultimately conserved in the form of ATP. This is the most important aspect of aerobic metabolism. In this terminal stage of oxidation, as electrons and protons reduce molecular oxygen to water, ADP and Pi combine to form ATP. Phosphorylation occurs at three sites along the respiratory chain in a series of complex reactions which include phosphorylated protein intermediates. The enzymes catalyzing the entire sequence of reactions of oxidative phosphorylation, which include the initial dehydrogenation, electron transport, and the coupled phosphorylations themselves, are all found in the mitochondria. Oxidative phosphorylation may be demonstrated quantitatively by incubating isolated mitochondria in the presence of oxygen and:

    a. an oxidizable substrate plus coenzymes

    b. inorganic phosphate

    c. a phosphate acceptor (ADP is the primary Pi acceptor)

    d. Mg^{+3} (necessary for the phosphorylation process)

    e. An ATP trapping system which regenerates ADP. (The enzyme hexokinase and glucose serve in this capacity via the ATP-dependent formation of glucose-6-phosphate and ADP.)

In order to demonstrate phosphorylation it is also necessary to minimize the effect of Mg-ATPases, which in essence uncouple phosphorylation by hydrolyzing ATP to ADP and Pi. The ATPase is found both in the mitochondria and in the crystalline hexokinase. There are several ways of minimizing the uncoupling action of Mg-ATPase: (a) Use very fresh preparations of mitochondria; (b) add a minimum amount of Mg^{+2} and a maximum

294

of K^+ to inhibit Mg-ATPases, particularly in the case of mitochondria obtained from liver; (c) add F^- to inhibit ATPase; or (d) use an excess of the trapping system to lessen exposure of ATP to hydrolysis.

As oxidation takes place, inorganic phosphate will disappear from the medium and ATP formation from ADP will occur. Usually, results of such experiments are recorded as the P/O ratio, which is defined as:

$$P/O = \frac{\text{micromoles of P disappearing}}{\text{microatoms of O taken up}}$$

The P/O ratio indicates the number of phosphorylations occurring for each pair of electrons passing down the chain. The maximum P/O ratios which have been observed in isolated mitochondria are about 3.0, depending on the substrate utilized. This ratio allows a calculation of the efficiency with which energy is stored or retained in the oxidation of a given substrate. A P/O ratio of 3.0 is an indication of 70 per cent efficient utilization of available energy.

The mechanism of the oxidative phosphorylation process is not known in detail and represents a major unsolved problem in contemporary biochemistry. The enzymes involved are very labile. Badly damaged mitochondria will still oxidize certain substrates (the electron transport enzymes themselves are relatively stable) but will no longer catalyze the coupled phosphorylations. Consequently, every precaution, with due speed, is necessary in order to preserve the coupled phosphorylation system.

It is also known that certain substances are capable of "uncoupling" phosphorylation; i.e., they permit oxidation of substrates to occur but inhibit phosphorylation. Certain nitro and halophenols, dyes such as methylene blue, Ca^{+2}, and the antibiotics, aureomycin and gramicidin, are all effective uncoupling agents. Thryoid hormone is also able to uncouple phosphorylation; this fact may underlie its mode of metabolic control.

The NAD^+-dependent oxidation of the substrate β-D-hydroxybutyrate to acetoacetate is a convenient reaction to study in liver mitochondria. No complicating additional oxidations of product occur because the liver cannot further oxidize acetoacetate. The oxidation of β-D-hydroxybutyrate to acetoacetate by the enzyme β-D-hydroxybutyrate dehydrogenase can be measured in the presence of the components described above, and the disappearance of inorganic phosphate from the medium as well as the oxidation of β-D-hydroxybutyrate can be analyzed spectrophotometrically. In the latter determination, the formation of acetoacetate is measured rather than the disappearance of substrate. The extent of oxidation, or the amount of oxygen taken up, could also be measured manometrically, a very involved process; or O_2 uptake could be measured potentiometrically with an oxygen electrode, a more direct and convenient method. Therefore, if an oxygen electrode is available, it may be useful to demonstrate the principles of this experiment using several substrates of the Krebs cycle. Another alternative method would be to measure the newly synthesized ATP directly by measuring the amount of glucose-6-phosphate formed in the presence of hexokinase and an excess of glucose and ADP.

EXPERIMENT 24

1. PREPARATION OF RAT LIVER MITOCHONDRIA

Work in groups of four:

As in previous experiments, one member of a pair of students will carry out the preparation of mitochondria while the other member of the pair sets up the flasks and tubes for analysis. Read ahead and set up tubes and reagents indicated in Tables 58, 59, and 60. In the case of the latter, add only buffer and water to the tubes. As each addition is made the students should check it off; this will eliminate errors of omission or addition. Since phosphorylation is extremely labile, it is important to see that all additions to the tubes are made prior to the final preparation of fresh mitochondria.

Table 58

Reagent							
	1	**2**	**3**	**4**	**5**	**6**	**7**
	Zero Time Control	*Complete*	*+Cytochrome*	*−Mg^{+2}*	*−ATP*	*+DNP*	*Blank*
A. 0.100 M standard K phosphate buffer, pH 7.4, ml	0.4	0.4	0.4	0.4	0.4	0.4	
B. 0.1 M MgCl$_2$, ml	0.1	0.1	0.1		0.1	0.1	0.1
C. 0.02 M ATP, ml	0.2	0.2	0.2	0.2		0.2	0.2
D. 0.01 M NAD$^+$ in 0.25 M nicotinamide, ml	0.2	0.2	0.2	0.2	0.2	0.2	0.2
E. 0.1 M β-D-hydroxybutyrate, ml	0.2	0.2	0.2	0.2	0.2	0.2	0.2
F. 0.5 M glucose in 10 mg per cent crystalline hexokinase, ml	0.2	0.2	0.2	0.2	0.2	0.2	0.2
G. 0.4 M NaF, ml	0.1	0.1	0.1	0.1	0.1	0.1	0.1
H. 4 × 10^{-5} M cytochrome c, ml			0.1				
I. 0.002 M 2,4-dinitrophenol, ml						0.1	
J. H$_2$O, ml	0.1	0.1		0.2	0.3		0.5
K. liver mitochondria, ml	0.5	0.5	0.5	0.5	0.5	0.5	0.5

Since the phosphorylation enzymes are very labile, the preparation of mitochondria must be carried out quickly and temperature held near 0° C throughout the preparation procedure. A freshly removed rat liver is placed in ice-cold 0.25 M sucrose and weighed. Make a 10 per cent homogenate and isolate mitochondria as in the previous experiment. This time, however, make the final suspension more concentrated, that is, from 5 grams of liver you should get 5 ml of mitochondria suspension. *Save the remainder of mitochondria in an ice bath.*

2. INCUBATION PROCEDURE

To flasks 1 and 7, first add 8.0 ml of 6 per cent trichloracetic acid, and then add 0.5 ml of mitochondria, and mix. Centrifuge flasks 1 and 7 immediately, remove 1 ml aliquots, and transfer to labeled flasks; dilute this with 3.0 ml of water. Vessel 1 is the zero time control and 7 is used as the reagent blank.

Add 0.5 ml of mitochondrial suspension at staggered 1.0 minute intervals to vessels 2 through 6. Place the vessels in the Dubnoff shaker for 40 minutes at room temperature.

At the end of 40 minutes, add 8.0 ml of 6 per cent TCA to vessels 2 through 6 at one minute intervals to stop the reaction, and mix. Transfer vessel contents to conical tubes and centrifuge 3 to 5 minutes using the table model centrifuge.

Transfer 1.0 ml aliquots of each supernatant to correspondingly numbered tubes; add 3.0 ml of H_2O and mix. Aliquots of these diluted supernatants will be used for the determination of acetoacetate and phosphate.

3. ESTIMATION OF PHOSPHATE

Prepare a series of tubes as shown in Table 59. Mix and read at 660 nm after 30 minutes.

Table 59

	Reagent	Tube Number						
		1	2	3	4	5	6	7
A.	0.1 M NaAc, ml	1.0	1.0	1.0	1.0	1.0	1.0	1.0
B.	0.1 M HAc buffer, pH 4.0, ml	7.0	7.0	7.0	7.0	7.0	7.0	7.0
C.	molybdate reagent, ml	0.5	0.5	0.5	0.5	0.5	0.5	0.5
D.	diluted aliquot, ml	1.0	1.0	1.0	1.0	1.0	1.0	1.0
E.	1 per cent fresh ascorbic acid, ml	0.5	0.5	0.5	0.5	0.5	0.5	0.5
F.	absorbance, 660 nm							
G.	μmoles of Pi/ml							
H.	total μmoles Pi remaining							
I.	μmoles of Pi taken up							

From the standard, tube 1, determine the amount of phosphate in each tube and calculate total amount of phosphate in each reaction vessel. By using tube 7 as the reagent blank, the endogenous phosphate or Pi in the reagent will be subtracted from the other tubes automatically.

4. ANALYSIS OF ACETOACETATE

Read the directions *carefully* and *completely* before beginning the determination.

(a) Acetoacetate is determined by measuring colorimetrically the diazo product of the reaction of acetoacetate and *p*-nitrobenzene diazo hydroxide. The diazotization is complete after 30 minutes if performed at pH 5 to 5.2. The colored compound is completely removed from the aqueous phase by shaking briefly with ethylacetate, and the intensity of the color of the organic phase is read in the spectrophotometer.

(b) Add sample to the 1.0 M acetate buffer according to Table 60. The buffer is necessary in order to neutralize the trichloroacetic acid. Use an extra tube, number 8, containing 0.1 μmoles of acetoacetate, as your standard. It should give an absorbance of 0.120.

(c) Prepare fresh 60 ml of diazotization mixture as follows: to 40.0 ml of *p*-nitroaniline add 6.0 ml of a 0.5 per cent sodium nitrite solution. The

Table 60

	Reagents	Tube Number							
		1	2	3	4	5	6	7	8
A.	1.0 M acetate buffer, pH 5.0, ml	1.0	1.0	1.0	1.0	1.0	1.0	1.0	1.0
B.	*diluted* supernatant, ml	1.0	1.0	1.0	1.0	1.0	1.0	1.0	
C.	water, ml	1.0	1.0	1.0	1.0	1.0	1.0	1.0	1.0
D.	standard acetoacetate, 0.1 μmole/ml, ml								1.0
E.	diazotization mixture, ml	6.0	6.0	6.0	6.0	6.0	6.0	6.0	6.0
F.	5 N HCl, ml	2.0	2.0	2.0	2.0	2.0	2.0	2.0	2.0
G.	ethylacetate, ml	10	10	10	10	10	10	10	10
H.	absorbance, 430 nm								
I.	μmoles of AcAc formed								
J.	P/O ratio								

colorless solution is chilled in an ice bath and 14.0 ml of 0.2 M sodium acetate is added.

(d) Add 6.0 ml of the reagent to each tube and shake, using parafilm as a stopper to mix by inversion.

(e) Allow to stand 30 to 35 minutes at room temperature; then add 2.0 ml of 5 N HCl to each tube and shake again. Using a propipette, add 10.0 ml of ethylacetate to each tube and shake. Remove the ethylacetate layer with disposable pipettes and place in a clean, dry, matched colorimeter tube.

(f) Read the sample at 430 nm in the spectrophotometer using tube 1 or 7 as a blank, which again serves to automatically correct for endogenous aceto-acetate in the reagent and liver preparation.

5. CALCULATION OF THE P/O RATIO

(a) The P/O ratio is determined by the ratio of the calculated μmoles of Pi *disappearing* in tubes 2 to 6 as compared to the zero time tube 1, and by the μmoles of AcAc formed in these same tubes, as compared to the zero time tube 1. In the latter case, it must be noted that the number of μmoles of AcAc formed is equivalent to the number of μmoles of β-hydroxybutyric acid dis-appearing; and for each μmole of AcAc formed, $\frac{1}{2}O_2$ or 1 μmole of monoatomic oxygen is taken up, in accordance with the overall reaction equation:

$$\underset{\overset{|}{OH}}{\overset{\overset{H}{|}}{CH_3-C-CH_2COOH}} + \tfrac{1}{2}O_2 \rightleftharpoons \overset{\overset{O}{\|}}{CH_3-C-CH_2COOH} + H_2O$$

(b) You started out with a reaction volume of 2.0 ml. At the end of the incubation time you made a dilution to 10 ml by adding 8 ml of 6 per cent TCA.

(c) You took a 1.0 ml aliquot and diluted it to 4.0 ml. You therefore have successively make a 1:5 and 1:4 dilution, or a total dilution of 1:20.

(d) The original concentration of Pi was 20 μmoles/ml or 40 μmoles in 2.0 ml. The final concentration in the zero time control therefore is 1 μmole/ml. This is the actual amount of Pi you use as your reference standard, since you take a one ml aliquot of the diluted TCA extract.

(e) In order to determine the total Pi remaining in each tube you may employ the formula

$$\text{Total Pi } remaining \quad (C_u) = \frac{A_s}{A_u} \times C_s \times 20 \times 2,$$

$$\text{or} \qquad\qquad = \frac{A_s}{A_u} \times 40,$$

where $C_s = 1.0$ μmole.

(f) In the case of the acetoacetate, the same considerations apply, except that $C_s = 0.1$ μmole.

$$\text{Total AcAc formed} \quad (C_u) = \frac{A_s}{A_u} \times C_s \times 20 \times 2 = \frac{A_s}{A_u} \times 4.$$

QUESTIONS

What is the action of 2,4-dinitrophenol?

Were Mg^{+2}, ATP, and cytochrome c necessary for demonstrating oxidative phosphorylation?

Explain what might have happened in those tubes where more Pi was formed to begin with.

Calculate the Pi which should have been formed in accordance with the amount of acetoacetate formed.

What would the P/O ratio then be, and what can you conclude about the fragility of this system?

REFERENCES

1. Chance, B., and Williams, G. R. *The Respiratory Chain and Oxidative Phosphorylation.* Adv. in Enz. **17** 65 (1956).
2. Lehninger, A. L., Wadkins, C. L., Cooper, C., Devlin, T. M., and Gamble, J. L., Jr. *Oxidative Phosphorylation.* Science **128** 450 (1958).
3. Lehninger, A. L. *Bioenergetics.* W. A. Benjamin, Inc., New York, 1965.
4. Lehninger, A. L. *Mitochondrion.* W. A. Benjamin, Inc., New York, 1964.
5. *Laboratory Manual of Physiological Chemistry.* Department of Physiological Chemistry, Johns Hopkins University, Baltimore, 1957.
6. Hardy, H. A., and Wellman, H. *Oxidative Phosphorylation: Role of Inorganic Phosphates and Acceptor Systems in Control of Metabolic Rates.* J. Biol. Chem. **195** 275 (1952).
7. Lowry, O. H., and Lopez, J. A. *The Determination of Inorganic Phosphates in the Presence of Labile Phosphate Esters.* J. Biol. Chem. **162** 421 (1946).
8. Walker, P. G. *A Colorimetric Method for the Estimation of Acetoacetate.* Biochem. J. **58** 699 (1954).

CHAPTER 25

THE CARBONIC ACID-BICARBONATE BUFFER SYSTEM AND THE REGULATION OF pH IN TISSUE FLUID

PRINCIPLES

One of the major end products of metabolism is carbon dioxide. This metabolite is also the primary constituent of the most important extracellular buffer system in the body, the carbonic acid-bicarbonate system. Its primary role concerns the regulation of the pH of blood, but it is also of considerable importance as a buffer in all extracellular and intracellular fluid.

In essence, the buffering action of the bicarbonate-carbonic acid system is no different from that of the buffers which you have previously considered. As a proton donor, carbonic acid ionizes to give a proton and a proton acceptor, the bicarbonate ion:

$$[H_2CO_3] \leftrightarrows [H^+] + [HCO_3{}^-]$$

This equilibrium relationship may also be expressed logarithmically in the form of the Henderson-Hasselbalch equation:

$$pH = pK_a' + \log \frac{[HCO_3{}^-]}{[H_2CO_3]}$$

However, this system differs from those previously considered; its components are in equilibrium with a *gas*, carbon dioxide, and the properties of this buffer system are dependent upon this fact. Moreover, the unique suitability of the bicarbonate-carbonic acid pair to serve as a biological buffer also stems from this fact.

Carbon dioxide is soluble in water, and the concentration of carbon dioxide dissolved in water is dependent upon the partial pressure of CO_2 in the gas phase which is in equilibrium with the aqueous phase. Expressed mathematically:

(1) $$[\text{Dissolved } CO_2] = k_0[P_{CO_2}]$$

This is an expression of *Henry's law;* k_0 is a proportionality constant and has dimensions of moles per liter per unit of pressure.

But since CO_2 also reacts chemically with water:

(2) $$\text{dissolved } CO_2 + H_2O \leftrightarrows H_2CO_3$$

and:

(3) $$\frac{[H_2CO_3]}{[\text{dissolved } CO_2][H_2O]} = K_{eq}.$$

Rearranging terms:

(4) $$[H_2CO_3] = K_{eq}[H_2O][\text{dissolved } CO_2].$$

However, since the concentration of water does not change appreciably during the reaction it may be regarded as a constant.

Therefore:

(5) $$[H_2CO_3] = K_E [\text{dissolved } CO_2]$$

where K_E is not the equilibrium constant, but is $K_{eq}[H_2O]$.

Equation (1) stated that $[\text{dissolved } CO_2] = k_0[P_{CO_2}]$. Hence we may substitute this value in Equation (5) and obtain:

(6) $$[H_2CO_3] = K_E k_0 P_{CO_2}$$

Furthermore, carbonic acid is a weak acid and dissociates in two steps:

(a) $$[H_2CO_3] \leftrightarrows [H^+] + [HCO_3^-]$$
(b) $$[HCO_3^=] \leftrightarrows [H^+] + [CO_3^=]$$

(K_{eq} and pK_a for each reaction is given in Table 2, page 26.) But since hydrogen ion concentration compatible with life only concerns the first dissociation, the dissociation of the second proton may be disregarded.

Expressing the first dissociation of H_2CO_3 in terms of the Henderson–Hasselbalch equation:

(7) $$pH = pK_a' + \log \frac{[HCO_3^-]}{[H_2CO_3]}$$

By substituting Equation (6) into Equation (7) and combining pK_a' with $-\log K_E$ we obtain:

$$(8) \qquad pH = pK_a' + \log \frac{[HCO_3^-]}{k_0 P_{CO_2}}.$$

pK_1' is now a conglomeration of constants and is *not* the same as pK_a'.

From Equation (8) it is possible to make several deductions. In the first place, if the partial pressure of CO_2 is kept constant, then the pH of the solution will rise with increasing HCO_3^- concentration and fall with decreasing HCO_3^- concentration. On the other hand, if HCO_3^- is held constant, then pH will decrease with rising P_{CO_2} and increase as P_{CO_2} falls. Both of these situations have counterparts in the physiological regulation of plasma pH, since, in the course of tissue respiration, CO_2 is passed into the blood and eventually is exhaled from the lungs.

These equations are useful for preparing a bicarbonate buffer of known pH and composition. However, in research or clinical work, there is a different problem. The question which frequently arises is, "What is the pH of tissue fluid?" This would be easy to answer if it were possible to measure tissue pH directly, but this is not easily done. However, from the Henderson–Hasselbalch equation it should be possible to *calculate* the pH of a plasma sample if the concentrations of H_2CO_3 and HCO_3^- in the sample could be measured. (Precise values for the necessary constants have been determined and are available.) Although it is not convenient to measure $[H_2CO_3]$ or $[HCO_3^-]$, two related quantities, $[\text{Total } CO_2]$ and P_{CO_2}, are easily measured. The purpose of the simple mathematical derivation given below is to show how measurement of these quantities can be used to arrive at a value for plasma or tissue fluid pH. First of all:

$$(9) \quad [\text{Total } CO_2] = [HCO_3^-] + [H_2CO_3] + [\text{Dissolved } CO_2].$$

Rearranging terms in Equation (9), we obtain:

$$(10) \quad [HCO_3^-] = [\text{Total } CO_2] - [H_2CO_3] - [\text{Dissolved } CO_2].$$

Substituting this value for $[HCO_3^-]$ in Equation (8), we obtain:

$$(11) \quad pH = pK_1' + \log \frac{[\text{Total } CO_2] - [H_2CO_3] - [\text{Dissolved } CO_2]}{k_0 P_{CO_2}}.$$

Henry's law states that $[\text{Dissolved } CO_2] = k_0 P_{CO_2}$. By substituting, we obtain:

$$(12) \qquad pH = pK_1' + \log \frac{[\text{Total } CO_2] - [H_2CO_3] - k_0 P_{CO_2}}{k_0 P_{CO_2}}.$$

According to Equations (1) and (5),

$$[\text{dissolved } CO_2] = k_0 P_{CO_2},$$

and

$$H_2CO_3 = K_E[\text{Dissolved } CO_2].$$

Therefore

$$[H_2CO_3] = K_E k_0 P_{CO_2}.$$

Substituting this value for $[H_2CO_3]$ in Equation (12), we obtain:

$$(13) \quad pH = pK_1' + \log\left[\frac{[\text{Total } CO_2] - K_E k_0 P_{CO_2} - k_0 P_{CO_2}}{k_0 P_{CO_2}}\right]$$

Collecting and simplifying terms, this becomes:

$$(14) \quad pH = pK_1' + \log\left[\frac{[\text{Total } CO_2]}{k_0 P_{CO_2}} - (K_E + 1)\right]$$

K_E is so small (its actual value is about 0.0001) that $(K_E + 1)$ is not appreciably different from unity; therefore it may be assumed that:

$$(15) \quad [K_E + 1] = 1.$$

We may therefore write:

$$(16) \quad pH = pK_1' + \log\left[\frac{[\text{Total } CO_2]}{k_0 P_{CO_2}} - 1\right]$$

or

$$(17) \quad pH = pK_1' + \log\frac{[\text{Total } CO_2] - k_0 P_{CO_2}}{k_0 P_{CO_2}}.$$

Both pK_1' and k_0 have been accurately determined experimentally, and Table 61 provides values for k_0 and pK_1 at an ionic strength of 0.1. A constant ionic strength must be kept in mind since there is variation of these constants with changing ionic strength. Concentrations are expressed in millimoles per liter and pressure in mm of Hg.

Total CO_2 in plasma can be measured by means of the Van Slyke apparatus, and the numerical value substituted in Equation (18). The partial pressure of CO_2 in alveolar air can be measured by a technique described in the following

Table 61

Temperature °C	k_0 (μmoles/mm Hg)	pK_1'
20	0.0507	6.219
22	0.0479	6.207
24	0.0453	6.196
26	0.0426	6.186
37	0.0300	6.110

section. Since alveolar air is the gas phase which is in equilibrium with arterial blood, the value for alveolar air can be substituted into Equation (18). Hence, by determining *both* (Total CO_2) and P_{CO_2} it is possible to calculate the pH of a plasma sample and to determine whether or not it is within the accepted normal limits of pH, 7.35 to 7.45. It cannot be emphasized too strongly that estimation of a single one of the variables—(Total CO_2) or P_{CO_2}—or (HCO_3^-) is not sufficient to indicate whether or not there is a disturbance of the pH.

EXPERIMENT 25

Part A.   The Role of Plasma CO_2 in the Regulation of pH

1. RENAL REGULATION OF PLASMA pH

Micturate a fresh sample of 20 to 30 ml of urine into a 100 ml beaker. Now, empty the bladder. Determine the pH of the urine sample. Take by mouth 5 grams of $NaHCO_3$. Determine the pH of urine samples collected every 30 minutes during the remainder of the period.

What are your observations and conclusions about the physiological effect of $NaHCO_3$?

2. DETERMINATION OF SIMULATED LUNG-ALVEOLAR CO_2 TENSION

CO_2 pressures between 40 and 50 mm of mercury (approximating alveolar lung pressure), over a 30 millimolar solution of $NaHCO_3$ (simulating plasma concentration) are conditions which result in a pH value in a measurable range. Now use Equation (8) and Table 61, page 304, to calculate the pH change when P_{CO_2} is 25, 30, 35, 40, 45, 50, 55, and 60 mm at a HCO_3^- concentration of 30 millimolar and an ionic strength of 0.1. Submit your calculations.

3. DETERMINATION OF THE CO_2 TENSION OF LABORATORY AIR

The CO_2 pressure in uncontaminated air is about 0.2 mm of mercury. Because the pressure is so low, a much lower bicarbonate concentration will be needed to give suitable pH values. Using Equation (8), together with the data in Table 61, calculate the values for pH at 24° when P_{CO_2} is 0.20, 0.30, 0.40, and 0.50 mm. *Plot* pH versus log P_{CO_2}.

Place 5 ml of a solution composed of 0.50 mM $NaHCO_3$ and 200 mM NaCl in a test tube. Add 1.0 ml of 1×10^{-4} M phenol red and 4 ml of H_2O to bring to a volume of 10 ml. By means of a large rubber bulb and a constricted tube, bubble air gently through this mixture till no color change is detected. Determine the absorbancy. The pH is determined from absorbance values obtained for phenol red at varying pH (cf. your data from Experiment 3). Read the CO_2 pressure from your graph.

What blank would you use?

Blow through the test solution. Determine the pH once again. This method is sometimes used to determine the CO_2 tension in submarines. What can you say is the CO_2 tension of alveolar air?

306

What effect would breathing through the mouth have on the pH of saliva? Suppose the blood plasma contains 28 mM per liter of HCO_3^-, the saliva contains 14 mM per liter and both are expressed to the same CO_2 tension. Calculate the difference in pH you would expect.

4. EFFECT OF HCO_3^- CONCENTRATION ON pH AT CONSTANT CO_2 TENSION

Place the components indicated in Table 62 into matched colorimeter tubes.

Table 62

Reagents	Tube Number			
	1	2	3	4
A. 1×10^{-4} M phenol red, ml	1.0	1.0	1.0	1.0
B. 0.1 M $NaHCO_3$, ml	0.5	1.0	1.5	2.0
C. distilled water, ml	8.5	8.0	7.5	7.0
D. absorbance (before gasing)				
E. pH (before gasing)				
F. absorbance (after gasing)				

Gas each tube for 2 minutes by allowing a fine stream of gas composed of 5 per cent CO_2 and 95 per cent N_2 to bubble through the solution. Stopper immediately with a tightly fitting rubber stopper. Determine the absorbance at 550 mμ on the spectrophotometer. Use a tube containing distilled water as the blank. Determine the pH of each solution from your data in Experiment 3 as above. Record these values in Table 62. On graph paper, plot pH as the ordinate versus the logarithm of the bicarbonate concentration as the abscissa. Assume that the total bicarbonate concentration is the quantity added as $NaHCO_3$, because the amount contributed by the dissociation of H_2CO_3 will be small by comparison. Verify this by a calculation.

The variation of pH with changing HCO_3 concentration should be apparent from the graph.

Part B. Hydrogen Ion and Oxygen Transport by Hemoglobin in the Regulation of pH

PRINCIPLES

Respiration in living cells may be described as the mechanism of O_2 release to tissue from an exchange medium, and of CO_2 uptake by the exchange medium with a minimum change in pH. When one considers blood as the exchange medium in mammalian tissue, it is possible to show that respiratory gaseous exchange takes place in a cyclical fashion. In the lungs there is a gain of 5.5 volumes of O_2 and a loss of 5 volumes of CO_2; the reverse takes place at the cell sites.

The oxygen is carried in blood in two ways: dissolved, and in combination with the protein hemoglobin (Hb) as the adduct form, oxyhemoglobin (HbO_2). There are four subunits of hemoglobin, each with a porphyrin ring. The molecular weight of this molecule is 64,450 including the heme groups, each subunit having a minimum molecular weight of 16,100. The oxygenation reaction may be written simply as:

$$Hb + 4O_2 \rightleftharpoons Hb(O_2)_4$$

A simple calculation will show that each gram of Hb will be able to carry 1.39 ml of O_2 under standard conditions.

$$\frac{22,400 \text{ ml}}{16,100 \text{ g}} = \frac{X}{1 \text{ g Hb}} ; \qquad X = 1.39 \text{ ml}$$

The average amount of hemoglobin found in 100 ml of blood is 15 g; then:

$$15 \times 1.39 = 20.9 \text{ ml of } O_2/100 \text{ ml of blood.}$$

Also, there are 150 gm Hb/liter; therefore, $\frac{150}{16,100} = 0.0093$ moles Hb per liter.

The contribution of dissolved oxygen to the total capacity of blood is 1.2 ml/100 ml of blood, thus totaling 22.1 volume per cent (1.2 + 20.9).

The extent to which Hb will combine with O_2 depends on P_{O_2}. When P_{O_2} is high, Hb is almost completely saturated with O_2, and vice versa. Oxygenation of Hb is also dependent upon P_{CO_2}. As P_{CO_2} increases, a faster dissociation of HbO_2 is favored.

The shape of the curve is sigmoidal, suggesting an enhancing allosteric effect on the intensity of association of O_2 with increased involvement of Hb subunits (Hill Reaction). As the HbO_2 adduct first forms, O_2 affinity is not

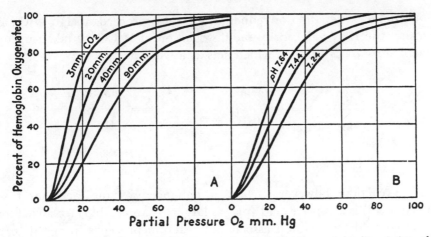

FIGURE 45. The effects of increased P_{CO_2} and decreased pH upon the dissociation of oxy-hemoglobin. Increased temperature also shifts the curve in the same direction as increased P_{CO_2}. (A after Barcroft, B after Peters and Van Slyke. From *Physiology and Biophysics*, T. C. Ruch and H. D. Patton, eds. Philadelphia, W. B. Saunders, 1965.)

great, but as association proceeds, there is a facilitation of further O_2 addition, as indicated in the area of steep rise. The rate of adduct formation of eventually tapers off, and reaches a plateau as full saturation is approached.

Under physiological conditions, arterial blood has a P_{CO_2} of 40 mm of Hg and a P_{O_2} of 80 mm. At the tissue site the exchange O_2 is formed by two factors: a lowered tissue P_{O_2} to $<$10 mm Hg and an elevated P_{CO_2} of 47 mm Hg. As a consequence of these facilitating effects, the P_{O_2} of venous blood drops to 40 mm Hg and the P_{CO_2} simultaneously is elevated to 40 mm Hg; the O_2 saturation of Hb drops from a high of 95 per cent in arterial blood to a low of 65 per cent in venous blood.

At the same time, hemoglobin plays a major role in maintaining a constant pH during this exchange, especially in view of the formation and subsequent dissociation of H_2CO_3. At a physiological pH of 7.4, all carboxyl groups of hemoglobin are protonated or undissociated, and hemoglobin has a net positive charge. The titration curves for the oxygenated and deoxygenated forms of hemoglobin would indicate that each has two major inflections, one at around pH 4.5 and the other around pH 6.8, the latter of which is attributed to the high histidine content of this molecule. At physiological pH, then, the imidazole group is mainly responsible for the coupled reduction and buffering action of hemoglobin during respiration. A simple calculation from the linear portion of the titration curve for hemoglobin would indicate that 0.245 μmoles of H^+ per millimole of HbO_2 will lower the pH 0.1 of a unit from pH 7.4. As an example of this buffering action, if 3 μmoles of H^+ per ml were added to 1 liter of blood at pH 7.4, and the volume change is neglected, the calculated drop would be 0.13 pH units, for blood containing 150 g or 9 millimoles of Hb per liter. In the absence of this buffering action, addition of the same amount of H^+ to one liter of water would be equivalent to a 0.003 M H^+ solution, and cause a pH drop to 2.52.

As the deoxygenation or reduction of HbO_2 proceeds to $Hb + O_2$ at the tissue site, the H^+ produced by the dissociation of H_2CO_3 is taken up by the imidazole group of histidine, which acts as a weak acid.

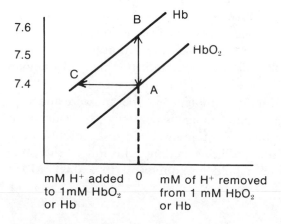

This reaction, however, is further enhanced by the fact that the Hb (reduced form) is a weaker acid (pK_a is larger) than is HbO_2 (oxidized form), and consequently the proton association reaction proceeds from left to right to a greater extent.

If we closely examine only the narrow linear portion of the titration curves of the two forms of hemoglobin, we see the situation shown in Figure 46.

In going from A to B, the curve indicates that if we take a solution of HbO_2 and deoxygenate it the pH will rise, because, as stated before, the reduced form of Hb is a weaker acid and H^+ is taken up from solution. The reverse is also true; as Hb becomes oxygenated, there is a release of H^+, with a concommitant drop in pH. This coupling of O_2 binding with proton release is called the Bohr effect. While this effect is not fully understood, the change in proton affinity appears to be related to change in direction of the valence attachment of Fe^{2+} in the imidazole ring, from the O_2 adduct to the unprotonated imidazole group of Hb. In this experiment, you will attempt to demonstrate the Bohr effect.

In order to close this discussion, however, it will be necessary to further explain the buffering action of Hb. In physiological circumstances, the pH is maintained as deoxygenation takes place; that is, upon going from A to C, the H_2CO_3 formed dissociates and the protons are taken up by Hb. Actually, for every millimole of HbO_2 reduced, 0.7 millimole of H^+ can be added to blood without changing pH. For instance, a solution of arterial blood at 90 per cent saturation will contain 9.3 mM $HbO_2 \times 0.9 = 8.4$ mM HbO_2.

FIGURE 46. Titration of hemoglobin.

This blood is reduced to 45 per cent saturation at the tissue site, thus leaving 9.3 × 0.45 or 4.2 mM of HbO_2 unchanged. This is equivalent to the reduction of 4.2 mM of HbO_2, and is also equivalent to an uptake of 2.94 μmoles of H^+ (4.2 × 0.7). This extraordinary buffering capacity of the hemoglobin in the red blood cells is the reason why respiration can take place with only a slight change in pH, from 7.4 to 7.35. The slight drop is due to the fact that not all the H^+ released is completely buffered by Hb.

DEMONSTRATION OF BOHR EFFECT

Spin 10 ml of whole blood for 5 minutes at 48,000 × g in a plastic centrifuge tube of suitable size.

Wash one volume of packed red blood cells with three volumes of 0.9 per cent NaCl, centrifuge, discard supernatant, repeat the wash, and hemolyze by shaking vigorously for 5 minutes with one volume of toluene and two volumes of 0.09 per cent NaCl per volume of packed cells. After toluene treatment, centrifuge the hemoglobin suspension for 5 minutes in the clinical centrifuge with conical glass tubes at 4000 rpm. Adjust the clarified supernatant to pH 7.5 with 0.1 M NaOH; then dilute to 10 times the original cell volume.

Place about 10 ml of this 1 per cent hemoglobin solution in a small filter flask and another 10 ml aliquot in a 50 ml Erlenmeyer flask. Stopper the filter flask and attach it (through a trap) to the water aspirator. Evacuate for 5 minutes with gentle swirling and then clamp off with a screw clamp and keep sealed until ready to measure the pH. Aerate the solution in the Erlenmeyer flask with a stream of CO_2-*free* air. After 5 minutes cease aeration and stopper tightly.

Compare the colors of the two solutions. Standardize a pH meter with pH 7 standard buffer and measure the pH of each of the hemoglobin solutions. The measurement must be made quickly in order to avoid changes in the O_2 and CO_2 content of the two solutions.

REFERENCES

1. *Laboratory Manual of Physiological Chemistry.* Department of Physiological Chemistry, Johns Hopkins University, Baltimore, 1957.
2. *Laboratory Biochemistry.* Department of Biological Chemistry, University of Michigan, 1954.
3. *Experimental Biochemistry Laboratory Manual.* Department of Biochemistry, University of Wisconsin, 1967.

CHAPTER 26

DEMONSTRATION OF TRANSAMINATION IN LIVER EXTRACTS

PRINCIPLE

Transaminases are enzymes which catalyze the general reaction indicated by the equation:

$$RCOCOOH + R'CHNH_2COOH \rightleftharpoons RCHNH_2COOH + R'COCOOH$$

These enzymes are widely distributed and are found in plants, bacteria and animal tissues. The transaminase enzymes require pyridoxal phosphate as a coenzyme. These reactions are readily reversible. The most important transaminases are the following:

(1)   α-ketoglutarate + amino acid $\rightleftharpoons$ glutamic acid + keto acid

(2)   oxaloactic acid + amino acid $\rightleftharpoons$ aspartic acid + keto acid

(3)   glutamic acid + oxalacetic acid $\rightleftharpoons$ α-ketoglutarate + aspartic acid

(4)   glutamic acid + pyruvic acid $\rightleftharpoons$ α-ketoglutarate + alanine

By means of these four reactions, the transaminases can account for transfer of all the alpha amino groups of amino acids to either glutamic or aspartic acid. The amine of glutamate eventually forms one-half of the amine in urea via glutamic dehydrogenase and carbamyl phosphate synthetase action. The amine of aspartate forms the other half of urea via the formation of arginosuccinate.

In this experiment you will demonstrate the occurrence of reaction (2) in extracts of rat liver. The reaction components and products will be identified by the use of paper chromatography. Authentic samples of the amino and keto acids will provide the R_f values required as standards in the identification of the enzymatic reaction products. This is the actual method employed by Feldman and Gunsalus to demonstrate transamination between α-KGA and a large number of amino acids, thus proving that this reaction is not confined to just a few amino acids, as was once thought.

EXPERIMENT 26

Work in pairs.

(1) Set up the reagents listed in Table 63 in 12 ml heavy-walled conical centrifuge tubes. These should be ready beforehand so that the enzyme preparations may be added *as soon as they are available.*

(2) One fresh rat liver is homogenized in a volume of cold potassium phosphate buffer (0.05 M, pH 7.4) equal to 3 times its wet weight. Seven times the wet weight of cold buffer is added to the homogenate, and mixed

Table 63

		Tube Number			
Reagents		1	2	3	4
A.	0.2 M alpha-ketoglutarate, ml			0.3	0.3
B.	0.2 M D,L-alanine,* ml			0.3	0.3
C.	0.2 M sodium pyruvate, ml	0.3	0.3		
D.	0.2 M L-glutamate, ml	0.3	0.3		
E.	0.1 M sodium arsenite,* ml	0.4	0.4	0.4	0.4
F.	liver enzyme, ml	1.0		1.0	
G.	heated liver enzyme, ml		1.0		1.0

* Sodium arsenite is added to prevent the decarboxylation of the alpha-keto acids by other enzymes present in the liver. L-alanine is the active form of the amino acid for transamination, but because D-alanine is non-inhibitory, the racemic mixture may be used.

well; the diluted homogenate is centrifuged in the cold for 1 minute at 8000 rpm. From 5 grams of liver, $3 \times 5 + 7 \times 5 = 50$ ml of supernatant will be obtained. The supernatant contains the enzyme and must be kept cold while other steps are performed. Use fresh preparations in tubes 1 and 3.

(3) Immerse a 12 ml centrifuge tube containing 3 ml of the cold liver supernatant preparation in a boiling water bath for 5 minutes. Use this preparation in tubes 2 and 4.

(4) Incubate all of the tubes at 37° C for 45 minutes in a constant temperature bath. At the prescribed time, add 6.0 ml of cold 95 per cent ethanol to each tube. Keep them in an ice bath. Permit the tubes to remain there until a flocculent protein precipitate appears; centrifuge for 3 to 5 minutes in the clinical centrifuge.

(5) During the incubation period, prepare sheets of chromatography paper. These preparations, including spotting and handling of the paper sheets, should be accomplished while wearing plastic gloves.

Cut two sheets of Whatman #1 filter paper with scissors into 40×40 cm squares. Next draw a line 5 cm above, and parallel to, the machine cut edge of the paper and divide each line into 6 equal parts with a pencil mark, starting 5 cm from either of the long edges. Cut a 2.5 cm square corner from the ends designated to be dipped into the solvent. Designate one sheet for amino acid analysis and one for keto acid analysis.

(6) Samples of supernatant solutions from each of the above experimental tubes will be spotted on each of the *two* sheets of paper; one will be used to determine the amino acids, and the other to determine the alpha-keto acids. One student will proceed with the preparation of the keto acid paper chromatogram while the other student will handle the amino acid chromatogram.

AMINO ACID DETECTION

With the aid of a piece of capillary tubing, apply a spot from each tube, each 1 cm in diameter. Apply sample from a single tube 5 times, one on top of the other, drying each spot prior to application of the next one. Apply the clear supernatant from each experimental tube to the spot designated for amino acid analysis. Be careful and methodical.

In addition to the supernatant solutions, there should be two control spots for alanine and glutamic acid. Make only one small application from the 0.2 M solutions of these amino acids used in the experiment.

KETO ACID DETECTION

The same procedure is repeated for the paper designated for keto acid analysis. Two single control spots of 0.2 M pyruvate and alpha keto glutamate should be included; however, *in this case*, 8 applications of these control acids should be made. As these spots are applied, a spot of half-saturated 2,4-dinitrophenylhydrazine in 0.1 N HCl should be applied on top of the supernatant solutions or control alpha keto acids. In this way a yellow 2,4-dinitrophenyl hydrosine derivative of the alpha keto acids is made which can be detected by eye.

The chromatogram containing the alpha keto acids is placed in a chromatography jar containing *n*-butanol saturated with 3 per cent ammonia. The other sheet, spotted with the amino acids, is placed in a similar jar containing 3 : 1 *n*-propanol-water as solvent. Both chromatograms are developed overnight, removed, and placed in a hood to dry.

SPRAYS

For detection of the amino acid spots the next day, the appropriate paper is sprayed with ninhydrin reagent. The paper is then heated at 150° in a chromatography oven. The 2,4-dinitrophenyl hydrazine spots of the keto acids on the other paper should be easily seen. If not, spray the chromatograms with a solution of 10 per cent NaOH in 95 per cent EtOH.

Interpret the results on the basis of the chromatographic data.

Calculate the R_f values of all spots. How well do the experimental spots agree with the controls?

REFERENCES

1. Feldman, L. J., and Gunsalus, I. C. *The Occurrence of a Wide Variety of Transaminases in Bacteria.* J. Biol. Chem. **187** 821 (1950).
2. Cammarata, P. S., and Cohen, P. P. *Scope of the Transamination Reaction in Animal Tissues.* J. Biol. Chem. **187** 439 (1950).
3. *Laboratory Manual of Physiological Chemistry.* Department of Physiological Chemistry, Johns Hopkins University, Baltimore, 1957.

CHAPTER 27

BIOSYNTHESIS OF UREA (THE KREBS-HENSELEIT UREA CYCLE IN LIVER)

PRINCIPLE

There is a phylogenetic variation in the end products of nitrogen metabolism. The lower forms such as bony fishes excrete ammonia; birds and reptiles excrete uric acid, while virtually all of the higher forms predominantly excrete urea along with other nitrogenous products. The mechanism by which urea is formed by the mammalian liver is a cyclic process (Fig. 47).

The conversion of ornithine to citrulline occurs in the mitochondria. The mechanism of this reaction involves several steps including the formation of carbamyl phosphate. Ornithine transcarbamylase, which catalyzes the condensation of carbamyl phosphate and ornithine to citrulline, is found in mammalian, but not avian, liver. The conversion of citrulline to arginine occurs in the cytoplasm by way of the intermediate formation of arginosuccinic acid. The final cleavage of arginine to urea and ornithine is catalyzed by the enzyme arginase which is found primarily in the cytoplasm.

There is a relationship between the urea cycle and the biosynthesis of pyrimidines. The formation of carbamyl phosphate is an initial step in both pathways. In birds, pyrimidines are formed from citrulline, presumably by the hydrolysis of arginosuccinate to ureidosuccinic acid and ornithine. The ureidosuccinic acid is known to be a pyrimidine precursor. In the rat, however, citrulline cannot form pyrimidines. There is evidence, also, that the same mechanism that contributes a carbamyl group to the ornithine to form citrulline, is employed to donate a carbamyl group to aspartic acid to form ureidosuccinic acid.

317

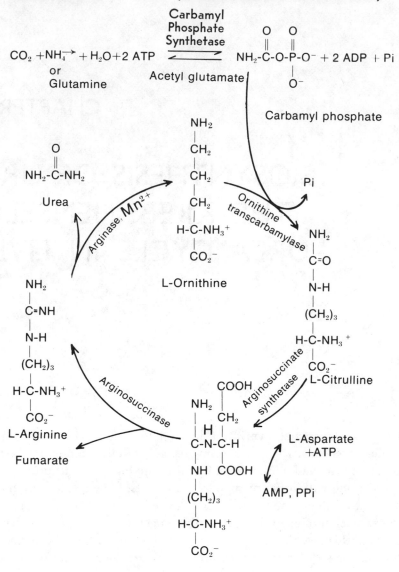

FIGURE 47.   The Krebs-Henseleit Urea Cycle.

In this experiment you will attempt to demonstrate the conversion of citrulline, ornithine, and arginine to urea using acetone powder extracts and subcellular particles. The phosphorylation accompanying the oxidation of the added phosphoglyceric acid serves as a source of ATP, which is required for urea synthesis.

EXPERIMENT 27

Two different types of tissue extracts will be used in order to demonstrate the presence of the enzymes of the urea cycle. The class will be divided in half, and each half will work as pairs. The enzyme extracts will already have been prepared. One half of the class will use an acetone powder extract of beef or pork liver, and the other half will use acetyl trimethyl ammonia bromide (CTAB) extract of fresh rat liver.

Table 64

Reagents		Tube Number									
		1	2	3	4	5	6	7	8	9	10
A.	medium cocktail,[1] ml	1.6	1.6	1.6	1.6	1.6	1.6	1.6	1.6	1.6	1.6
B.	0.1 M D,L-citrulline, ml	0.2		0.2							
C.	0.1 M D,L-arginine, ml		0.2		0.2						
D.	0.1 M D,L-ornithine, ml					0.2		0.2	0.2	0.2	
E.	0.06 M carbamyl phosphate (dilithium), ml						0.2	0.2			
F.	0.4 M NH$_4$HCO$_3$, ml								0.2	0.2	
G.	0.05 M acetyl L-glutamate									0.2	
H.	enzyme extract[2], ml			1.5	1.5	1.5	1.5	1.5	1.5	1.5	1.5
I.	H$_2$O, ml	2.2	2.2	0.7	0.7	0.7	0.7	0.5	0.5	0.3	0.9

[1] Medium Cocktail: each 1.6 ml contains 0.6 ml of 0.025 M potassium phosphate buffer, pH 7.8; 0.3 ml of 0.1 M L-aspartic acid; 0.4 ml of 0.03 M MgSO$_4$; 0.1 ml of 0.05 M ATP; 0.2 ml of 0.1 M phosphoglyceric acid.

2 See directions in the Instructor's Manual for this experiment.

(a) Set up tubes with all reagents except enzyme as indicated in Table 64. Equilibrate the tubes and the enzyme solution separately for 5 minutes at 37°. At zero time, add enzyme to appropriate tubes and incubate for 30 minutes. While this incubation is being carried out, preparations for a urea standard curve should be made.

(b) Pipette 0, 1, 2, 3, and 4 ml of urea standard into five separate test tubes and dilute the volume of each tube to 7 ml with water. To each tube add 5 ml of sulfuric-phosphoric acid mixture. *Do not pipette.* Use a burette. Mix well.

Add 0.4 ml of 4 per cent α-isonitrosopropiophenone to each tube. Mix by inversion with the aid of parawax squares. Mark the level of fluid with rubber bands. Cover the tubes with aluminum foil and put them in a boiling water bath for 45 minutes. *This entire procedure should be done in the dark.*

(c) At the end of this time cool the tubes for 10 minutes in cold water. Restore the original volume to each tube with water. Transfer contents to spectrophotometer tubes and read at 540 nm. Use the tube which contains no urea as blank.

(d) Plot absorbance vs. urea concentration on graph paper.

(e) At the end of the 30 minute incubation time, the experimental tubes are removed from the bath, and 2 ml of 15 per cent metaphosphoric acid is added to each tube. Centrifuge the contents of each tube for 5 minutes and transfer 2 ml of the supernatant to dry test tubes.

(f) Dilute all tubes to 7 ml with water and proceed as in step c for determining the standard curve. *Use the same blank.*

(g) Estimate the μmoles of urea produced per mg crude enzyme per hour, after determining the protein concentration in 0.1 ml of the enzyme extract by the Biuret method as was done in previous experiments. Remember that you have taken a 2 ml aliquot out of a total volume of 6 ml after adding the metaphosphoric acid.

Explain why you are able to observe urea formation in each of those tubes in which it is formed.

Are all of the enzymes of the urea cycle found in this acetone powder extract or in the CTAB extract?

What does the color formed in tube 1 indicate to you?

Do the results from tubes 3, 4, and 5 unequivocably show urea formation? If not, why not?

How can you interpret the results in tubes 7, 8, and 9?

Have you shown an ATP dependence for urea formation?

REFERENCES

1. Burnett, G. H., and Cohen, P. P. *Carbamyl Phosphate-Ornithine Transcarbamylase.* J. Biol. Chem. **229** 337 (1957).
2. Ratner, S., and Pappas, A. *Biosynthesis of Urea.* J. Biol. Chem. **179** 1182 (1949).
3. Archibald, P. M. *Colorimetric Determination of Urea.* J. Biol. Chem. **157** 507 (1945).
4. Cohen, P. P., and Brown, G. W., Jr. "Ammonia Metabolism and Urea Biosynthesis," in *Comparative Biochemistry, Vol. II* (M. Florkin and H. S. Mason, eds.) Academic Press, New York, 1960.
5. Ratner, S. *Urea Synthesis and Metabolism of Arginine and Citrulline.* Adv. in Enz. **15** 319 (1954).

CHAPTER 28

THE INCORPORATION OF UC¹⁴ TYROSINE INTO BRAIN MICROSOMAL PROTEINS

The biological significance of enzymes and proteins in tissue metabolism by this time has been made obvious. Of even greater interest, however, is the relationship between DNA, the genes contained therein, RNA, and proteins. In brain tissue this relationship may be of particular importance because this process in neuronal cells has been related to memory function. There is ample evidence to indicate that brain tissue is very active in the synthesis and turnover of proteins. There is, at the same time, sufficient evidence to indicate that the mechanism in brain is essentially identical to that found in liver or any other organ. The questions remaining concern identification of the kinds of proteins being synthesized in point of time, and associating them with a specific kind of behavior, such as memory.

In this experiment an attempt will be made to understand the basic requirements and techniques for demonstrating *in vitro* protein synthesis in rat brain microsomes.

In order to demonstrate *in vitro* protein synthesis it is necessary for a number of components to be present: (a) at least 20 amino acid activating enzymes, (b) at least 20 transfer RNAs specific for each of these amino acids, (c) ribosomes, (d) at least one messenger RNA, (e) a number of coenzymes, such as GTP and ATP, (f) the cations K^+ and Mg^{+2}, and (g) a single C^{14}-labeled amino acid, which is used to demonstrate *de novo* incorporation of an amino acid into protein.

In order to assure the presence of each of these components, a microsomal preparation will be used to provide a source of (a), (b), and (c). The coenzymes, cations and radioactive amino acid are supplemented in the medium. After completion of the experiment, an extraction procedure similar to the one used for isolating DNA and RNA is employed; however, in this case rather than throwing away the precipitated proteins, it is the RNA and DNA which are of no interest, and the purified proteins are analyzed for isotope incorporation.

PREPARATION OF BRAIN HOMOGENATES

Work in pairs.

(1) Decapitate a rat and immediately excise the brain, including the cerebellar hemispheres.

(2) Place the brain in a previously tared 25 ml beaker containing 10 ml of the phosphate buffer, sucrose and salt medium. This mixture is diluted with an equal volume of 0.008 M $MgCl_2$ just prior to use. Determine the wet brain weight and wash it once with fresh medium.

(3) Homogenize the brain in 20 ml of medium. Make 7 up and down passes at 1000 rpm of the homogenizing motor. Use 12 ml polyethylene centrifuge tubes and centrifuge for 1 minute at 8000 rpm, including acceleration time. Decant and centrifuge the supernatant, S_1, for 5 minutes at 48,000 g.

(4) Recover the supernatant (S_2) and use it to initiate the reaction. Save 0.1 ml for a biuret determination.

(5) Add the reagents indicated in Table 65 to heavy-walled conical centrifuge tubes. *Remember* you are handling a radioactive amino acid. Exercise all precautions previously outlined for the use of isotopes in a laboratory. *Wear gloves and use a propipette.* Be sure that you do not add isotope to tube 4. See directions in step concerning tube 1.

Table 65

	Reagents	Tube Number 1	2	3	4
A.	UC^{14} tyrosine, 2 μc/ml, ml	0.1	0.1	0.1	
B.	energy systems[1]	0.2		0.2	0.2
C.	water		0.2		0.1
D.	microsomal fraction[2]	0.7	0.7	0.7	0.7

[1] Energy System: Make up fresh: 0.005 M ATP, 0.00125 M GTP, 0.10 M phosphocreatine, and 50 units per ml of creatine kinase. Thus each tube contains 1 μmole ATP, 0.25 μmoles GTP, 20 μmoles of phosphocreatine, and 10 units of creatine kinase.

[2] Phosphate buffer-sucrose, salt medium, pH 7.5: 0.04 M KH_2PO_4, 0.07 M $KHCO_3$, 0.05 M KCl, 0.7 M Sucrose, pH 7.5. *This mixture is diluted with an equal volume of 0.008 M $MgCl_2$ just prior to use.*

(6) Initiate the reaction by adding microsomes to tubes 2, 3, 4. Incubate with shaking for 30 minutes at 37°, and terminate the reaction by addition of 1.0 ml of 20 per cent *ice cold* TCA. An endogenous control (tube 1) should contain 1 ml 20 per cent TCA prior to the addition of homogenate, and be

allowed to stand with the rest of the tubes for 15 minutes in ice. This should be done after a 30 minute incubation of tubes 2, 3, and 4 at 37°.

(7) While the tubes are standing in ice for 15 minutes, agitate with a stirring rod. Use four separate stirring rods. *Do not interchange them for the rest of the procedure.* Keep the rods in the conical centrifuge tubes. *Do not contaminate the table top.*

(8) After this period all tubes are centrifuged for 3 minutes on the table model centrifuge. The TCA supernatant is removed by aspiration with the aid of a capillary pipette.

(9) Add 5 ml of 5 per cent TCA to all tubes. Break up the pellet first with the stirring rod, then with the aid of a vortex mixer. Immerse the tubes in a 90° bath for 10 minutes; agitate every 2 to 3 minutes. Centrifuge and discard the supernatant.

(10) Add 5.0 ml of cold 5 per cent TCA. Place in a bucket of ice, and extract the precipitate by resuspension as above. Centrifuge again and discard the supernatant.

(11) Extract with 5 ml of 95 per cent EtOH; resuspend and warm to 60 to 70° for 3 minutes. Centrifuge and discard the supernatant.

(12) Extract with 5 ml of EtOH/ether/CHCl$_3$ (2:2:1); warm to 55 to 60° for 3 minutes. Centrifuge and discard the supernatant.

(13) Extract with 5 ml of anhydrous ether/acetone (2:1); mix for 3 minutes. Centrifuge and discard the supernatant.

(14) Suspend the pellet in 1 ml of ether:acetone (2:1). With the aid of a capillary pipette, transfer the extracted protein to tared planchets. Allow the solvent to dry and determine the amount of radioactivity incorporated. See directions in Experiment 22.

If a liquid scintillation counter is available, proceed as follows:

(15) Allow the pellet to dry in *warm air*. Weigh out 1 to 3 mg of protein (to the nearest 1/100 of mg) directly into tared scintillation vials. Add 1 ml of 2 N NaOH and cap loosely. Heat the vials in a sand bath at approximately 130° C, which is near the upper surface, for 30 to 60 minutes.

(16) Allow the vials to cool, and then add 2 to 2.5 ml of BBS-2 Beckman Biosolve reagent. Mix gently until a clear solution is obtained. Allow 10 minutes for the reaction to take place. If the solution retains color, add 2 drops of 3 per cent H$_2$O$_2$ or 4 per cent SnCl$_2$ solution and mix again.

(17) Add 15 ml of scintillation cocktail containing 5 g of PPO and 100 mg of POPOP in one liter of scintillation or spectrophotometer grade toluene.

(18) Determine the DPM per mg protein, corrected for background and for non-enzymic absorption. Obtain K_e, the efficiency factor for the instrument used, from your instructor. Record and interpret your results.

REFERENCES

1. Stenzel, K. H., Aronson, R. F., and Rubin, A. L. *In Vitro Synthesis of Brain Proteins II. Properties of the Complete System.* Biochem. **5** 930 (1966).
2. Lajtha, A. *Instability of Cerebral Proteins.* Biochem. Biophys. Res. Comm. **23** 294 (1966).

3. Campbell, M. K., Mahler, H. R., Moore, W. J., and Tewari, S. *Protein Synthesis from Rat Brain.* Biochem. **5** 1174 (1966).
4. Rubin, A. L., and Stenzel, K. H. *In Vitro Synthesis of Brain Proteins.* Proc. Nat. Acad. Sci. **53** 963 (1965).
5. Murthy, R. V., and Rappoport, D. H. *Biochemistry of the Developing Brain VI. Preparation and Properties of Ribosomes.* Biochem. Biophys. Acta **95** 132 (1965).
6. Clouet, D. H., and Waelsch, H. *Amino Acid and Protein Metabolism of the Brain IX.* J. Neurochem. **10** 51 (1963).
7. Siekevitz, P. *Uptake of Radioactive Alanine in Vitro into Proteins of Rat Liver Fractions.* J. Biol. Chem. **195** 549 (1952).

TABLE OF FOUR-PLACE LOGARITHMS

x	0	1	2	3	4	5	6	7	8	9	1	2	3	4	5	6	7	8	9
															Average differences				
10	0000	0043	0086	0128	0170	0212	0253	0294	0334	0374	4	8	12	17	21	25	29	33	37
11	0414	0453	0492	0531	0569	0607	0645	0682	0719	0755	4	8	11	15	19	23	26	30	34
12	0792	0828	0864	0899	0934	0969	1004	1038	1072	1106	3	7	10	14	17	21	24	28	31
13	1139	1173	1206	1239	1271	1303	1335	1367	1399	1430	3	6	10	13	16	19	23	26	29
14	1461	1492	1523	1553	1584	1614	1644	1673	1703	1732	3	6	9	12	15	18	21	24	27
15	1761	1790	1818	1847	1875	1903	1931	1959	1987	2014	3	6	8	11	14	17	20	22	25
16	2041	2068	2095	2122	2148	2175	2201	2227	2253	2279	3	5	8	11	13	16	18	21	24
17	2304	2330	2355	2380	2405	2430	2455	2480	2504	2529	2	5	7	10	12	15	17	20	22
18	2553	2577	2601	2625	2648	2672	2695	2718	2742	2765	2	5	7	9	12	14	16	19	21
19	2788	2810	2833	2856	2878	2900	2923	2945	2967	2989	2	4	7	9	11	13	16	18	20
20	3010	3032	3054	3075	3096	3118	3139	3160	3181	3201	2	4	6	8	11	13	15	17	19
21	3222	3243	3263	3284	3304	3324	3345	3365	3385	3404	2	4	6	8	10	12	14	16	18
22	3424	3444	3464	3483	3502	3522	3541	3560	3579	3598	2	4	6	8	10	12	14	15	17
23	3617	3636	3655	3674	3692	3711	3729	3747	3766	3784	2	4	6	7	9	11	13	15	17
24	3802	3820	3838	3856	3874	3892	3909	3927	3945	3962	2	4	5	7	9	11	12	14	16
25	3979	3997	4014	4031	4048	4065	4082	4099	4116	4133	2	3	5	7	9	10	12	14	15
26	4150	4166	4183	4200	4216	4232	4249	4265	4281	4298	2	3	5	7	8	10	11	13	15
27	4314	4330	4346	4362	4378	4393	4409	4425	4440	4456	2	3	5	6	8	9	11	13	14
28	4472	4487	4502	4518	4533	4548	4564	4579	4594	4609	2	3	5	6	8	9	11	12	14
29	4624	4639	4654	4669	4683	4698	4713	4728	4742	4757	1	3	4	6	7	9	10	12	13
30	4771	4786	4800	4814	4829	4843	4857	4871	4886	4900	1	3	4	6	7	9	10	11	13
31	4914	4928	4942	4955	4969	4983	4997	5011	5024	5038	1	3	4	6	7	8	10	11	12
32	5051	5065	5079	5092	5105	5119	5132	5145	5159	5172	1	3	4	5	7	8	9	11	12
33	5185	5198	5211	5224	5237	5250	5263	5276	5289	5302	1	3	4	5	6	8	9	10	12
34	5315	5328	5340	5353	5366	5378	5391	5403	5416	5428	1	3	4	5	6	8	9	10	11
35	5441	5453	5465	5478	5490	5502	5514	5527	5539	5551	1	2	4	5	6	7	9	10	11
36	5563	5575	5587	5599	5611	5623	5635	5647	5658	5670	1	2	4	5	6	7	8	10	11
37	5682	5694	5705	5717	5729	5740	5752	5763	5775	5786	1	2	3	5	6	7	8	9	10
38	5798	5809	5821	5832	5843	5855	5866	5877	5888	5899	1	2	3	5	6	7	8	9	10
39	5911	5922	5933	5944	5955	5966	5977	5988	5999	6010	1	2	3	4	5	7	8	9	10
40	6021	6031	6042	6053	6064	6075	6085	6096	6107	6117	1	2	3	4	5	6	8	9	10
41	6128	6138	6149	6160	6170	6180	6191	6201	6212	6222	1	2	3	4	5	6	7	8	9
42	6232	6243	6253	6263	6274	6284	6294	6304	6314	6325	1	2	3	4	5	6	7	8	9
43	6335	6345	6355	6365	6375	6385	6395	6405	6415	6425	1	2	3	4	5	6	7	8	9
44	6435	6444	6454	6464	6474	6484	6493	6503	6513	6522	1	2	3	4	5	6	7	8	9
45	6532	6542	6551	6561	6571	6580	6590	6599	6609	6618	1	2	3	4	5	6	7	8	9
46	6628	6637	6646	6656	6665	6675	6684	6693	6702	6712	1	2	3	4	5	6	7	7	8
47	6721	6730	6739	6749	6758	6767	6776	6785	6794	6803	1	2	3	4	5	5	6	7	8
48	6812	6821	6830	6839	6848	6857	6866	6875	6884	6893	1	2	3	4	5	5	6	7	8
49	6902	6911	6920	6928	6937	6946	6955	6964	6972	6981	1	2	3	4	4	5	6	7	8
50	6990	6998	7007	7016	7024	7033	7042	7050	7059	7067	1	2	3	3	4	5	6	7	8
51	7076	7084	7093	7101	7110	7118	7126	7135	7143	7152	1	2	3	3	4	5	6	7	8
52	7160	7168	7177	7185	7193	7202	7210	7218	7226	7235	1	2	2	3	4	5	6	7	7
53	7243	7251	7259	7267	7275	7284	7292	7300	7308	7316	1	2	2	3	4	5	6	6	7
54	7324	7332	7340	7348	7356	7364	7372	7380	7388	7396	1	2	2	3	4	5	6	6	7
x	0	1	2	3	4	5	6	7	8	9	1	2	3	4	5	6	7	8	9

TABLE OF FOUR-PLACE LOGARITHMS (CONTINUED)

x	0	1	2	3	4	5	6	7	8	9	1	2	3	4	5	6	7	8	9
														Average differences					
55	7404	7412	7419	7427	7435	7443	7451	7459	7466	7474	1	2	2	3	4	5	5	6	7
56	7482	7490	7497	7505	7513	7520	7528	7536	7543	7551	1	2	2	3	4	5	5	6	7
57	7559	7566	7574	7582	7589	7597	7604	7612	7619	7627	1	2	2	3	4	5	5	6	7
58	7634	7642	7649	7657	7664	7672	7679	7686	7694	7701	1	1	2	3	4	4	5	6	7
59	7709	7716	7723	7731	7738	7745	7752	7760	7767	7774	1	1	2	3	4	4	5	6	7
60	7782	7789	7796	7803	7810	7818	7825	7832	7839	7846	1	1	2	3	4	4	5	6	6
61	7853	7860	7868	7875	7882	7889	7896	7903	7910	7917	1	1	2	3	4	4	5	6	6
62	7924	7931	7938	7945	7952	7959	7966	7973	7980	7987	1	1	2	3	3	4	5	6	6
63	7993	8000	8007	8014	8021	8028	8035	8041	8048	8055	1	1	2	3	3	4	5	5	6
64	8062	8069	8075	8082	8089	8096	8102	8109	8116	8122	1	1	2	3	3	4	5	5	6
65	8129	8136	8142	8149	8156	8162	8169	8176	8182	8189	1	1	2	3	3	4	5	5	6
66	8195	8202	8209	8215	8222	8228	8235	8241	8248	8254	1	1	2	3	3	4	5	5	6
67	8261	8267	8274	8280	8287	8293	8299	8306	8312	8319	1	1	2	3	3	4	5	5	6
68	8325	8331	8338	8344	8351	8357	8363	8370	8376	8382	1	1	2	3	3	4	4	5	6
69	8388	8395	8401	8407	8414	8420	8426	8432	8439	8445	1	1	2	2	3	4	4	5	6
70	8451	8457	8463	8470	8476	8482	8488	8494	8500	8506	1	1	2	2	3	4	4	5	6
71	8513	8519	8525	8531	8537	8543	8549	8555	8561	8567	1	1	2	2	3	4	4	5	5
72	8573	8579	8585	8591	8597	8603	8609	8615	8621	8627	1	1	2	2	3	4	4	5	5
73	8633	8639	8645	8651	8657	8663	8669	8675	8681	8686	1	1	2	2	3	4	4	5	5
74	8692	8698	8704	8710	8716	8722	8727	8733	8739	8745	1	1	2	2	3	4	4	5	5
75	8751	8756	8762	8768	8774	8779	8785	8791	8797	8802	1	1	2	2	3	3	4	5	5
76	8808	8814	8820	8825	8831	8837	8842	8848	8854	8859	1	1	2	2	3	3	4	5	5
77	8865	8871	8876	8882	8887	8893	8899	8904	8910	8915	1	1	2	2	3	3	4	4	5
78	8921	8927	8932	8938	8943	8949	8954	8960	8965	8971	1	1	2	2	3	3	4	4	5
79	8976	8982	8987	8993	8998	9004	9009	9015	9020	9025	1	1	2	2	3	3	4	4	5
80	9031	9036	9042	9047	9053	9058	9063	9069	9074	9079	1	1	2	2	3	3	4	4	5
81	9085	9090	9096	9101	9106	9112	9117	9122	9128	9133	1	1	2	2	3	3	4	4	5
82	9138	9143	9149	9154	9159	9165	9170	9175	9180	9186	1	1	2	2	3	3	4	4	5
83	9191	9196	9201	9206	9212	9217	9222	9227	9232	9238	1	1	2	2	3	3	4	4	5
84	9243	9248	9253	9258	9263	9269	9274	9279	9284	9289	1	1	2	2	3	3	4	4	5
85	9294	9299	9304	9309	9315	9320	9325	9330	9335	9340	1	1	2	2	3	3	4	4	5
86	9345	9350	9355	9360	9365	9370	9375	9380	9385	9390	1	1	2	2	3	3	4	4	5
87	9395	9400	9405	9410	9415	9420	9425	9430	9435	9440	0	1	1	2	2	3	3	4	4
88	9445	9450	9455	9460	9465	9469	9474	9479	9484	9489	0	1	1	2	2	3	3	4	4
89	9494	9499	9504	9509	9513	9518	9523	9528	9533	9538	0	1	1	2	2	3	3	4	4
90	9542	9547	9552	9557	9562	9566	9571	9576	9581	9586	0	1	1	2	2	3	3	4	4
91	9590	9595	9600	9605	9609	9614	9619	9624	9628	9633	0	1	1	2	2	3	3	4	4
92	9638	9643	9647	9652	9657	9661	9666	9671	9675	9680	0	1	1	2	2	3	3	4	4
93	9685	9689	9694	9699	9703	9708	9713	9717	9722	9727	0	1	1	2	2	3	3	4	4
94	9731	9736	9741	9745	9750	9754	9759	9763	9768	9773	0	1	1	2	2	3	3	4	4
95	9777	9782	9786	9791	9795	9800	9805	9809	9814	9818	0	1	1	2	2	3	3	4	4
96	9823	9827	9832	9836	9841	9845	9850	9854	9859	9863	0	1	1	2	2	3	3	4	4
97	9868	9872	9877	9881	9886	9890	9894	9899	9903	9908	0	1	1	2	2	3	3	4	4
98	9912	9917	9921	9926	9930	9934	9939	9943	9948	9952	0	1	1	2	2	3	3	4	4
99	9956	9961	9965	9969	9974	9978	9983	9987	9991	9996	0	1	1	2	2	3	3	3	4
x	0	1	2	3	4	5	6	7	8	9	1	2	3	4	5	6	7	8	9

INDEX